Fruit Production

— Minor Fruits —

The Authors

Dr. Hare Krishna graduated from the IAS, B.H.U., Varanasi and completed M.Sc. and Ph.D. from IARI, New Delhi. Presently, he is working as a Senior Scientist (Horticulture) at ICAR- Central Institute for Arid Horticulture, Bikaner (Rajasthan). He has been associated with the development of *ber* variety Thar Malti and mulberry varieties Thar Lohit and Thar Harit. In his initial years of career, so far he has published research papers in 45 international and national journals of repute; authored 05 books and edited 02 books. He has developed guidelines for the conduct of DUS on Indian jujube (*Ber*) as Nodal Officer (nominated by PPV&FRA, New Delhi). He has been awarded 'Certificate of Excellence in Reviewing' by 'Scientia Horticulturae' and also an Editorial Board Member of it.

Dr. R.R. Sharma completed Post Graduation from IARI, New Delhi in 1994 and joined as Scientist in the Division of Fruits and Horticultural Technology, IARI, New Delhi and currently Dr. Sharma is working as Principal Scientist. During his scientific career, he has served several research projects and has been associated with release of mango hybrids like *Pusa Arunima, Pusa Surya, Pusa Lalima, Pusa Shresth, Pusa Pitamber etc.,* and standardized several fruit production technologies. He is a prolific writer and has published 50 research articles in international journals, 65 in national journals, 160 popular articles in magazines, and 25 book chapters. He has also authored 10 books on different aspects of fruit production.

Fruit Production

— Minor Fruits —

Hare Krishna

R.R. Sharma

2017

Daya Publishing House®

A Division of

Astral International Pvt. Ltd.

New Delhi – 110 002

Cataloging in Publication Data--DK
  Courtesy: D.K. Agencies (P) Ltd. <docinfo@dkagencies.com>

  Hare Krishna, 1981- author.
Fruit production : minor fruits / authors, Hare Krishna, R.R. Sharma.
      pages cm
      Includes bibliographical references and index.
      ISBN 978-93-86071-74-3 (International Edition)

      1. Fruit-culture--India. 2. Fruit-culture. 3. Fruit trade--India. 4. Fruit trade.   I. Sharma, R. R., Dr., author. II. Title.

SB354.6.I4H37 2017 DDC 634.0954   23

Published by   : **Daya Publishing House®**
                            A Division of
                            Astral International Pvt. Ltd.
                            – ISO 9001:2015 Certified Company –
                            4736/23, Ansari Road, Darya Ganj
                            New Delhi-110 002
                            Ph. 011-43549197, 23278134
                            E-mail: info@astralint.com
                            Website: www.astralint.com

भारत सरकार
कृषि अनुसंधान और शिक्षा विभाग एवं
भारतीय कृषि अनुसंधान परिषद
कृषि मंत्रालय, कृषि भवन, नई दिल्ली 110 001

GOVERNMENT OF INDIA
DEPARTMENT OF AGRICULTURAL RESEARCH & EDUCATION
AND
INDIAN COUNCIL OF AGRICULTURAL RESEARCH
MINISTRY OF AGRICULTURE, KRISHI BHAVAN, NEW DELHI 110 001
Tel.: 23382629; 23386711 Fax: 91-11-23384773
E-mail: dg.icar@nic.in

डा. एस. अय्यप्पन
सचिव एवं महानिदेशक
Dr. S. AYYAPPAN
Ex. SECRETARY & DIRECTOR GENERAL

Foreword

The cultivation of fruit crops is, presently, one of the most important and remunerative enterprises of agriculture. Growing of fruits translate into realization of more yield and income from a unit area of land than agricultural crops. Besides social and nutritive values, fruits also contribute in strengthening national economy by fetching foreign exchange. Annually, our country earns Rs. 240830 Lakhs of foreign exchange by export of major fruits and nuts like mango, grapes, banana, guava, pineapple, sapota, walnut *etc.* On the other hand, minor fruits are not so extensively cultivated and their consumption and trade tend to be more limited, geographically and quantitatively, than those of the major fruits, despite having high nutritive and medicinal attributes and also their ability to grow under adverse soil and climatic conditions. Further, if exploited properly, these fruits have the potential of transforming the economy of rural or disadvantaged areas because the crops are of explicit quality with great nutritional, medicinal, organoleptic, economic and traditional importance. Therefore, it is important to include these fruits in our health promotion campaigns. This is the need of the hour to focus on this important and economically rewarding area of horticulture.

The need for intensifying research on minor fruits and syllabus taught at University level necessitate requirement for a textbook, which covers all contents of the production of minor fruit crops. I convey my good wishes to the authors of this book for having designed and delivered such University-level textbook for the undergraduate and post-graduate students.

I am confident that this book will be very useful for teachers, scientists, researchers, UG and PG students, extension personnel, and even growers of our country.

S. Ayyappan

Preface

The history of fruit cultivation dates back to 11,000 years ago, while in India, the specific reference of cultivation of fruits and trees was found in the text *Upavanavinoda*, a small section of *Sarangadharapaddhati* (13th cent. A.D.) and the *Kallola* 10 and 11 of the *Ivatattvaratneikara* (18th C.A.D). And since then, the odyssey of horticulture in India had been expanded and evolved greatly. Today, India holds a prominent position in horticulture and has emerged as the second largest producer of fruits in world. Fruits like mango, banana, citrus, apple, guava, papaya, pomegranate, grapes, sapota, pineapple, litchi *etc.* are well known in both local and international markets and are being grown in different parts of the country as commercial fruit crops in organized orchards. Contrarily, the cultivation and trade of minor fruits are confined to specific areas, particularly in rural and tribal belts; though, these fruits are unique in their own taste, palatability and nutritive value and are incomparable with any other food commodity. In general, they have low requirement for external inputs for production and are well adapted to a wide range of growing conditions including marginal lands or marginal climatic conditions.

Ample literatures are available on various aspects of cultivation of such fruit crops, which have been published from time to time. However keeping in view the enormously increasing and updated scientific knowledge, an immediate need was felt to bring out a book on this topic in new attire so as to draw out a positive and unswerving response in students, teachers, researchers and growers. This is the second book of our proposed 'Fruit Production' trilogy.

For writing this book, authors have freely consulted several books, research periodicals, research reports, web content *etc.* in order to bring in the most recent and relevant information together. Utmost care has been taken to make the book comprehensive to the readers supported with the ample numbers of tables and plates. However, there are chances of errors and omissions, we, therefore, invite constructive suggestions from the readers to make this more comprehensive.

This book could not have come in its present form without the helps received from various colleagues and students in different stages during the preparation of the manuscript and for which they are profusely thanked.

We also acknowledge our family members who were ever there to extend unconditional support during the preparation of this book and we thank all of them from the core of our hearts.

In the last but not the least, we thank Astral Publishing House, New Delhi, who had faith in us and gave the responsibility of writing this proposal.

Hare Krishna

R.R. Sharma

Contents

1
Almond

INTRODUCTION

Almond is one of the major nut fruit crops in the world. It is primarily cultivated for its energy rich kernels. The almonds do not fall under the category of nuts because the stone is endocarp and exocarp and mesocarp of the fruit form the leathery husk, covering the stone. There is no involvement of involucre in the formation of outer leathery hull as in nuts.

ORIGIN, HISTORY AND DISTRIBUTION

Almond is native to hot arid region of Western Asia and Mediterranean region. It is believed that from these regions, it was carried to Greece and North Africa and several other countries of the world for domestication. Almond is mainly cultivated in countries falling between 36° and 45°N latitude. More than 50 per cent of almond production is from USA and Spain. The other leading almond producing countries are China, Italy, Iran, Morocco, Portugal, Turkey, France, Algeria, Greece, Afghanistan, Syria and Persia (Table 1.1). In Southern hemisphere, the almond cultivation is limited to South Africa, Australia and Argentina. In India, almonds are grown in Jammu and Kashmir and dry temperate zone of Himachal Pradesh (Kinnaur and Lahaul Valley). However, green almond cultivation is done in wet temperate zone of Himachal Pradesh (Shimla and Kullu districts), Uttarakhand and sub-mountainous areas of Punjab. The commercial cultivation of almond could not pick up in India due to problems like damage by rains, winds, frost and hail storms during flowering time, lack of suitable varieties and self incompatibility.

Table 1.1: Major Almond Producing Countries of the World (2012)

Country	Production (Tonnes)	Country	Production (MT)
United States of America	720000	China, mainland	43000
Spain	215100	Algeria	33996
Australia	142680	Libya	32000
Iran (Islamic Republic of)	100000	Greece	29000
Morocco	99067	Lebanon	26000
Italy	89865	Chile	24500
Syrian Arab Republic	86271	Pakistan	21000
Turkey	75055	Uzbekistan	20000
Tunisia	70000	Israel	9450
Afghanistan	62000	Occupied Palestinian Territory	8500
Australia	142680	Portugal	7200

Source: www. http://faostat.fao.org.

COMPOSITION AND USES

Composition

Almonds are considered as world's healthiest food as its kernels are a very good source of proteins (glutamic acid = 6, 810/100 g), fats, dietary fiber, vitamin E, biotin, and several essential minerals such as calcium, magnesium, phosphorus, potassium and copper (Table 1.2). Almonds contain high amount of monounsaturated fats, which have been associated with reduced risk of heart disease. In addition to their cholesterol-lowering effects, almonds' ability to reduce heart disease risk may also be partly due to the antioxidant action of the vitamin E found in the almonds. Almonds also reduce after meal blood sugar level, hence protects us from diabetes.

Uses

The nuts are usually eaten when they are ripe. The kernels are highly delicious and in great demand. Almonds are a rich source of oil ranging from 36 to 60 per cent, of which, 62 per cent is monounsaturated oleic acid, 29 per cent is linoleic acid and only 9 per cent is saturated fatty acid. Almond oil is called as *badam rogani,* which is used in confectionary, pharmaceuticals and cosmetics preparations. Nowadays, green almonds are also preferred by some consumers. The fruit of the wild forms (bitter almond) contains the glycoside *amygdalin*, which is transformed into hydrogen cyanide when kernels are either crushed, chewed, or any other injury, and can be fatal.

Almonds, like other nuts, are susceptible to aflatoxin-producing molds such as *Aspergillus flavus* and *Aspergillus parasiticus*. It is unsafe to eat mold infected nuts. Hence, some countries have strict limits on allowable levels of aflatoxin contamination of almonds and require adequate testing before the nuts can be marketed to their citizens.

Table 1.2: Nutritional Value of Almond per 100 g

Attribute	Content	Attribute	Content
Energy	576 cal	Protein	21.22 g
Carbohydrates	21.68 g	Vitamin E	26.2 mg
Dietary fibre	12.2 g	Calcium	264 mg
Sugars	3.89 g	Magnesium	268 mg
Monounsaturated fatty acids	30.889 g	Phosphorus	484 mg
Polyunsaturated fatty acids	12.070 g	Potassium	705 mg

TAXONOMY AND BOTANICAL DESCRIPTIONS

Taxonomy

Almonds belong to family Rosaceae, sub-family Prunoideae, subgenus Amygdalus, section Euamygdalsu and genera *Prunus*. The cultivated almond is *Prunus amygdalus* Batsch but several prefer to name it as *Prunus dulcis* Mill. The chromosome number of almond is, n= 8 and most of the cultivars are diploids with somatic number, 2n = 16. It is also believed that cultivated almond is a hybrid between *P. fenzliana* and *P. bucharia* because both are closely related to it. The bitter almond is *Prunus dulcis* var. *amara*.

Botany

The almond is a small- to-medium sized deciduous tree with a spreading, open canopy, usually growing 4-10 m in height. Its young twigs are green at first, becoming purplish where exposed to sunlight, then grey in their second year. The leaves are 3-5 inches long, with acute tips and finely serrated margins. Almond flowers are nearly identical to peach and other *Prunus* flowers in structure, but light pink or white in colour, and fragrant. The flowers usually appear before the leaves in the early spring. Flowers are borne laterally on spurs or short lateral branches, or sometimes laterally on long shoots.

Almonds are self-incompatible, and require cross pollination. Pollinators (honeybees) are absolutely essential, especially since cool, wet weather can occur at the relatively early blooming period. The entire fruit including the hull is a drupe; however, the hull dries and splits prior to harvest.

SOIL AND CLIMATIC REQUIREMENTS

Soil

Almonds often grow in infertile soil. However, light sandy soils are most suitable for almond cultivation. Although, almond is more resistant to dry soils as compared to most other temperate fruits, it shows definite response to summer irrigation. Thus, deep, fertile, well drained light loam soils are most suitable for profitable almond cultivation. Very heavy clay soils are not usually desirable for almonds.

Climate

Almond grows best in Mediterranean climates with warm, dry summers and mild, wet winters. Almond can be grown in temperate and subtropical climatic conditions; however, temperature is the most limiting factor for commercial production. It requires moderate winter for good bud burst in spring. Spring frost is another limiting factor. The chilling requirement depends on cultivar and varies from 100-700 hours at or below 7.2°C to break dormancy. Further, it is susceptible to injury by rainy weather in spring and summer. Similarly, almonds prefer dry and low humid conditions during fruit ripening.

IMPORTANT VARIETIES

The commercial varieties of almond grown throughout the world are Non Pareil, Mission, Ne Plus Ultra, Peerless, Eureka, Kapreil, Thompson, Ballico and Merced. The promising varieties for Indian conditions are:

- ✰ Dry temperate zone: Ne Plus Ultra, Texas (Mission).
- ✰ High and Mid-hills (Wet temperate zone): Merced, Non Pareil, IXL, Nauni Selection, Nikitskyi, White Brandis.
- ✰ Low- hills and valley areas: Drake, Katha, Peerless, Ne Plus Ultra.
- ✰ Mid sub-tropical areas: Hybrid No. 15, NB 258, JK 39, JK 55, JK 57 and JK 75.

On the basis of availability of fruits, almond varieties have been classified as under:

- ✰ **Early varieities:** IXL, Jordanolo, Ne Plus Ultra, Peerless.
- ✰ **Mid-iseason varieities:** Cressey, Davey, Drake, Kapareil, Merced, Nonpareil, Norman, Paxman, Price Cluster, Profuse, Vesta.
- ✰ **Late varieities**: Ballico, Butte, Emerald, Empire, Mission (Texas), Ripon, Ruby, Thompson, Tioga, Wawonal, Yosemite.
- ✰ **Very Late:** Tardy, Non Pareil.

Chief characteristics of some of the important varieties of almond are as under:

- ✰ **Non Pareil:** This is the most important variety of almond. It is a regular bearer, mid bloomer (3rd week of March) and early maturing variety. Its canopy is upright to spreading. It bears flower and nuts on both spurs as well as on long shoots having good ability to renew fruiting wood, It is relatively resistant to frost. The nut has an extra light colour, papery shell, and have high shelling percentage (60 per cent) and medium sized kernel. The variety is suitable for export market.

- ✰ **California Paper Shell:** This is another important variety of almond. It is also a regular bearer, mid-bloomer variety (2nd week of March) but matures late than Non Pareil. Its tree has upright growth habit and thus suitable for high density plantations. It bears flower and nuts on both spurs as well as on long shoots with good ability to renew fruiting wood. The nut and kernel are longer in size with extra light colour, papery shell, having high shelling percentage (50 per cent.) This variety is also suitable for export market.

☆ **Merced:** It is a regular bearer variety, which blooms during 3rd week of March and ready to harvest after 152 days from full bloom. The tree is upright and thus suitable for high density plantations. It bears flower and nuts on both spurs as well as on long shoots with good ability to renew fruiting wood. Shell is papery shell with high shelling percentage (56 per cent). The variety is also suitable for export of kernels.

☆ **IXL:** It is a regular bearer variety, which blooms during 3rd week of March and ready to harvest after 151 days from full bloom. The tree is spreading type and of intermediate vigour. It bears flower and nuts on both spurs as well as on long shoots with good ability to renew fruiting wood. The nut and kernels are medium and shells are soft having high shelling percentage (55 per cent).

☆ **Shalimar:** It is a regular bearer, which blooms during 2nd week of March and ready to harvest after 143 days from full bloom. The tree growth habit is spreading or drooping type. It bears flower and nuts on both long shoots and spurs. The shell colour is light, papery with high shelling percentage (50 per cent), highly suitable for export.

☆ **Makhdoom:** A regular bearer type, which blooms during 1st week of March and ready to harvest after 141 days from full bloom. The tree growth habit is spreading and drooping type. It bears flowers and nuts on long shoots and spurs. The shell colour is medium, soft type with a shelling percentage of about 42 per cent.

☆ **Waris:** It is a regular bearer variety, which blooms during 3rd week of March and ready to harvest after 145 days from full bloom. The tree is upright in growth habit and thus suitable to grow under high density plantations. It bears flower and nuts on long shoots and spurs. The shell colour is medium, and produces plump kernels with high shelling percentage (48 per cent).

☆ **Drake:** Its trees are low in vigour, spreading, and mid-season variety. Its nuts are small to medium in size, bold and roundish with pointed apex and light creamy whitish brown in colour and semi soft shelled.

☆ **Ne-Plus Ultra:** It is a mid-season variety. Its trees are vigourous and spreading. Its nuts are medium to large flattened, bold and light brown in colour, and paper shelled.

☆ **Pranyaj:** It is also a mid-season variety. Its trees are moderate in vigour and spreading and nuts are medium, brown and flattened to bulge. Its kernel is medium to large in size and papery shelled.

☆ **Primorskij:** It is late season variety. Its trees are spreading and moderately vigorous. Nuts are medium to large, bold, slightly flattened and brown in colour. Its kernel is medium to large. It is a soft paper shelled variety.

ROOTSTOCKS AND PROPAGATION

Almond seedlings, bitter almond and wild peach seedlings can be used as rootstock. Bitter almond is better than sweet almond as rootstock. Almond seedlings are still preferred as rootstock because of their tolerance to drought, limy soils, iron chlorosis and longevity. Almond trees on peach rootstock have short life especially in soils high in lime and sodium. Some of the other important rootstocks of almond are *behmi* (suitable for cool high hills), Marianna 2624 (resistant to oak root fungus) and Nemaguard (suitable for nematode infested areas). Peach x almond hybrids (GF 677 and GF 557) are now used in intensive almond orchards.

Almond can be propagated by seed or budding and grafting. Rootstocks are primarily propagated by seeds. Budding can be done in autumn, spring and summer. T budding is usually preferred over other methods of vegetative propagation.

PLANTING AND ORCHARD ESTABLISHMENT

The planting of almond is done is December-January, when it is dormant. Although planting distance depends on cultivar, rootstock use, soil fertility or irrigation system but usually, the planting distance is 6 m in flat fertile soils. However, it can be reduced to 4-5 m in less fertile shallow soils. Pits of 1 x 1 x1 m are dug and filled with the mixture of soil and FYM about one month prior to planting. In flat areas, the square or hexagonal system of planting should be followed, however, in hills the planting is done in terraces or contours.

FLOWERING, POLLINATION AND FRUIT SET

In most part of the world, almond flowers during spring seasons. Most of the commercial cultivars of almond are self-incompatible and some combinations are even cross-incompatible. Hence, it is necessary to interplant pollenizers in the commercial plantations. Pollination is entirely dependent on honeybee's activity. Unfavourable weather for honeybee's activity adversely affects normal fruit set. Provision of 5-6 bee colonies per hectare is necessary for assured cross pollination and adequate fruit-set. The following combination of varieties should be followed in commercial orchards:

a. Non Pareil – Ne Plus Ultra – Mission (Texas)

b. Peerless – Non Pareil – Mission

c. Ne Plus Ultra – Peerless – Non Pareil

d. Davey – Non Pareil – Mission

e. Merced – Non Pareil – Mission

f. Ne Plus Ultra – Non Pareil – Davey

Ne Plus Ultra, Drake and Dhebar are considered good pollinizers.

Pollination Recommendations and Practices

The literature on almond pollination leaves no doubt about the need for an ample supply of bees to pollinate the flowers. There is no other choice than to have honeybees perform this task. There should be either one pollenizer row of trees for every three rows of the main variety, or two rows of pollenizer trees for each two of the main variety. As per literature, the pollination of California's almonds is the largest annual managed pollination event in the world, with close to one million hives (nearly half of all beehives in the USA) being trucked in February to the almond groves. Much of the pollination is managed by pollination brokers, who contract with migratory beekeepers from different states.

INTERCULTURAL OPERATIONS

Training and Pruning

Almond plants can be trained either by modified central leader system or by open centre system. After first year of establishment, head back the tree to 1m above ground level. Afterwards, three primary scaffold branches are retained along with few temporary branches below trunk, removing only those which compete strongly with primary scaffolds and the tree is trained to a shape of modified leader system. One year old wood is pruned each year in December/January when tree is completely dormant.

Almond bears fruits on short spurs or one year old shoots. Almond produces most of their fruit on short spurs which remain fruitful for about 5 years. Almonds do not require heavy pruning. Thinning of branches rather than heading back is recommended. Therefore, pruning should be done in such a manner that 1/5th of fruiting wood is replaced each year. Prune in such a way that new wood with new spur growth is constantly replacing spur that are no longer fruitful. To achieve this, only prune or remove older branches that are 1.2 to 3.7 cm in diameter. Thin out very little of the smaller wood, except to remove unwanted water sprouts or suckers. Larger fewer cuts are more desirable than numerous small cuts, which encourages the problem of gummosis. This practice is called bulk pruning.

Nutrition Management

Almond is a heavy feeder. Therefore, it requires substantial amount of manures and fertilizers for proper growth, development and production of quality nuts. In general, almond has high nitrogen requirement similar to peach, however, it can tolerate a lower available potassium content of soil than apple and plum. Before applying any manure of fertilizer, it is advisable to get soil test and leaf nutrient analysis done, as fertilizer doses depend largely on available nutrients in the soil, previous fertilizer use, age of the plant and type of fertilizer used. A mature bearing tree of almond above the age of 7 years requires 40 kg FYM, 2.0 kg CAN, 1.5 kg single super phosphate and 1.2 kg muriate of potash annually.

It is recommended to apply well rotten farm yard manure (FYM) during winter (Dec.-Jan.). CAN/Urea may be applied in 2-3 split doses. 1st half dose should be applied along with DAP (full) and MOP (full) at fortnight before expected bloom, 2nd dose (1/4th) of CAN/urea may be applied about 3 weeks after fruit set and third dose of Urea (1/4th) should be applied in May- June. Foliar spray of urea about 1.5-

2.0 per cent may also be given for promoting fruiting bud formation and subsequent growth in the next season.

Almond is found to be sensitive to boron deficiency which affects flowering and fruit set and may cause stem gummosis. Spray boric acid (0.1 per cent) before flowering and after petal fall to overcome boron deficiency.

Water Management

In India, almond is largely grown as rainfed fruit crop. However, for commercial cultivation and fruitful production, ample supply of irrigation water is required. Immediately after planting, apply sufficient water to settle the soil around the roots and give irrigation at an interval of 10-15 days afterwards. However, avoid heavy irrigation during flowering as it will result in complete drop of flowers.

The critical stages of almond which are most sensitive to water shortages are flowering (Feb to March) and fruit development (April, May and June). Therefore, irrigations must be provided during these stages for getting higher yield of quality nuts. The drip irrigation is highly effective in almond. It not only helps in increasing water efficiency but also increases yield of quality nut with a saving of about 40-50 per cent water over conventional irrigation systems. Mulching of tree basins with straw or black polythene further conserve moisture and reduces irrigation intervals. In dry rainfed areas where there is no source of irrigation water, rain water harvesting methods depending upon the topography combined with straw or black polythene mulching can be very useful for conservation of moisture, enhancing water availability and increasing plant growth, fruit set and nut yield.

Intercropping

During the pre-bearing period, intercropping with suitable crops is highly suitable and beneficial in almond orchards. It helps in utilization of interspaces and available resources and additional income to the orchardist. Pea, pulses, knol khol, carrot, mustard, onion and garlic *etc.*, can be grown as intercrops. Some legume crops (*e.g.*, peas and pulses) helps in increasing soil fertility, and some (*e.g.*, mustard) helps in augmenting pollination by attracting bees.

PLANT PROTECTION

Major Insect-Pests and their Management

Several insect-pests cause damage to almond but the major insect-pests and their control measures are given hereunder:

Almond Weevil (*Myllocerus lactivirens*)

This is very serious pest of almonds. Weevils feed on young and tender leaves and eat away whole leaves. For their control, shake the tree after sunset and then destroy them. Spray carbaryl (0.1 per cent) or Dimethoate (0.02 per cent) in the period of severe infestation.

Bark Eating Caterpillars

Caterpillars feed upon the bark as well as on stem under the cover of their webbing. Therefore, remove webbing and treat the main limbs and trunk of the attacked tree during March to October with Dimethoate (0.02 per cent) and inject it into holes excavated by the caterpillar with syringe, to get rid-off such pests.

Borers

Several types of borers cause damage to almond by boring into the stem or shoots. In case of severe infestation, the plant becomes weak or sometimes may die. The above ground borers are evident from the frass and faecal pellets or drying up of the terminal shoots or dead bark. The flat headed borer generally attacks the portion of the stem and branches exposed to the sun. For the management of borers, remove the dead bark and apply water proof paint on hard wood and spray the main trunk upto 1.5 m with Rogor (0.02 per cent) or swab the main trunk and limbs with Rogor (0.2 per cent). Covering of the exposed part of the stem to sun with dry grass or gunny bags soaked with Rogor (0.02 per cent) once a month from in March-October is quite effective.

Termites

Termites feed on roots and stem portion near the ground level of almost all types of fruit trees in warm localities. The severely affected trees often dry or the fruit bearing capacity is greatly reduced. Trees infested with stem borers are more vulnerable to the attack of termites. Destroy the termitaria in the vicinity of the orchards and drench the soil with Chlorpyriphos (0.1 per cent).

Root-knot Nematode (*Meloidogyne incognita*)

The root knot nematode makes knot like galls on roots, resulting in malfunctioning of infested plants. In addition, the infested plants show stunting and yellowing of leaves. Plant growth is often patchy in nurseries. The other nematode species, which attack almond are *Pratylenchus pruni*, *Macroporthenia xenoplax*. Their infestation results in the development of fibrous roots and stubby fasciculation causing stunting of plants and reduction in yield. Use Phorate 10G granules @ 25kg/ha nursery beds and @ 10-20 g/tree in the orchards depending upon tree age. Nematode resistant rootstocks like Shalin, Nemaguard and Yunnan *etc.*, can also be used in nematode infested vicinities. Trap crops such as marigold, berseem and mustard can be planted in between lines of plantation to reduce the nematode population in infested orchards.

Insect-pests like leaf curl aphid, sanjose scale *etc.*, can be effectively controlled by the measures suggested for their control in fruits like peach, cherry and/or plum *etc.*

Major Diseases and their Management

Several diseases infest almond and cause severe losses but the major diseases and their control measures are given below:

Bacterial Gummosis

It is the most destructive disease of almond trees and fruits. It is caused by a bacterium, *Pseudomonas syringae*. The affected portions such as bark, outer sap-wood and fruits develop circular to elongated soaked gumming lesions. For its management, spray streptocycline @ 10g/100 L water before the onset of rainy season or alternatively spray copper oxychloride and Bordeaux mixture @ 0.3 per cent after leaf fall. In addition, cut and burn the affected plant parts immediately to restrict the further spread of the disease.

Brown Rot

It is caused by a fungus, *Monilinia fructicola*. This fungus causes blossom blight, fruit rot, twig blight, and branch canker. Brown rot of ripening fruit is very common, and it generally occurs as the fruit approaches maturity. The first evidence of fruit infection is the appearance of a small brown spot, frequently originating in a slight wound caused by insect feeding or egg-laying activities. The rotted area rapidly expands and eventually becomes covered with tan-gray fungal fruiting tufts. Fruit rotted by brown rot usually retain their form and usually remain attached to the tree for some time after being completely rotted. Later they may drop off the trees. For it management, remove all rotted fruits after harvest to reduce the amount of fungus over-wintering in orchards. Adequate pruning will increase air circulation, allowing faster drying and fewer fruit infections. Apply fungicide sprays during bloom and as fruit ripens or as suggested for brown rot of plums.

Collar Rot

This disease is caused by a fungus, *Phytophthora cactorum*, and occurs primarily in poorly drained soils like root rot but in this disease, rotting starts from the collar portion of stem and spreads on both sides from the site of infection. There is a development of brown, soft, spongy cankered area around collar region of the plants. Affected plant starts declining and finally may die. The removal of soil around collar portion during November-December and exposure to sun, removal of affected bark and application of Chaubattia paste and soil drenching around trunk with mancozeb (0.3-0.4 per cent), copper oxychloride (0.5 per cent) or ridomil MZ (0.3 per cent) has shown appreciable control. Clonal rootstocks like M 2, M 4, M 9 and MM 113 have been found to be resistant to collar rot whereas MM 106 is susceptible to the disease.

Crown Gall

It occurs worldwide and cause huge losses especially to nursery stock. It is caused by bacterium, *Agrobacterium tumefaciens*. In this disease, there is formation of small galls on the stem and roots. It is very difficult to control this disease. However, the affected plants should be destroyed in order to contain the spread of the disease.

Almond Rust

It is a widespread disease and attack leaves and fruits late in the season. It is caused by a fungus, *Tranzschelia pruni-spinosae*, which causes chlorotic spots containing red-brownish pustules in the underside of leaves and sometimes on fruits, usually after heavy rainfall in the spring. It can be controlled by spraying wettable sulphur and Zineb (0.2 per cent) before the expected attack of the disease.

Shot Hole

It is caused by a fungus, *Stigmina carpophila*, which causes shot holes or lesions on young shoots, flowers, fruits and leaves and small purple area surrounded by a yellow-halo. Later these areas turn brown and the dead tissue may fall or whole leaf may drop off. It can be controlled by a spray of captan, or captafol (0.02 per cent) after flowering or by a spary Bordeaux mixture (2 per cent) during autumn.

Anthracnose

In the recent years, anthracnose has appreaed as pre- and postharvest disease of almond. Previously, *Colletotrichum gloeosporioides* was regarded as its causal agent, but now it is established that *Colletotrichum acutatum* is its causal fungus. This disease attacks the blossoms, leaves, fruit, and limbs of an almond tree. Infected leaves tend to develop water-soaked lesions that eventually fade in colour. Defoliation can occur, but leaves often remain attached to the branches. Infected nuts often have a crater-like lesion in which the affected area turns a reddish orange. Often, the fruit may gum profusely as the fungus is able to penetrate into the kernel, killing the embryo. Affected nuts often remained attached to the spur. Shoots and spurs that bear infected nuts often become infected and die.

Viral Diseases

Viral diseases like Yellow Bud Mosaic, caused by tomato ring spot virus; Almond Calico, caused by Prunus Necrotic Ring Spot Virus (PNSRV); Infectious Bud Failure, caused by strain of PNRSV and non-infectious bud failure *etc.*, may also infect almond. It is very difficult to control such diseases once the infection has occurred. However, it is advisable to use healthy bud wood for grafting or budding and destroy the infected plants from the nursery or orchards as and when they are noticed.

MATURITY, HARVESTING AND YIELD

Maturity

Almonds can be harvested green or dry, depending on the market demand and purpose of the grower. However, the mature and well-filled nuts should only be harvested. In general, almonds are ready for harvest from early August to late-September. Harvesting should begin when about 90-95 per cent of the nuts have hulls that have split open. Splitting of hulls begins in the top of the tree and progresses downward. It is also important to keep the tree well watered up to full harvesting, since the hulls will not split well if the trees are under water stress. Delay in harvest increases risk of navel orange worm infestation and damage caused by direct picking of fruits.

In wet temperate zone mid and high hills of India, the green almonds are harvested before onset of rains. The kernels of green almonds are used for confectionery, healthy foods and fresh consumption as green almonds can't be stored for long.

Harvesting Method

The best way to knock almonds from trees is to strike the small branches with a pole or to strike the major branches with a rubber mallet made for that purpose. In India, farmers usually climb the tree and shake the branches. In California, almonds are harvested by mechanical tree shakers. However, young trees may be damaged by shakers, so are harvested by hand knocking in the first few years. After harvesting, nuts are picked up promptly and piled/stacked for few days for easy dehulling.

Yield

Almond tress start bearing at the age of 4-5 years and commercial bearing starts only after 18-20 years, and bear profitably up to 50 years. A well-managed tree can yield 25-30 kg almonds per year.

POSHARVEST HANDLING

Hulling and Drying

After harvesting, remove hulls promptly from the nuts. The hulls are usually removed by hand. Almonds harvested at the proper time usually require additional drying to prevent mold growth in storage. To dry the nuts, spread them in a thin layer on a tray or screen to allow good air circulation. Birds commonly steal almonds while they are drying, hence cover the drying nuts with screen or plastic netting to prevent loss. If the chances of rain is there, cover the nuts. Periodically check the nuts for dryness. Remove shells from several nuts and break the kernels. Rubbery kernels indicate that additional drying is necessary. Almonds are ready for storage when their kernels are crisp to brittle when broken. Nuts can be dried by forced hot air until their moisture content reaches 5-7 per cent.

Storage

When properly dried, in-shell almonds can be stored for 8 months at room temperature (20.0°C), and for a year or more at (0° to 7.2°C). Shelled almonds will retain quality for 1 year or so at 0°C. Before storing almonds at room temperature, freezing the nuts at 0°F for 48 hours is necessary to kills insect pests and eggs. Almond kernels can absorb objectionable odours in storage. Hence, always store them in airtight containers away from strong-smelling materials such as onions or garlic.

Processing

Shelled nuts are sorted for size and appearance. The nuts are bleached for color improvement, then salted, roasted, and/or flavored before packaging.

Pasteurization of Almonds

Due to problem caused by *Salmonella* bacteria in stored almonds, now pasteurization of almond kernels has been recommended by several countries of the world. To achieve this dramatic reduction in *Salmonella* bacteria, two major alternatives have been suggested: (1) exposure of almonds to steam heat sufficient to raise the surface temperature of the almond kernels to about 93°C or (2) insertion of the kernels into a closed chamber where they could be exposed to propylene oxide gas. However, due to carcinogenic effects, fumigation with propylene oxide (PPO) is not allowed in the processing of almonds.

2
Aonla

INTRODUCTION

Aonla or Indian gooseberry (*Emblica officinalis* Gaertn.) is an ancient, indigenous fruit of India. It is cultivated in India since the *Vedic* era and is associated with country's tradition, culture and heritage. Its importance to ancient Indian civilization can be well understood with the fact that it is known as *'Divya'* and *'Amrut'* or *'Amrit Phala'* in Sanskrit, which literally means fruit of heaven or nectar fruit. The Sanskrit name, *Amlaki*, translates as the Sustainer or The Fruit where the Goddess of Prosperity resides. In Hindu religious mythology the tree is worshipped as the Mother Earth as its fruits are considered to nurture humankind as they are very nourishing. It is considered as a wonder fruit for health conscious population. The popularity and nutritional value of *aonla* can be explained from the change of the western proverb "an apple a day keeps the doctor away" to "an *aonla* a day keeps the doctor at bay". *Aonla* is known by different names in different parts of the world for example, Emblic myrobalan, Indian gooseberry, malacca tree (English), *Aonla*, *amla* (India), *mak kham bom* (Lao/Thai), *melaka* (Malay), *zee phyu thee* (Myanmar), *ganlanshu* (Chinese), *kimalaka* (Indonesian). This crop is quite hardy, prolific bearer and is remunerative for the farmers having marginal land. Traditionally, *aonla* has been regarded as a crop of forest or household. However during the last two decades, substantial expansion in the area under *aonla* cultivation across the country, utilizing the wasteland, has been noticed. As a result, there has been increase in efficient utilization of natural resources, rehabilitation of wastelands and improved economic return to farmers coupled with enhanced employment.

COMPOSITION AND USES

Composition

Aonla is well known for its nutritional attributes. The edible *aonla* fruit tissue has 3 times the protein concentration and 160 times the ascorbic acid concentration of an apple. It is rich in polyphenols, tannins, minerals and is regarded as one of the richest source of vitamin C and its content of ascorbic acid is next to only to that of Barbados cherry (*Malpighia glabra* L.). The fruit also contains a high amount of salt, carbohydrates, phosphorus, calcium, iron and vitamins (Table 2.1).

Table 2.1: Nutritional Value of Fruit of *Aonla* (per 100g edible portion)

Edible Portion (per cent)	Moisture (per cent)	Protein (g)	Fat (g)	Minerals (g)	Fiber (g)
89	81.8	0.5	0.1	0.7	3.4

Carbohydrates (g)	Calcium (per cent)	Phosphorus (per cent)	Iron (mg)	Vitamin C (mg)	Nicotinic Acid (mg)
14.1	0.05	0.02	1.2	600	0.2

Uses

All parts of the plant including the fruits of *aonla* are medicinally rich and are used in the preparation of various *Ayurvedic* medicines. The medicinal properties of *aonla* have been mentioned in old *Ayurvedic* texts, such as *Charaksamhita* and *Sushrutsamhita*. *Aonla* fruits are used as an ingredient in more than 175 formulations of *Ayurveda*. The active ingredient that has significant pharmacological action in *aonla* is designated by Indian scientist as 'Phyllemblin'. Fruits are commercially used for preparation of *chayanprash* and *triphala*, which is used in the treatment of chronic dysentery, biliousness and other disorders. *Aonla* is effective in the treatment of dyspepsia, diabetes *etc.* Fruits are acrid, cooling, refrigerant, diuretic and laxative and dried fruits have been reported to be effective in cure of haemorrhages, diarrhea, dysentery, anemia, jaundice, peptic ulcer, dyspepsia and cough. The fruits have been found to have hypolipidaemic, antiathero-sclerotic, hepatoprotective, antioxidant, antimutagenic, cytoprotective, antitumour and antimicrobial properties. *Aonla* fruit ash contains chromium, zinc and copper. It is considered as adaptogenic, which improves immunity. The plant is consider an effective antiseptic for cleaning wounds and it is one of the many plant palliatives for snakebite and scorpion stinging. The leaves of *aonla* are use as a mouthwash and as a lotion for sore eyes. An ointment made from the oil obtained from the burnt seeds is used to cure skin infections by some rural communities. The leaves of *aonla* are employed for dyeing matting, bamboo wickerwork, silk and wool into brown colours. Grey and black colours are obtained when iron salts are used as mordants. Matting can be dyed dark colours with a decoction of the bark. The fruits are used to prepare a black ink and a hair dye. Immature fruits, bark and leaves are used for tanning in India and Thailand. Leaves and fruits are used for animal fodder, whereas leaves can also be applied as green manure. Although the wood may warp and split, it is used for the construction of furniture and implements and it is very durable when submerged in water. The wood is also suitable as firewood and produces charcoal of good quality.

Fruits have good demand in the industries for the preparation of various health care products also like hair oils, dyes, shampoos, face creams and toothpowders. The fruits are edible but rarely eaten fresh because they are astringent and sour; therefore, they are commonly used for preparation of pickles, candy, jelly, jam *etc*. The *aonla* preserve (*murabbas*) is one of the specialties of the Indian fruit-preservation industry. These days, several *aonla* based health beverages are available in the market.

Aonla also has religious importance as well in India as it has been regarded as the sacred tree under Hindu mythology. Under certain religious discourses, the worship of the tree is mandated and leaves are offered to the Hindu Gods and Goddesses. In Himachal Pradesh, *aonla* tree is worshipped in the month of *Kartik* as a mark of auspicious and chaste.

ORIGIN, HISTORY AND DISTRIBUTION

Its geographical area of origin ranges from the Northern India and Southern Himalayas of Nepal, to the South of the Indian sub continent, Bhutan, Bangladesh, Myanmar, Pakistan, Sri Lanka, Thailand and Indo-China to Southern China (Fujian, Guangdong, Guangxi, Sichuan, Taiwan, Yunnan), and Malesia (in Peninsular Malaysia, Singapore, Sumatra, Borneo, Java, the Lesser Sunda Islands and Ambon). The tree is found growing in the plains and sub-mountain on tracts all over the Indian subcontinent from 200 m and ascending to 1450 m in the Himalayas. Its natural habitat like other members of its family starts from Burma in the East and extends to Afghanistan in the West. Its exotic distribution includes Bermuda, Caribbean Islands, Cuba, Puerto Rico, the Philippines, Trinidad and Tobago, Central America, Panama, North America (USA; Florida, Hawaii, Indiana), Iran, Iraq and Japan. At present, *aonla* is widely distributed in south Asia and to a limited extent in the American regions. It is very rarely reported in the African continent.

In India, seedling trees are of common occurrence in the mixed deciduous dry forests, ascending from sea level (Western and Eastern ghats, Aravali and Vindhyan hills) to 1500 m asl, from North-West Himalayas (Jammu and Kashmir, Himachal Pradesh, Uttrakhand) to Eastern Himalayas in Assam, Meghalaya, Mizoram, Manipur and Tripura. The natural distribution of wild *aonla* is found on the Himalayas, Chota Nagpur, Bihar, Odisha, West Bengal, North Circars, Deccan, Karnataka and in Western Ghats. A quite distinct strain of *aonla* grows wild in the mid-hill regions of the Western Himalayas even at places which experience mild snowfall during winter months. It bears smaller fruits and is a very heavy cropper. The major difference between these wild trees and the large-fruited types cultivated in the plains is of winter-hardiness. Whereas the improved types are susceptible to frost injury, the wild *aonla* is not damaged. Domestication of *aonla* in India was first started in Varanasi district of Uttar Pradesh with the initiative of Maharaja (King) of Kashi (Varanasi). Later, a superior genotype 'Banarasi' was selected from the wild *aonla* trees available in large number in the nearby Vindhya hills. Authentic information on the cultivation of *aonla* dates back to 1881-82 in the Pratapgarh district of Uttar Pradesh. The ailing state owner of the district (King) was advised for regular consumption of *aonla* fruit in one way or other. As per information available, few *aonla* trees were introduced from Vanarasi and few from Gujarat. Those brought from Varanasi were named as Banarasi and those brought from Gujarat were known as Francis and later on as Hathijhool (because of its drooping branches). A seedling of Banarasi, with prolific bearing and flat fruits was named as Chakla and now it is known as Chakaiya. Shri W.P.A.R. Nagarajan and his son Shri Arun Nagrajan, Planters, Pattiveeranpatti are the main contributors in extension of *aonla* cultivation in Southern India.

Table 2.2: Major *Aonla* Growing Belts in India

State	Districts
Andhra Pradesh	Anantapur, Prakasam, Nellore, Medak, West Godavari, Nalgonda, Guntur, Rangareddy, Adilabad, East Godavari, Chittoor
Assam	Kamrup, Tinsukia, Nagaon, Jorhat, Lakhimpur, Goalpara
Bihar	Aurangabad, Gaya, Jamui
Chhattisgarh	Raipur, Baloda Bazar, Gariaband, Mahasamund, Dhamtari, Kabirdham (Kawardha), Baster (Jagdalpur), Kondagaon, Kanker, Korba, Raigarh, Koriya, Bilaspur, Janjgir-Champa, Rajnandgaon
Gujarat	Sabarkantha, Banaskantha, Bharuch, Vadodara
Haryana	Bhiwani, Gurgaon, Hisar, Sirsa, Mewat
H. P.	Shimla (Rampur, Kumarsain), Kullu (Anni, Nirmad), Mandi (Karsog, Gohar, Suder Nagar), Kangra (Dehra, Gopipur, Pragpur, Indora, Nurpur), Sirmour (Rajgarh, Nauradhar, Haripurdhar, Pacchad), Bilaspur (Ghumarwin, Jandutta), Una (Amb, Bangana, Gagret), Hamirpur (Nadaun, Bhoranj)
J&K	Kathua, Udhampur, Jammu, Samba, Reasi, Doda, Kishtwar, Rajouri
Jharkhand	Palamu, Latehar, Dhanbad, Girdih, Bokaro, Gumla, Hazaribagh, Ranchi
Karnataka	Bangaluru (Urban), Shimoga, Bidar, Koppal, Chamarajnagar, Mandya, Mysore
Madhya Pradesh	Satna, Panna, Seoni, Umaria, Katni, Jabalpur, Narsinghpur, Hoshangabad, Harda, Bhopal, Sehore, Raisen, Guna, Vidisha, Sagar, Damoh, Bhind, Morena, Sheopur, Gwalior, Shivpuri, Chhatarpur, Datia, Tikamgarh, Jhabua
Maharashtra	Solapur, Amravati, Akola, Washim, Satara, Osmanabad
Punjab	Gurdaspur, Hoshiarpur, Ropar
Rajasthan	Jaipur, Alwar, Ajmer, Baran, Kota, Jhalawar, Pali, Chittorgarh, Jodhpur, Nagaur,Sawai Madhopur, Tonk, Jalore, Karauli, Barmer, Banswara, Sriganganagar
Tamil Nadu	Coimbatore, Dindigul, Madurai, Salem, Sivaganga, Theni, Thoothukudi,Tirupur, Tirunelveli, Virudhunagar
Uttar Pradesh	Pratapgrah, Sultanpur, Allahabad, Mathura, Fatehpur, Faizabad, Jhansi,Lalitpur, Chitrakoot
Uttarakhand	Haridwar, Dehradun, Udham Singh Nagar
West Bengal	Purulia, Bankura, Birbhum, Pashchim (West) Midnapore

In India, commercial cultivation of *aonla* is common in Pratapgarh, Faizabad, Rai Bareilly, Varanasi, Jaunpur, Sultanpur, Kanpur, Azamgarh, Banda, Agra and Mathura districts of Uttar Pradesh. Its intensive plantation is being done in the salt affected areas of the state of Uttar Pradesh, including ravenous areas in Agra,

Mathura, Etawah, Fatehpur and semi arid tract of Bundelkhand. *Aonla* cultivation is also spreading rapidly in the semi arid regions of Maharashtra, Gujarat, Rajasthan, Andhra Pradesh, Karnataka, Tamil Nadu, Aravali ranges in Haryana and Kandi area in Punjab and Himachal Pradesh (Table 2.2).

TAXONOMY AND BOTANICAL DESCRIPTION

Taxonomy

The genus *Emblica* (earlier *Phyllanthus*) belongs to the family Euphorbiaceae and order Euphorbiales. *Phyllanthus* is derived from two Greek words: *phullon*, meaning a leaf, and *anthos*, a flower, referring to the bearing of flowers on the axils of leaves. The basic chromosome number of *aonla* is x=7. The cultivated form of *aonla* have been identified as tetraploid (2n=28).

Botany

Small to medium sized deciduous tree, sometimes up to 25 m tall but usually much shorter, up to 7.5 m; trunk often crooked and gnarled, up to 35 cm in diameter; bark thin, smooth, grey, peeling in patches, with numerous knobs. *Aonla* tree is characterized by phyllanthoid branching habit with two types of shoots. On the basis of growth characteristics, these have been characterized as long (indeterminate) and short (determinate) shoots. These are also referred as branch and branchlet. The indeterminate shoots are longer and continue to put new growth in the season. These shoots do not fall from the tree and also do not bear flowers, irrespective of period of their emergence. While on the other hand, determinate shoots appear on the nodes of indeterminate shoots and their number at each node may vary from 3 to 5 in different cultivars. These determinate shoots bear small sized (10-13 mm length, 2-3 mm width) leaves, arranged so closely that apparently appears to be a pinnately compound leaf. First few proximal nodes on the determinate shoots are barren (without leaves), which are reduced to dark brown scarious cataphylls. Succeeding nodes are with green but reduced leaves. Leaves alternate, distichous and densely crowded along lateral twigs, simple and entire, glabrous, sessile; stipules triangular; blade narrowly oblong, 5–25 mm × 1–5 mm, rounded and more or less oblique at base, acute or obtuse and mucronate at apex. Flowers fascicled in axils of leaves or fallen leaves, unisexual, the male flowers numerous at base of young twigs, the female flowers solitary and higher up the twig; male flowers pedicellate, with 6 pale green perianth lobes 1.5–2.5 mm long and 3 entirely connate stamens; female flowers sessile, with 6 somewhat larger perianth lobes, a cup-like disk, and a superior 3-celled ovary crowned by 3 styles, connate for more than half of their length and deeply bifid at apex. Fruit a depressed globose drupe up to 4 cm in diameter, pale green changing to yellow when mature; stone with 3 slightly dehiscent compartments, each usually containing 2 seeds. Seeds trigonous, 4–5 mm × 2–3 mm.

FLOWERING AND FRUIT DEVELOPMENT

Aonla trees are marginally leafless in the dry season, usually during winters. Shoot growth may occur during spring *i.e.* February-March and continue to May.

However, leaves develop completely only after fruit set. The flowers are produced at the leaf base of the newly developed leaves of determinate shoots. *Aonla* flowers open in the late afternoon, mainly from 6.00-7.00 p.m. and dehiscence of anthers occurs soon, or about 10-15 minutes after anthesis. They last only 2-3 days. The female flowers take about 72 hours to open completely and the stigmatic receptivity occurs between the 2nd and 5th day of anthesis. In Northern India, flowering takes place from March to May. In Chennai, Tamil Nadu and Western states the trees bloom in June-July and again in February-March, the second flowering producing only a small crop. Flowers are pollinated by insects, particularly by *Melipona* spp., *Apis* spp. and coccinelid beetles. Common visiting hours of honey bees are late evening and early morning. Flowering and fruit set takes place in spring to early summer and soon after the fruits enter dormancy throughout summer season and till monsoon. Fertilization takes place within 36 hours after pollination. However, the zygote and the endosperm nuclei remain in the uni-nucleate stage for periods of 120 days *i.e.*, until the rainy season. It is suggested that the auxins translocated from the shoot-tip to the fruit cause dormancy of fruits by a mechanism similar to that operative in apical dominance.

Only few female flowers of *aonla* produce fruits. Initial fruit set varies from 12-18 per cent. Initial low fruit set is due to lack of female flowers and/or lack of pollinators, or both. *Aonla* is an out-crossing species with some level of selfing. Fruit set is very low following selfing, and in self-pollinated plants most of the fruits drop before attaining maturity. Artificial cross pollination increase initial fruit set as well as final retention. Flower and fruit drop in *aonla* is divided in three stages; (i) the 'first drop', which accounts for 70 per cent shedding of flowers in the first 3 weeks after onset of blooming is attributed to lack of pollination; (ii) the 'second drop' occurs from June to September, consists of drop of young fruit lets at the time of dormancy break may be due to lack of pollination and fertilization and (iii) the 'third drop' is spread over a period of rapid growth from August to October may be due to lack of auxins *i.e.*, embryological and physiological factors.

After fruit set, the fruits remain dormant through summer and start growing with the onset of monsoon and are ready for harvest in November to January, depending on the location. The fruit takes about 5-7 months to mature, usually in December to March. Therefore, this crop is one of the most ideal crop for arid conditions.

SOIL AND CLIMATIC REQUIREMENTS

Soil

Aonla can be grown in a wide range of soils but well drained deep sandy loam soil having good water holding capacity is considered the best. Heavy soils or high water table areas are not suited for cultivation. In sandy soils, *aonla* plants can be successfully grown if irrigation facility is available. Calcareous soils are usually not suitable for its growth. However, if some amendments are used, *aonla* plantation can be raised on saline and sodic wastelands. *Aonla* has good tolerance to salinity, alkalinity, sodicity and also has ability to withstand drought conditions.

Climate

Aonla is cultivated in arid and semi-arid parts of India in the States of Rajasthan, UP, Haryana, Maharashtra, Gujarat and Tamil Nadu. It can be successfully grown in hot arid climate. The trees shed their leaves and become dormant during winters. Heavy frost during winter is not conducive to its cultivation. A mature *aonla* tree can tolerate freezing as well as high temperature of 46°C. Though it can bear temperature upto 46 °C during summer months, but the temperature should not be high at the time of flowering. It affects fruit setting and may sometime lead to complete unfruitfulness, if there are hot and dry winds too. Warm temperature seems conducive for the initiation of floral buds. Ample humidity is essential for initiation of fruit growth of dormant fruitlets during July–August. Annual rainfall of 630-800 mm is ideal for its growth. Dry spells result in heavy dropping and delay in initiation of fruit growth.

IMPORTANT VARIETIES

Cultivars like Chakaiya, Banarasi and Francis have been grown in India. Shy bearing in Banarasi and predominance of internal necrosis in Francis are, however, serious demerits of these two cultivars. Now, some promising selections have been made from the great genetic variability found in Vindhayan and Arawali region of India. Salient characters of *aonla* cultivars are given below:

- ☆ **Chakaiya**: This is a late maturing cultivar with small to medium fruit size (30-35 g). Flattened fruit apex, deep stem end cavity and leveled style end are some typical characters of this cultivar. Some fibre is present in the fruit pulp which favours longer shelf life.

- ☆ **Kanchan (Narendra *Aonla* 4)**: This is a late maturing cultivar selected from Chakaiya seedlings. Owing to more number of female flowers, it gives higher fruit yield. Fruits are small to medium in size with light green colour and are suitable for making pickle.

- ☆ **Krishna (Narendra *Aonla* 5)**: This is a high yielding early maturing cultivar selected from population of Banarsi trees. The fruit is large (40-50 g) with less fibrous pulp. Fruits can be used for the preparation of preserve, candy and juices.

- ☆ **Narendra *Aonla* 6**: This is also a selection from Chakaiya. Fruit is small to medium in size and oval-round in shape and has less fibrous and semi-hard flesh suitable for preserve making.

- ☆ **Narendra *Aonla* 7**: This is a mid-season, necrosis-free cultivar selected from seedling population of Francis. The tree is dwarf and has heavy bearing capacity. The fruits are large (40-50 g), round, smooth with light yellow pulp generally free from fibre. The fruits are suitable for making preserve and other processed products.

- ☆ **Narendra *Aonla* 9**: It is a seedling selection from open pollinated seedling of cultivar Banarasi. Tree is tall with semi-spreading growth habit. Fruits are large, flat end round with smooth, semi-translucent, light green peel colour.

☆ **Narendra *Aonla* 10:** It has originated from a seedling selection developed from the cv. Banarasi and NA-10 is locally known in India as Agra Bold. The tree is semi tall and semi spreading. It is an earliest maturing genotype of *aonla* and bears moderately. Fruit moderate in size, flattened round; skin rough, colour yellowish green with pink tinge. Fruits are mildly susceptible to necrosis. It has tendency to bear in alternate years.

☆ **Narendra *Aonla* 20:** This is clone of Chakaiya. Trees are spreading in nature, medium to tall, fruits are bold and shining. It has very less fiber content.

☆ **Goma Aishwariya:** It is a selection from NA-7 and has been released by Central Horticultural Experiment Station, Godhra. It is an early and drought tolerant. It has low fiber content and is suitable for processing and export.

☆ **Lakshmi-52:** A seedling selection of Francis, Lakshmi-52, was identified in the village Bhadausi, Garwara, district Pratapgrah, Uttar Pradesh. Its tree growth is semi-erect and branches are semi-spreading which do not droop. The fruit matures during November-December and is free from necrosis having a yield potential of 2-2.5 q/tree (10 years onwards). The fruit weighs 40-60 g containing 90.4 per cent pulp, 82.5 per cent moisture, 11° Brix TSS, 1.7 per cent acids, 512 mg/100g vitamin C and 3.3 per cent tannins. The fruits have great processing potential due to bigger size and high nutritive quality.

☆ **Balwant:** This is an early maturing cultivar selected from Banarasi population at Agra. The plants are tolerant to sodic soil. Fruits are large (40-45 g/fruit) with round shape and flattened stem end. Pulp has less fibre. Its fruits are suitable for the preparation of products.

☆ **BSR-1:** It is a high-yielding (155 kg/tree, 42 952 kg/ha), self-fruitful, and late-maturing cultivar of *aonla* selected from a large number of germplasm from Bhavanisagar, Tamil Nadu, India. Its fruits are flattened at the base and round at the apex, with an average weight of 27.30 g/fruit. The fruits contain high total soluble solids (18.1 degrees brix) and vitamin C (620 mg/100 g of flesh), low phenol (29.75 mg/g of flesh), and high crude fibre content (4.31 per cent).

Besides, Anand 1, Anand 2 and Anand 3 have been selected as promising strains in Gujarat. Pant Aonla 1 has been released from GBPUA&T for *tarai* and *bhabhar* areas of Uttarakhand.

PLANT PROPAGATION

Aonla has long been raised through seeds and budding. From seed propagation, there is prolonged juvenility and wide variability. Seed propagation is not advisable except to raise rootstocks for vegetative propagation. For propagation through seeds, ripened fruits should be harvested from vigorous, heavily and regularly bearing trees. To ensure high quality seedling production, only fully matured fruits should be harvested. Only medium-sized fruits with good shape and high specific gravity should be selected as they give a higher germination percentage and rate of growth. Harvested fruits are usually depulped and sun dried. Later, seeds are extracted after breaking the stone and are sown in polythene tubes (40 x 15 cm) filled with mixture of FYM + clay + sand in 1: 1: 1 proportion. *Aonla* seeds remain viable for 3-4 years. February-March is the best time for sowing seeds in the nursery. Sowing can also be done during July-August in 15-20 cm raised nursery beds of 3 x 1 m size. Before sowing, the seeds should be treated with fungicide (Bavistin @ 2 g/kg seed). The seeds germinate within 8-10 days. Seed germination in *aonla* is epigeous. Newly germinated seedlings are shifted to flat beds at 20 x 20cm distance after one month so that they become ready for budding during next July. Spray of 0.3 per cent Dithane M-45 has been found to protect the plants from diseases in the nursery.

Aonla is commercially propagated by inarching/budding/grafting on seedling rootstock; however, star gooseberry (*Phyllanthus acidus*) as rootstock has also been attempted for *aonla* propagation. Patch and modified ring budding on rootstocks of pencil thickness are the most successful methods for *aonla* propagation. July-August is the best time for budding in *aonla*. Budding period can be extended up to September-October in North India but in that case bud union occurs but sprouting would commence during February-March. Plants can also be propagated by softwood grafting and inarching. Propagation of *aonla* in polybag, polytube, "root trainer" or *in situ* orchards needs to be standardized and commercialized. Old seedling trees can also be converted into improved cultivars by top working. T budding also gives high success rates in rejuvenation of old orchards and inferior trees. Softwood grafting can also be used successfully.

A high percentage of rooting (87 per cent) has been reported from semi-hardwood cuttings collected from the middle portions of invigorated shoots of young trees and planted in beds with 1500 ppm of GA_3 at a temperature about 33 °C. It is also important to select branches with good number of female flowers as branches with more number of male flowers are likely to bear a large number of male flowers later in their life time.

PLANTING AND ORCHARD ESTABLISHMENT

Grafted or budded *aonla* plants are planted 7–10m apart under square system during July–August or February. Pits of 1 x 1 x 1 m size are dug during May-June and are filled after 15 days with FYM and top soil mixture in 1: 1 ratio. To protect the plants from termite, chlorpyriphos can be applied in each pit. Plantation of *aonla* saplings is done during July-August. If assured irrigation and protection facilities are available, planting can be done during February and October. Since *aonla* has self incompatibility, ten per cent population (15 plants/ha) should consist of pollinizer trees to ensure good fruiting. The best combination is NA 6 and NA 7 or Kanchan. In any case, planting of more than two cultivars in a block would take care of the pollinizer requirement. In *aonla* orchards, *ber*, guava and lemon are ideal filler plants. These are planted in the centre of each square of *aonla* plants. Hedge-row planting is also being tried keeping line-to-line distance of 8 m, while plant-to-plant distance is reduced to 4-5 m. Besides, triangular planting system is also becoming common practice for better utilization of space and solar radiation.

Plantation of seedling trees on the borders to serve as wind break also provides seeds for raising rootstock plants.

For shelter belt plantation, of 2-3 staggard rows of fast and growing tall tree species should be provided. *Sheesham (Dalbergia sissoo), neem, gonda (Cordia myxa), ardu (Ailanthus excelsa), etc.* are good species for this purpose in arid region. Temporary shelter can be provided by planting fast going bushes of castor, *dhaincha, etc.*

INTERCULTURAL OPERATIONS

Training and Pruning

Due to the brittle nature of the wood, fruit bearing branches of *aonla* frequently break off during heavy fruiting. This suggests that the plants should be trained to develop canopy at manageble heights with strong foundation framework. To provide a desirable frame, the trees are trained at initial stages. *Aonla* plant should be encouraged to develop a medium-headed tree. Plants should be trained to modified central leader system with a single stem up to the height of 0.75-1.0 m from the ground. Two to four branches with wide crotch angle, appearing in the opposite directions should be encouraged in early years. The unwanted branches are pinched off during March–April. In the subsequent years, 4-6 branches should be allowed to develop. As per growth habit, shedding of all determinate shoots encourages new growth in coming season. Regular pruning of a bearing *aonla* tree is not required; however, it should be done only after harvest of the crop. Further, dead, infested, broken, weak or overlapping branches should be removed regularly.

Nutrition Management

Judicious application of manures and fertilizers increases productivity of *aonla* trees. The quantity of manures and fertilizers depends on the fertility of the soil. Tissue analysis of leaf samples taken from the middle portion of 3-4 months old intermediate shoots gives fair idea of nutrient status of the tree. Application of 10 kg FYM, 100 g N, 50 g P_2O_5 and 50 g K_2O per plant per year has been considered sufficient in arid region soils. The doses should be increased every year by the same quantity up to 10 years and then stabilized. The best time for manure and fertilizer application is January-February and June-July when the plants are in floriferous and fruit development stages, respectively. Full dose of FYM and phosphorus and half dose of nitrogen and potash are applied during spring season and the remaining half dose is applied during the rainy season.

Deficiencies of micronutrients cause poor growth in *aonla*. Application of 250-500 g zinc sulphate/plant is found to be beneficial in sandy soils of arid region. Three foliar sprays with 0.06 per cent borax at 10-15 days interval has been found to reduce the malady of necrosis.

Water Management

Aonla is a drought hardy plant but responds well to irrigation. Plants in early years of establishment need assured irrigation in the absence of which, the growth is poor. Frequent irrigations at short intervals of 3-4 days are done for about one month after planting to ensure essential establishment of *aonla* plants in light soils of arid region. Afterwards, irrigation interval can be increased to 25-30 days. Since *aonla* is a deciduous fruit tree, in which flowering and fruit set takes place in spring following which the fruits enter dormancy throughout summer season and till monsoon. Hence, plants do not require much irrigation during summer, when most crops would require it frequently. However, light irrigations during March-July at 10-15 days interval are most beneficial. Drip irrigation on 20 per cent wetted area basis on alternate days has been found to improve growth and yield of *aonla*. In water scarcity areas, pitcher irrigation can also be successfully utilized. In a normal monsoon year, irrigation during rainy season may not be required but in late maturing cultivars, irrigation is essential during September-October to avoid moisture stress. During winter, light irrigations are done particularly before the suspected time of frost occurrence to save the young plants from damage.

Weeding

Aonla plants need weeding, especially at the early stages, as the canopy has not developed properly. Regular weeding of tree basin should be done. Hoeing in the basin around the tree trunk also breaks soil crust and ensures good root development.

Mulching

Mulching with locally available materials such as paddy straw, local grasses, banana leaves or sugarcane trash @20 kg per basin can be done to check moisture loss during summer. Organic mulches should be evenly distributed with a covering of 10-15 cm depth. During rainy season, the basin should be cleared and the partly decomposed mulched material should be racked in the basin for its further incorporation in the soil. Mulching with organic materials, over a number of years, shall be helpful in improving the organic-matter content, infiltration rate, and restricting the upward movement of soluble salts and thus escaping their toxicity menace in salt-affected soils. In case of polyethylene mulching, 100-micron thick LLDPE/LDPE mulch film is used. The size of mulch should be increased as the tree grows.

Intercropping

Aonla tree starts fruit production after 3-4 years of planting and the commercial yield is achieved after 8-10 years. During the initial years, inter space between tree rows can be utilized for growing intercrops which not only provide additional income but also improve the physico-chemical properties of the soil and check weed. Drought hardy leguminous, solanaceous and cucurbitaceous vegetables are good intercrops in rainfed as well as irrigated *aonla* orchards. Some models are: *aonla* + *ber* (2-tier), *aonla* + guava (2-tier), *aonla* + *ber* + *phalsa* (3-tier), *aonla* + *dhaincha* + wheat or barely, *aonla* + *dhaincha* + onion/garlic/fenugreek or brinjal and *aonla* + *dhaincha* + German chamomile (3-tier). Moth bean as *kharif* season crop can also be taken in rotation with *rabi* crops like fenugreek, chickpea, mustard and cumin.

PLANT PROTECTION

For good growth and production, *aonla* trees should be protected from diseases and pests. Bark eating caterpillar, shoot capsule borer and termite are the main pests and rust is a serious disease of *aonla*. Clean cultivation and maintenance of health and vigour of trees, integrated pest and disease management, destruction of infected parts, appropriate nutrients, water and weed management are suggested as good agricultural practices in *aonla* cultivation.

Major Insect-Pests and their Management

Bark Eating Caterpillar (*Indarbela tetraonis* Moore)

Bark eating caterpillar feeds on the stem and branches and making holes. The attack of this pest may be identified by the presence of irregular tunnels and patches covered with silken-web consisting of excreta and chewed up wood particles, on the shoots, branches, and trunk. For its management, avoid overcrowding of branches, keep the orchards clean to prevent infestation, periodically monitoring for early detection, killing the caterpillars mechanically by inserting iron spike in shelter holes made by these borers at early stage of infestation, plugging the holes with cotton swab soaked in kerosene/petrol/Monocrotophos @0.05 per cent followed by external plastering with moist clay.

Shoot Gall Maker (*Betousa stylophora* Swinhoe)

This pest attacks primarily on nursery plants and old bearing trees. The insect is active from June to December and causes gall formation on stem and shoot. The larva of this moth tunnels in the apical portion of the shoot and infested portion bulge into gall, which check growth of the shoots. When larva is active, reddish frass is extruded through a hole at one end. Fresh galls are generally formed from June to August.

Overcrowding of branches should be discouraged so as to avoid development of congenial environment for pest resurgence. Removal and destruction of the affected portions followed by 2-3 sprays of 0.05 per cent monocrotophos at 15 days interval provide effective control.

Mealy Bug (*Nipaecoccus vestatar* Newstead)

Incidence of mealy bugs is noticed from March to July with the peak population in April-May. Eggs are laid in ovisacs formed by the females and nymphs get settled on plant parts (leaf, shoot and inflorescence) and suck the sap. Fully matured larvae jump out of the fruits, drop down in the soil and pupate there. Full life cycle is completed in 20 to 28 days, with the last larval stage diapausing within the soil.

For its management, deep ploughing of the orchard after harvesting for exposing the diapausing larvae and spray of 0.2 per cent carbaryl or 0.04 per cent monocrotophos at the beginning of the fruiting is recommended.

Termites

Termites cause serious damage in light soils of arid region. Mixing of chlorpyriphos @ 10 ml/10 liters/plant at the time of pit filling and later application of 5-10 ml chlorphyriphos with irrigation water give effective control. Painting of the trunk up to 30 cm height from ground level with slaked lime + chlorpyriphos mixture checks crawling of termites upwards on the tree trunk.

Major Diseases and their Management

Among the diseases, rust is the most important disease of *aonla*. It causes fruit and leaf infection and causes significant loss. The other important disease problems are fruit rots caused by number of fungi.

Rust (*Ravenelia emblicae* Styd.)

Rust is a serious disease of *aonla* infecting the leaves and fruits and cause severe economic losses. On fruits, initially few black pustules appear, which later develop in a ring. The pustules join together and cover big area of the fruit. The black spores are exposed after rupturing a papery covering. Fruits give dirty look and lose its market value. On leaves, pinkish brown pustules develop which may be arranged in groups or scattered as isolated pustule.

It can be controlled by 3-4 sprays of Blitox-50 or Diathane M-45 at 15 days interval.

Fruit Rot

There are number of rot such as soft fruit rot, *Pestalatia* fruit rot, *Cladosporium* rot of fruit and *Aspergillus aculeatus* and *Alternaria* and *Phoma* fruit rots. For their management, one spray of carbendazim 0.1 per cent 15 days prior to fruit harvest, careful harvesting to avoid any injury to the fruits, storage of fruits in clean containers, adoption of proper sanitary measures during storage and transit and treatment of fruits with borax or sodium chloride may be done.

Physiological Disorders and their Management

Internal necrosis a physiological disorder has been observed in *aonla* fruits. Francis variety is highly susceptible followed by *Banarasi*. Incidence initiates with browning of mesocarp which extends towards the epicarp resulting into brownish-black appearance of flesh. Infection has not been noticed on other *aonla* cultivars like *Chakaiya*, NA6 and NA7. For its management combined spray of zinc sulphate (0.4 per cent) + copper sulphate (0.4 per cent) and borax (0.4 per cent) during September-October has been found effective.

HARVESTING AND YIELD

Vegetatively propagated trees start bearing 3-4 years after planting while the seedling trees take more than 8 years to start flowering and fruiting. In vegetatively propagated trees, commercial yield of 150-200 kg starts after 8-10 years. In Northern India, *aonla* trees flower during March-April and fruits are available from November to January. In Southern and coastal region, *aonla* trees flower twice, therefore, fruits are available for most of the parts of a year. Year round availability of *aonla* fruit can be realized by growing cultivars with varying maturity groups under different geographical regions, manipulation of agro-techniques, storage and their efficient transportation. The fruits should be harvested in the morning or late in the evening and collected in plastic trays. In any case, fruits should not be allowed to fall on the ground as the injured fruits cause spoilage to other healthy fruits as well during packaging, transit and storage. Maturity of fruit is indicated by the presence of shining green colour and reduction in acridity and increase in vitamin C content. Change in seed colour from creamy-white to brown is also an indication of fruit maturity. Fully developed fruits are harvested. In general, harvestings should be carried out 2-4 times for complete harvesting of fruits on a tree. This helps in size gain of the fruits, which set later during their growth. Delay in harvesting results in heavy dropping of fruits particularly in *Banarasi* and Francis. *Aonla* cultivars have been classified in to three groups, based on their maturity, *i.e.* early, mid-season and late (Table 2.3).

Table 2.3: Maturity Season of *Aonla* Cultivars

Early	Mid-season	Late
Banarasi	Francis	Kanchan
Krishna	NA-7	Chakaiya
NA-10	NA-6	–

POSTHARVEST TECHNOLOGY

After harvest, the fruits are graded into 'A' 'B' and 'C' grades. The fruits are used after processing. 'A' grade fruits are used for preparation of preserve (*murabba*), candy and pickle, 'B' grade fruits are used for the preparation of *chyavanprash, triphala,* jam, syrup, squash, *etc.* and the small size 'C' grade fruits are used for making products like shreds, dried segments, powder and shampoo making. Prematurely dropped fruits can also be used for the preparation of mouth freshening and digestive products in combination with spices such as *harar, bahera,* asophoetida, *etc. Aonla* fruit is perishable in nature and has a short shelf life of 5–6 days at ambient conditions as fruit is sensitive to bruises, browning, desiccation and various postharvest diseases. Value addition through processing would be the only effective tool for economic utilization of increased production of *aonla* in the future. Processing not only reduces the postharvest losses but also provides higher returns to the growers.

3
Apricot

INTRODUCTION

Apricot is an attractive, delicious and nutritious temperate fruit. In India, it ranks next to plum among stone fruits in area and production. It is drought resistant, salt tolerant and hardy plant and thus requires less care and management than other temperate fruits.

ORIGIN, HISTORY AND DISTRIBUTION

The origin of the apricot is disputed. It was known in Armenia during ancient times, and has been cultivated there for so long, it is often thought to have originated there. Its scientific name *Prunus armeniaca* (Armenian plum) derives from that assumption only. Despite the great number of varieties of apricots that are grown in Armenia today (about 50), according to the Soviet botanist Nikolai Vavilov, its center of origin would be the Chinese region, where the domestication of apricot would have taken place. Other sources say that the apricot was first cultivated in India in about 3000 BC. However, now it is established that the primary centre of origin of apricot is Western China and secondary centre of origin is Western Asia. Apricots were first cultivated in western Asia a few centuries BC and introduced in Europe during Roman Era. Wild Indian apricots like zardalu and chulli are considered indigenous to Western Himalayan ranges of India. Today, apricot cultivation has spread to all parts of the globe with climates that support it, *i.e.,* apricots are being grown in temperate climate of all the continents of the world, Asia and Europe being the largest producers. Major apricot producing countries are Turkey, Iran Uzbekistan, Algeria, Italy, Pakistan, France, Morocco, Spain and Egypt (Table 3.1). Apricots are also grown commercially in Greece, Hungary, Australia,

Morocco, Syria, Iran, Afghanistan, Bulgaria, Rumania, China, Iraq, Israel and India In India, apricots are grown in J&K, Himachal Pradesh, Uttrakhand and to some extent in NE states.

Table 3.1: Apricot Producing Top-10 Countries of the World

Country	Production (Tonnes)	Country	Production (MT)
Turkey	795768	Greece	90200
Iran (Islamic Republic of)	460000	Japan	90000
Uzbekistan	365000	Afghanistan	83500
Algeria	269308	Syrian Arab Republic	72000
Italy	247146	Ukraine	62900
Pakistan	192500	Russian Federation	57000
France	189711	United States of America	55157
Morocco	122405	South Africa	52504
Spain	119400	China, mainland	50000
Egypt	98772	Turkmenistan	36091

Source: www. http://faostat.fao.org

COMPOSITION AND USES

Composition

It is a rich source of β-carotene, carbohydrates, proteins, and phosphorus and niacin than many other fruits. Fruit pulp contains 1 per cent pectin as calcium pectate (Table 3.2). The distinct aroma of the fruit is due to the presence of volatile compounds like benzylaldehyde, linderol, 4-terpineol and 2-phynylethanol.

Table 3.2: Nutritional Value of Apricot Fruit per 100 g

Attribute	Contents	
	Raw Apricots	Dried Apricots
Energy	48 kcal	241 kcal
Carbohydrates	11 g	63 g
Dietary fiber	2 g	7 g
Fat	0.4 g	0.6 g
Protein	1.4 g	0.5 g
β-carotene	1094 μg	2163 μg
Calcium	13 mg	55 mg
Phosphorus	23 mg	71 mg
Potassium	259 mg	1162 mg

Uses

Apricots are mainly used as table fruit. Although fruits can be canned, candied, frozen and dried. A number of processed products like jam, nectar, *papad*, leather *etc.*, can also be prepared from it. They are also used for making liquor in some countries. Apricot kernels yield good quality cooking oil. Consumption of apricots is beneficial in preventing age-related Macular Degeneration. Apricots are also a rich source of antioxidants, which are required by our body to aid the body's natural functioning. Dried and fresh apricots are one of the best sources of soluble dietary fiber, that dissolves easily in the body and thereby help in absorbing essential nutrients and reducing the LDL or bad cholesterol content in the body, thus lowering your risk of developing cardiovascular diseases. In Europe, apricots were long considered an aphrodisiac, and were used in this context in William Shakespeare's A Mid-summer Night's Dream. Apricot juice, topically applied, has tonic properties.

TAXONOMY AND BOTANICAL DESCRIPTIONS

Taxonomy

Apricot belongs to family Rosaceae, sub family Prunoideae, sub-genus Prunophora, section Armeniaca and genus *Prunus*. The domesticated apricot is *Prunus armeniaca* L. with basic chromosome number 8 and somatic number 16. Usually, an apricot tree is from the tree species *Prunus armeniaca*, but the species such as *Prunus brigantina*, *Prunus mandshurica*, *Prunus mume*, and *Prunus sibirica* are closely related, have similar fruit, and are also called apricots. Apricots hybridize with plums to produce plumcots. *Prunus dasycarpa* is a plumcot resembling purple apricot produced naturally by the cross of apricot (*Prunus armeniaca*) and myrobalan plum (*P. cerasifera*). Apricot–plum hybrids are called as plumcots, apriplums, pluots, or apriums in different parts of the world. A 'Plumcot' is 50 per cent plum, 50 per cent apricot; an 'Aprium' is 75 per cent apricot, 25 per cent plum; and the most popular hybrid, the 'Pluot' is 75 per cent plum, 25 per cent apricot.

Several other species of apricot are also cultivated but primarily for rootstock or ornamental purposes. For example, *P. ansu* is adapted to humid conditions, and mostly use as rootstock. *P. mume* is grown in Japan and its fruits are used for making pickle and liquor. *P. sibirica* can tolerate as low as -50°C temperature and bears almond like inedible fruits.

Botany

The apricots are small-to-medium sized trees, usually 8–12 m tall, with a dense, and spreading canopy. The leaves are ovate with a rounded base, a pointed tip and a finely serrated margin. The flowers have five white to pinkish petals which appear singly or in pairs in early spring before the leaves. The fruit is a drupe and is similar to a small peach. The fruit is yellow to orange in colur, often tinged red on the side most exposed to the sun. Fruits may be smooth (glabrous) or velvety with very short hairs (pubescent). The flesh is usually firm and not very juicy. Its taste can range from sweet to tart. The single seed is enclosed in a hard, stony shell, called as 'stone'.

SOIL AND CLIMATIC REQUIREMENTS

Soil

Apricots are quite hardy to grow and can be grown in all types of soils. However, it grows well in deep fertile clay loam and well drained soils. It can tolerate slight drought or slight waterlogged conditions. Similarly, it can also tolerate high lime content of the soil.

Climate

The apricots are more specific in climatic requirements but by and large its climatic requirements are similar to peach. It needs slightly more winter chilling (800-1000 chilling hours). Apricot can be successfully grown at an altitude between 900 and 2,000m above mean sea level. White-fleshed, sweet kernelled apricots require cooler climate and are grown in dry temperate region up to 3,000 m above mean sea level, whereas yellow fleshed, bitter-kernelled ones thrive better under the warmer climate of mid hills (900-1,500 m) the long cool winter (300-900 chilling hours below 7°C), and frost free and warm spring are favourable for good fruiting. Average summer temperature (16.6°-32.2°C) is suitable for better growth and quality fruit production. The sites located in North Eastern India at lower elevations and on South-Western at higher elevations are suitable for its cultivation. Spring frost causes extensive damage to the blossoms which are killed when temperature falls below 4°C. Apricots thrive better under low humidity as high humid conditions in summer increase the incidence of brown rot. An annual rainfall of about 50-100 cm well distributed throughout the season is good for normal growth and fruiting.

IMPORTANT VARIETIES

For commercial cultivation, the selection of cultivar is of paramount importance and crucial because cultivars have only limited climatic adaptability and a particular cultivar can't be grown on large tract in the hills. The most important varieties of apricot are New Castle, Blenheim, Royal, Tilton, Early Montgamet and Moorpark. The following varieties of apricot have been recommended for different states (Table 3.3).

The varieties like Shakarpara, Suffeda, Charmagaz, Afghanistan, Parchinar have the sweet kernels like almonds. Benazeer is a low chilling apricot which can be grown even in mid subtropical areas of Punjab.

In India, 3 hybrids have also been developed at Regional Horticulture Research Station, Chaubatia (Uttrakhand). These are Chaubatia Alankar, Chaubatia Kesar and Chaubatia Madhu, which usually mature by the last week of May or 1st week of June.

Chief characteristics of some of the important varieties are hereunder.

☆ **Kaisha:** Its trees are vigourous and spreading, and bear medium sized fruits with roundish flattened shape and prominent suture, skin pale lemon yellow with red blush, free stone and early season maturity.

Table 3.3: Recommended Apricot Varieties Recommended for different States of India

State	Districts
Himachal Pradesh	
Mid hills	New Castle, Early Shipley and Shakarpara
High hills	Kaisha, Nugget, Royal, Suffaida, Charmagaz and Nari
Dry temperate	Charmagaz, Safaida, Shakarpara and Kaisha
Uttarakhand	Charmgaz, Kaisha, Moorpark, Turkey, Ambroise, Early Shipley, Chaubattia Alankar, Chaubattia Madhu, Chaubattia Kesri and Bebeco
Jammu and Kashmir	
Ladakh	Halman, Rakchakarpa, Tokpopa, Margulam, Narmu and Khante
Kashmir	Turkey, Australian, Charmagaz, Rogan and Shakarapara
New promising varieties for midhills	Early maturing - Baiti, Beladi Late-maturing - Farmingdale, Alfred

☆ **New Castle:** Its trees are vigourous and spreading. The fruits are medium to large size with roundish shape, skin is lemon/barium yellow in colour and early season maturity.

☆ **Australian:** It ripens in end-July to late-August. Fruit size is extra large and highest among the cold arid cultivars. Fruits are round in shape with acute apex, structure distinct, mid flavour and medium sweet and acidic. Not liked for table purpose and good for processing. It is highly suitable for Leh and Ladakh region (J&K) of our country.

☆ **Charmagz:** It is a self incompatible cultivar and needs a pollinizer. The fruits are medium in size and roundish flat in shape. The skin is straw yellow with a light yellow flesh which is very sweet and highly flavoured. It is suitable for dessert and drying purposes. Highly suitable for dry temperate zones such as Lahaul and Spiti (H.P.) and Leh and Ladakh region (J&K) of our country.

☆ **Harcot:** It is an early to mid-season maturity. Its trees are upright to spreading and vigourous bearing medium to large with roundish heart shaped fruits. Peel is yellow-orange with red blush, having sweet kernel.

☆ **Turkey:** It is a mid-season variety. Its trees are vigourous and spreading. Fruits are medium in size and almost round shape, with deep yellow peel and brownish orange dots, free stone and sweet kernel.

☆ **Halman:** Spreading and vigourous tree, fruits are large with roundish is shape, skin deep yellow golden in colour, sweet kernel, suitable for drying and early to mid-season maturity in cold arid zones such as Lahaul and Spiti (H.P.) and Leh and Ladakh region (J&K) of our country.

☆ **Rokchey Karpo:** It is an early cultivar, ripens in end July to mid August. Fruits are medium to large in size, round with compressed pedicel end. Pulp light pale, juicy, sweet and mild acidic with pleasant flavour *etc.*

☆ **Tokpopa:** The fruits are ripe late and available in the month of August. Fruits are medium in size, round in shape, compressed with smooth skin, dull yellow in colour, juicy and acidic to sweet in taste. Highly suitable for Leh region (J&K) of our country.

☆ **Rogan:** Fruits are small, highly juicy, round in shape, glossy skin straw yellow, very soft, juicy and slightly acidic sweet pulp. It bears smallest fruit among all cultivated varieties.

☆ **Safaida:** Fruits are large in size, round in shape, skin is glossy, smooth, light yellow in colour, flesh soft maize yellow in colour, very sweet, less acidic with pleasing flavour. Highly suitable for Leh region (J&K) of our country.

☆ **Nari:** It is a late ripening variety available after mid August. Fruits are medium in size, oblong in shape, elliptical with truncate base, skin greenish to light yellow with red and blush towards the sun exposed surface. Pulp is light yellow in colour, sweet/acid ratio is good and pleasant flavour.

☆ **Shakar Para:** Fruits are sweet with pleasant aroma. It can be used for table purposes. Ripening period is late July-mid August. Fruits are medium in size, round in shape, skin is glossy, creamish yellow with rosy blush, pulp soft, light yellow in colour, sweet and less acidic and pleasant flavour. Highly suitable for dry temperate zones such as Lahaul and Spiti (H.P.) and Leh and Ladakh region (J&K) of our country.

☆ **CITH Apricot-1:** It is a self fertile and mid-season blooming, fruits are very large (79 g), round symmetrical and smooth distal end, yellowish orange and reddish blemishes, early maturing and tolerant to major pests and diseases.

☆ **CITH Apricot-2:** It is a self fertile and early to mid-season blooming type, fruits are very large in size, oblate, asymmetrical with slightly pointed beak, yellowish orange with reddish on exposed surface, early maturing and tolerant to leaf curl and *Stigmina* blight.

☆ **CITH Apricot-3:** It is also a self fertile and early to mid-season blooming type, fruits are medium in size, oblate in shape, asymmetrical with slightly pointed beak, yellowish orange with very little reddish tinge, early mid-season maturing and good quality and tolerant to major pests and diseases.

ROOTSTOCKS AND PROPAGATION

In general, seedlings of apricots, plums and peaches are the commonly used rootstocks for commercial apricot plantation. The apricot and peach seedlings are suitable for sandy soils and dry conditions whereas plum seedlings are suitable for heavy and wet conditions. However, the wild apricot such as *zardalu* and *chulli* seedlings are most suitable as they form the strong graft union with most of scion varieties and can withstand the advance soil and climatic conditions. Myrobalan plum seedling should be used as rootstock for apricots for heavy impervious soils.

Rootstocks resistant to nematodes like Lovel and Nemaguard should be used in nematode affected areas.

For raising the footstock, seeds are collected from fully ripe fruits of wild apricots. Apricots seeds require stratification for a period of 45-50 days at 4°C to break dormancy. The germination of seeds can also be hastened by soaking the seeds for 24 hours in 500 ppm GA_3 or 5ppm kinetin solution before sowing. The stratified seeds are sown 6-10 cm deep in well prepared nursery beds at a distance of 15-20 cm from seed to seed in rows 25-30 cm apart. After sowing, the beds are mulched with 6-10 cm thick grass and light irrigation is applied. The seedlings attain graftable size one year after sowing.

Tongue grafting, T-budding and chip budding are generally adopted for its multiplication. The seedlings of pencil thickness are grafted with tongue method in February, while the seedlings also give very good success. After one month of bud take, the tying material (polythene) should be removed. Aftercare of grafted plants like single stemming, staking, weeding, watering and plant-protection measures should be adopted at regular intervals. Application of farmyard manure @80 tons/ha and 30 kg/ha of P_2O_5 is recommended for better growth of grafted/budded plants.

PLANTING AND ORCHARD ESTABLISHMENT

Apricots are usually planted in winter when the plants are dormant but early planting gives better establishment of plants. Pits of 1m x 1m x 1m size are dug about a month before planting. They are filled with a mixture of soil and 50-60 kg well decomposed farmyard manure. About 1 kg single super phosphate and 10 ml chlorpyriphos solution (10ml/10 liters of water) is also added to each pit. Contour planting or terrace planting is most desirable. The plants on vigorous seedling stocks can be planted at a planting distance of 6 x 6 m in well prepared pits. The pits of 1 x 1 x 1 m size are dug about one month before planting. The grafting point should be kept at least 15-20 cm above the ground level at the time of planting.

On flat land, a regular layout system such as square and triangular is followed, while on the hill slopes, contour system is generally practiced. The spacing of plants varies with the soil, climate and vigour of cultivar. The plants are generally planted at a distance of 6m x 6m. Due to the absence of the dwarfing rootstock, high density planting is still to be standardized with proper training and pruning system, and with the use of growth retardants.

In general, two-year-old healthy and disease-free saplings are planted to achieve good success. Saplings are planted in the middle of the pit, and the surrounding soil is pressed gently so that roots are set properly. Watering should be done immediately to establish close contact between roots and soil. After planting, tree basin is mulched with 10 cm thick hay mulch to conserve soil moisture. In summer, watering should be done as and when required.

FLOWERING, POLLINATION AND FRUIT SET

Apricots start bearing at the age of 5-6 years. Flowering commences in the month of February and terminates in March. The duration of flowering varies with the cultivar and prevailing weather conditions. High temperature and low relative humidity shorten the flowering duration. Early flowering is usually damaged by spring frost. Apricot produces several thousand flowers, but about 15-20 per cent of the flowers set fruits. However, 10-15 per cent fruit set gives sufficient crop.

Most of the apricot varieties are self-fruitful and set fruits without pollenizer However, the varieties like Perfect, Charmagz and Riland are self-incompatible. The pollination is done by bees or wind. The fruit set and yield in apricot can be enhanced by a single spray of GA_3 (10 ppm) at flowering. Furthermore, a spray of 10 ppm NAA or 50 ppm 2,4,5-T at the beginning of pit hardening stage reduces the fruit drop significantly. Similarly, over-bearing in apricots should be reduced by chemical thinning of fruits.

Fruit set in apricot is rather heavy which results in under sized fruits, and increases the tendency of biennial bearing. Fruit thinning improves fruit size, promotes regular bearing, decreases limb breakage (due to heavy crop load) and maintains the tree vigour. Fruit thinning should be done within 40 days after full bloom (last week of April or first week of May). Both hand and chemical thinning methods are employed. Depending upon the crop load, the fruit may be thinned till the fruits are 6- 10cm apart. A spur should not have more than 2 fruits. For fruit thinning a large number of chemicals like DNOC, 3CPA, NAA, GA_3 and thiourea have been used, but ethephon is the most widely used chemical for fruit thinning. Thiourea (40-100 ppm) and 2,4-D and 2,4,5-T (40-100 ppm) are also good fruit thinner.

CULTURAL OPPERATIONS

Training and Pruning

Apricots trees develop large canopy if not pruned correctly. Therefore, it should be trained to achievable height by modified leader system of training. However, open centre system is also satisfactory for most areas. At the time of planting, one year old whip is headed back at about 60-70 cm above the ground and 3-5 well spaced shoots are allowed to grow in all directions. Pruning is more important in first dormant season because the framework developed in this period gives ultimate shape to the tree. In first dormant season, 3-5 primary scaffold branches arising at proper angles (45′), well spaced (10- 15 cm apart) and spirally arranged around the tree trunk are selected. The lowest branch should be 40-45 cm above the ground level. All the primary scaffold branches are headed back to half of their growth to get the secondary branches on them. During second dormant pruning, 5-7 well spaced secondary scaffold branches are selected on each primary branch and other are removed. At the end of third year, pruning is confined to the thinning of branches which are either over- crowding or crossing each other, for proper development of the framework and to admit adequate sunlight in the tree canopy.

Apricot bears on spurs and laterally on one year old shoots. The spurs have a short life of 3-4 years. Many of them are also broken during fruit plucking. The production of young growth is, therefore, essential for the initiation of new spurs which generally takes place at the bases of the growing laterals. In young bearing trees, pruning should be light and of corrective type but in older trees, heavy pruning should be done to maintain balance between growth and fruiting. In apricot, 25-30 per cent thinning of one year old shoots or one third heading back is recommended to improve fruit size and quality. After pruning Chaubattia paste (Copper Carbamate+Red Lead+Linseed oil::800g+800g+1 L) should be applied on the cut ends of the shoots.

Nutrition Management

Apricot trees require fairly good amount of manures and fertilizers for proper growth and production and thus must be manured and fertilized judiciously. The need for fertilizers or nutrients is affected by several factors, and it is best guided by tissue analysis. A mature apricot tree require 60 kg FYM, 2.0 kg calcium ammonium nitrate, 1.6 kg single super phosphate and 1 kg muriate of potash annually. Apricots respond well to foliar nitrogen spraying in the form of urea at the rate of 0.5 to 1.0 per cent. A prefall spray of 2.5 per cent urea improves fruits set and yield. About 60 kg well rotten FYM should be applied during winter along with P_2O_5 and K_2O at the time of basin preparation. Half dose of N fertilizer should be applied in spring before flowering and the rest half dose a month later. Foliar application of nitrogen in the form of urea at 1 per cent is quite effective and 2 to 3 sprays during fruit development stage can improve the fruit size as well as the plant growth.

Water Management

Apricot trees are deep-rooted and hence do not require regular irrigation except in dry periods. However, moisture stress during fruit development periods adversely affects fruit yield and quality. Yet, newly established plants require regular irrigations. The peak water use period is from April end to mid June, which coincides with fruit development period. Irrigation during this period improves fruit size and yield. It should be irrigated at 10 days intervals during May and 6-8days during June. Mulching with black polythene or hay (10-15cm thick) also helps in conserving soil moisture and temperature.

Weed Control

The tree basins should be kept weed free by manual weeding or herbicidal treatments. Atrazine or diuron at the rate of 4 kg/ha is effective for control of weeds for 4-5 months. Mulching of tree basins with 10-15 cm thick layer of grass is useful not only to conserve soil moisture but also keeps the weed growth in check.

PLANT PROTECTION

Major Insect-Pests and their Management

Several insect-pests cause serious damage to apricots but the major pests and their control measures are as under:

Indian Gypsy Moth (*Lymantria obfuscata*)

This is the most destructive pest of apricots and found in all apricot growing areas of the world. Its caterpillars feed on foliage during night and defoliate the whole tree. Caterpillars eat away the whole lamina and leave behind only the hard veins. It is better to collect its eggs and destroy them. Spray endosulfan (0.05 per cent) during March-April.

European Earwig (*Forficula auricularia*)

Mature trees generally tolerate damage well; if damage is caused to shoot tips of young trees then growth may be stunted; shallow, irregularly shaped areas may be present on fruit surface where insect has fed. Remove all weeds from around tree bases; remove all pruning debris and loose bark around trees; wrapping trunks tightly with plastic wrap before nymphs emerge can stop them climbing up the tree; if using insecticide, apply early in Spring when earwigs begin to be active.

Leaf Roller (*Archips argyrospila*)

In some parts of the world, leaf roller causes significant loss to apricots. Its caterpillar cuts the leaf and petiole and rolls and tie them together with silk webbing and then feed inside the roll on soft tissues. After this, it enters the fruit and feed on its pulp. The affected fruits become unfit for human consumption. It may cause defoliation of plant, and fruits may have substantial scarring from feeding damage. The caterpillars wriggle vigorously when disturbed and may drop from plant on a silken thread. For its control, monitor plants regularly for signs of infestation; remove weeds from plant bases as they can act as hosts for leaf rollers; avoid planting in areas where sugarbeet or alfalfa are grown nearby; *Bacillus thuringiensis* or Entrust SC may be applied to control insects on organically grown plants; spray 0.02 per cent decamethrin; apply sprays carefully to ensure that treatment reaches inside rolled leaves, remove the rolled and webbed flowers to check its further spread.

Twig Borer (*Anarsia lineatella*)

Feeding of insect may cause death of shoot tips; fruits at stem end. Most effective method of treatment is well-timed applications of insecticide around time of bloom; organically acceptable insecticides include *Bacillus thuringiensis* or Entrust; infestations can also be treated with appropriate organophosphate or pyrethroid insecticides.

Apricot Chalcid (*Furytoma samsonavi*)

This is also a serious pest of apricot, which lays eggs inside the tender fruits. Grubs on hatching feed inside the kernel and cause drop of the immature fruits. After feeding, grubs pupate inside the fruit. For its control, the infected and fallen fruits should be collected and destroyed.

Fruit Fly (*Dacus* sp.)

The females lay eggs on fruits, which on hatching enter the fruit and feed on its internal contents, thus rendering it unfit for human consumption. For its control, give 2 sprays of bait consisting of 0.1 per cent malathion and 1 per cent sugar during April-May. Spray of decamethrin (0.02 per cent) during this period is also useful.

Major Diseases and their Management

Major diseases of apricot are described briefly hereunder:

Bacterial Canker or Gummosis

This is the most serious disease of apricots. It is caused by a bacterium, *Pseudomonas syringae.* Its symptoms included the development of elongated cankers at the base of buds and randomly on the trunk and scaffold limbs. Damaged areas are slightly sunken and somewhat darker in colour than the surrounding bark. At both the upper and lower margins of the canker, narrow brown streaks extend into healthy tissue. As the trees break dormancy in the spring, gum is formed by the surrounding tissue and may exert enough pressure to break through the bark and flow. The area beneath the canker has a soured odor. Individual scaffolds or the entire tree usually dies shortly after leafing out in the spring. Roots are not affected. Extensive suckering often occurs at the tree base. The bacterium is a weak pathogen and causes serious damage only when a tree is in a dormant condition or weakened due to unfavorable growing conditions. Trees up to 7 years old, growing on deep sandy soil are most susceptible. Avoid using high nitrogen fertilizer rates in mid to late-summer. Do not encourage late fall growth. Prune when the trees are fully dormant (January and February). Spray streptocycline @ 10g/100 L water before the onset of rainy season or alternatively spray copper oxychloride/Bordeaux mixture @ 0.3 per cent after leaf fall. In addition, cut and burn the affected plant parts immediately to restrict the further spread of the disease.

Bacterial Spot

This is also a serious disease of apricot, which is caused by a bacterium, *Xanthomonas campestris* pv. *pruni.* Its symptoms appear on leaves as small, circular, or irregularly shaped, pale green lesions. During early development, lesions almost always are concentrated near the leaf tip. In advanced stages, the inner portion of the lesion falls out, giving the leaf a 'ragged' or 'shot hole' appearance. Leaves heavily infected with bacterial spot turn yellow and fall. Repeated infection can occur throughout the growing season as long as the environment is favorable. Symptoms first appear on fruit as small, olive brown, circular spots. Spots become slightly darker and depressed as the bacteria develops. Lesions are scattered over the fruit surface and tiny cracks develop in the center of the spots. Sometimes symptoms resemble peach scab. Leaf infection is more common than fruit infections. Apparently, more specific climatic conditions are necessary for fruit to become infected. Chemical control during the season is difficult. Dormant sprays have been somewhat effective if the spray is timed to protect stems during the fall infection period. Copper containing fungicides should be applied just as the leaves begin to shed. Resistant varieties should be grown.

Brown Rot

This is also considered as very destructive disease of apricots. It is caused the fungus *Monilinia fructicola*, which causes blossom blight, fruit rot, twig blight, and branch canker. Brown rot of ripening fruit is very common, and it generally occurs as the fruit approaches maturity. The brown rot fungus causes blossom blight and fruit rot, but fruit rot is the most common. Surface moisture and moderately warm temperatures favor disease development. With blossom blight, flowers turn brown and are water-soaked. The fungus grows down the pedicel into the stem resulting in dark brown, sunken areas. Young stems are often girdled causing twig dieback. In some instances, young fruit may become infected but not show symptoms until the fruit matures. Generally, fruit are resistant to infection during the hard green stages of development. Fruit are most susceptible near maturity. The fungus enters fruit directly or through natural openings or wounds. A brown, water-soaked lesion rapidly develops. The brown rot fungus overwinters in mummies, stem cankers and on infected fruit peduncles. Beetles or other insects can be vectors for the fungus. For its management, remove all rotten fruits after harvest to reduce the amount of fungus over-wintering in orchards. Adequate pruning will increase air circulation, allowing faster drying and fewer fruit infections. Apply fungicide sprays during bloom and as fruit ripens or as suggested for brown rot of plums.

Stigmina Blight

It is a fungal diseases caused by *Stigmina carpophylla*. In some parts of the world, blight is a serious disease of apricots. On infected leaves light yellow to reddish coloured spots appear, which subsequently fall down to form shot holes. It occurs mainly on apricot and peach. Fruits develop dish brown measles on the outer surface. Spray tree with copper oxychloride (300 g/1 00 L water) before leaf fall and bud swell.

Armillaria Root Rot

It is caused by a fungus, *Armillaria mellea*. The infected trees have poor terminal growth and small leaves; around midsummer the whole tree suddenly collapses; in orchards trees usually die in a circular pattern; infected trees often have a fan-shaped white fungal mat growing between the bark and wood of the crown. Once a tree is infected, there is no treatment and it should be removed, fumigants do not control fungi in soil adequately; do not plant apricot in newly cleared forest or on the site of old orchards with a history of *Armillaria*.

Eutypa Dieback

It is a common disease in some parts, caused by a fungus, *Eutypa lata*. The symptoms appear as cankers on branches, usually associated with a pruning wound which is several years old; discolored sapwood may extend above and below canker; leaves on branches around canker may suddenly wilt as branch dies; leaves remain attached to branches; discoloured bark and inner wood; gummy amber exudate may be present. Infected limbs should be removed 1 ft below any internal symptoms before harvest; if pruning is conducted out with this time, a fungicide should be applied to the pruning wounds.

Phytophthora Root and Crown Rot

It is fungal disease, caused by *Phytophthora* spp. Its symptoms include poor new growth; chlorotic leaves, small in size and sparse; fruit may be small, brightly colored and susceptible to sunburn; shoots may suffer from dieback and tree will often die within weeks or months of first signs of infection or decline gradually over several seasons; root crown may show signs of decay which develops into a canker; bark of infected crown tissue turns dark brown; cankers may occur on aerial parts of plant. Plant trees on a small mound to promote drainage; avoid over-watering trees in spring; treat soil around newly planted trees with fungicide; minimize the frequency and duration of water saturated soil; trees should be propagated from resistant rootstock and application of appropriate systemic fungicides may provide some protection from the disease.

Apoplexy

Sometimes apricot trees starts declining suddenly, which has been named as apoplexy or premature dieback. Initially, there is bark cracking from the limbs followed by pathogen infection resulting in the production of resin droplets in the xylem, phloem and bark tissues. Badly affected plants die within one or two years. Several causes like fluctuating temperature have been related to this problem, but *P. syringae* and *Cytospora cinta* are principal pathogens associated with it. It can be controlled by pruning of affected portion, spraying plants with Bordeaux mixture and planting of resistant varieties.

Leaf Spot (*Phyllosticta cerasicola*)

Small, angular, ash coloured spots appear on the underside of the leaf which results in pre-mature defoliation. Spray mancozeb (0.02 per cent) for its control.

Powdery Midew

Powdery mildew also causes heavy losses to apricot in some areas. It is caused by a fungus, *Sphaerotheca pannosa* or *Podosphaera tridactyla*. Whitish powdery white patches of fungal growth appear on fruits and leaves; rusty patches on fruits which turn brown and leathery and may crack. Apply fungicide during bloom and fruit development

Verticillium Wilt

It is caused by a fubngus, *Verticillium dahlia*. Its symptoms include withering of leaves on one or more spurs on 1 year old wood; leaves are dull and stunted; fruit small; older trees do not recover from disease. Plant apricot in soil with no history of disease; keep trees adequately fertilized and watered.

Shot Hole (*Cercospora* spp.)

The symptoms include appearance of small, circular spots on leaves, which later turn necrotic and subsequently the dead part is blown leaving a hole behind. Spray with copper oxychloride (300 g/100 L water) or mancozeb (250-300 g/100 L water) before the rainy season.

Rhizopus Rot

It is majot postharvest diseases of apricot, which is caused by caused by a fungus, *Rhizopus stolonifer* It occurs frequently in ripe or near-ripe apricot fruits held at 20 to 25°C. Cooling the fruit and keeping them below 5°C is very effective against this fungus.

Physiological Disorders and their Management

Gel Breakdown or Chilling Injury

This physiological problem is characterized in the earlier stages by the formation of water-soaked areas that subsequently turn brown. Breakdown of tissue is sometimes accompanied by sponginess and gel formation. Fruit stored between 2.0-5.0°C have short market life and less flavour. It is advised to store apricots below 5°C or more to avoid the problem of chilling injury.

MATURITY, HARVESTING AND YIELD

Maturity

The apricots should be harvested when they fully mature because immature fruits don't develop good color, taste and flavour. The best harvesting indices for apricots are the fruit colour, TSS and days from full bloom (DFFB). Normally, fruits are picked when their colour changes to yellow. TSS around 10 per cent is also considered good for harvesting most of the varieties. However, usually harvesting date is determined by skin ground color changes from green to yellow. The exact yellowish-green colour depends on the cultivar. Apricots should be picked when still firm because of their high bruising susceptibility when soft. Most apricot cultivars soften very fast making them very susceptible to bruising and subsequent decay.

Harvesting Method

Apricots for fresh consumption or processing are picked by hand and carefully handled. Trees are usually picked over 2-3 times each, when fruit are firm. Trunk shaking can be used for processed fruit, although apricots are said to be more susceptible to trunk damage than other stone fruits.

Yield

Apricot trees start fruiting at the age of 5 years and continue up to 30-35 years. They attain full bearing age at about 7-10 years, yielding 50-80 kg/tree or 15-22 tons/ha.

POSTHARVEST HANDLING

Grading and Packing

Before packing, fruits are graded according to their size. Fruits are packed in wooden boxes or CFB cartons for transportation and marketing. Each box is lined inside with newspaper sheets keeping the margins for overhanging the flaps. The boxes are initially padded with pine needles at the bottom to avoid the bruising of fruits. Wrapping of individual fruits is not done in apricot. Fruits are arranged in layers and top layer is covered with paper by bringing together overhanging flaps. Then top of the box is nailed.

Small sized CFB cartons are also used for packing apricots. The CFB cartons are lighter in weight, easy to handle and in packing. The fruits fetch better price because of lesser bruising damage. However, they are slightly more expensive than wooden boxes and need protection from direct rains.

Drying

Dried apricots are harvested later (fully ripe) than those for shipping, and exposed to SO_2 to avoid postharvest diseases. The drying ratio is 5.5:1 (lbs fresh fruit: lb dry fruit). Drying is either natural, in the sun, or in large dehydrators as with prunes. Canned apricots are immersed in syrup, at a ratio of 0.7 lbs fresh = 1 lb canned. In the high altitude dry areas in Kinnaur district of Himachal Pradesh, apricots especially wild apricots (*chulli*) are not harvested fresh but allowed to dry on the tree itself. Due to very low relative humidity, the apricots dry rather well on the tree. The dried product is of excellent quality, not obtained even after adopting modem techniques of drying (*e.g.* checking, sulphuring and mechanical dehydration). The apricots dried on the trees are approximately two to three times more expensive than sun dried apricots.

Storage

Apricots have an extremely short shelf-life of only 1-2 weeks at 0° C and 90 per cent relative humidity. They are susceptible to all postharvest diseases (Brown rot or *Rhizopus* fruit rot) to which other stone fruits are susceptible.

4

Avocado

INTRODUCTION

The Avocado (*Persea americana* Miller) is the only important edible fruit of the laurel family, Lauraceae to which other important crops like cinnamon, camphor and bay laurel also belong. It is also known as alligator pear, midshipman's butter, vegetable butter, or sometimes as butter pear, and called by Spanish-speaking people *aguacate, cura, cupandra,* or *palta*; in Portuguese, *abacate*; in French, *avocatier* and *Makhanphal* or *Kulu Naspati* in Hindi. Over the years, it has gained importance and its cultivation is distributed widely in the tropics and subtropics. The areas under avocado are fast growing in several countries over the globe as it has enormous potential in commercial trade.

COMPOSITION AND USES

Composition

Avocado is considered to be the most important contribution to human diet in the New World for being more nutritious than any other fruit in the New World. Compared to other fruits, avocados are highly nutritious and do not contain cholesterol. The avocado contains about twice of our daily needs for vitamins C, E, and beta-carotene as its calorie proportion. Rich in copper and iron, two mineral constituents of antioxidant enzymes, avocados again prove their nutritional quality. Potassium is also high in avocados, as it is has one of the highest potassium rates in tropical and non-tropical fruits and vegetables. The avocado is associated with lower blood pressure because it is high in monounsaturated fat. Since the fruit contains not more than 1 per cent sugar, it is recommended as high energy food for diabetics. The nutritive value of fruit is presented in Table 4.1.

Table 4.1: Nutrient Value of Avocado Fruit (100 g of edible fruit portion)

Constituent	Approximate Value	Constituent	Approximate Value	Constituent	Approximate Value
Water content	73 per cent	Sugars	0.7 per cent	Phosphorus	52 mg
Calories	170 kcal	Fibre	6.7 per cent	Potassium	485 mg
Protein	2 per cent	Calcium	12 mg	Sodium	7 g
Fat	14.7 per cent	Iron	0.6 mg	Vitamin C	10 mg
Starch	0.1 per cent	Magnesium	29 mg	Vitamin A	146 IU

Source: *USDA National Nutrient Database for Standard Reference, Release 25 (2013), http://ndb.nal. usda.gov/ndb/search/list

Uses

Avocado is usually eaten fresh and is neither sweet nor sour in taste. The edible pulp has a nutty flavour with a buttery texture. It can also be used in ice creams and milk shakes. Cooking impairs flavour and appearance of avocados; however, a variety of satisfactory frozen products can be prepared. The most popular ways of serving the avocado are in salads, in sandwich filling, as appetizers, dips and as "guacamole"; however, in India people prefer to eat it after mixing the pulp with sugar. The avocado has a gamut of culinary uses and the delicate flavor appeals to the gourmet. Avocado oil is used in preparation of cosmetics.

TAXONOMY AND BOTANICAL DESCRIPTIONS

Taxonomy

Avocado belongs to genus *Persea*, which contains some 50 species. One of these is the true avocado, "*americana*." So, adding the abbreviated name of its describer, Miller, the avocado is botanically designated *Persea americana* Mill. It belongs to the family Lauraceae and order Laurales. Avocado contains 24 chromosomes with bivalent pairing at meiosis indicating that n = 12. It is botanically classified in three groups: a) *Persea americana* Mill. var. *americana* (*P. gratissima* Gaertn.), West Indian Avocado; b) *P. americana* Mill. var. *drymifolia* Blake (*P. drymifolia* Schlecht. and Cham.), the Mexican Avocado; c) *P. nubigena* var. *guatemalensis* L. Wms., the Guatemalan Avocado (Table 4.2).

Botany

The avocado tree may be erect, usually to 9 m but sometimes to 18 m or more, with a trunk 30-60 cm in diameter, (greater in very old trees) or it may be short and spreading with branches beginning close to the ground. Almost evergreen, being shed briefly in dry seasons at blooming time, the leaves are alternate, dark-green and glossy on the upper surface, whitish on the underside; variable in shape (lanceolate, elliptic, oval, ovate or obovate), 7.5-40 cm long. Those of the Mexican

Table 4.2: Comparison of the Three main Avocado Races

Trait	Mexican	Guatemalan	West Indian
TREE*			
Climatic adaptation	Semitropical	Subtropical	Tropical
Cold tolerance	Most	Intermediate	Least
Salt tolerance	Least	Intermediate	Most
Hairiness	Most	Less	Less
Leaf anisette	Present	Absent	Absent
Leaf colour	Medium	Often redder	Paler
FRUIT*			
Months to mature	6	12 or more	5
Size	Small	Variable	Variable
Pedicel (stem)	Slender	Thick	Nail-head
Skin thickness	Very thin	Thick	Medium
Skin surface	Waxy bloom	Rough	Shiny
Seed size	Large	Small	Variable
Oil content	Highest	High	Low
Pulp flavor	Spicy	Often nutty	Mild*
Tree response to freezing temperatures#			
Young trees	−2 to −1°C	−3 to −2°C	−4 to −3°C
Mature trees	−4 to −1°C	−4 to −2°C	−8 to −3°C

* After Bergh and Ellstrand, 1989. (Bergh, B. and Ellstrand, N. 1989. Taxonomy of the Avocado. California Avocado Society Yearbook. 135-145.)

#http://edis.ifas.ufl.edu.

race are strongly anise-scented. Small, pale-green or yellow-green flowers are borne profusely in racemes near the branch tips. They lack petals but have 2 whorls of 3 perianth lobes, more or less pubescent, and 9 stamens with 2 basal orange nectar glands. The fruit, botanically a berry of one carpel containing a single seed, pear-shaped, often more or less necked, oval, or nearly round, may be 7.5-33 cm long and up to 15 cm wide. The skin may be yellow-green, deep-green or very dark-green, reddish-purple, or so dark a purple as to appear almost black, and is sometimes speckled with tiny yellow dots, it may be smooth or pebbled, glossy or dull, thin or leathery and up to 6 mm thick, pliable or granular and brittle. In some fruits, immediately beneath the skin there is a thin layer of soft, bright-green flesh, but generally the flesh is entirely pale to rich-yellow, buttery and bland or nut like in flavor. The single seed is oblate, round, conical or ovoid, 5-6.4 cm long, hard and heavy, ivory in color but enclosed in two brown, thin, papery seed coats often adhering to the flesh cavity, while the seed slips out readily. Some fruits are seedless because of lack of pollination or other factors.

ORIGIN, HISTORY AND DISTRIBUTION

Avocados are indigenous to tropical Mexico, Guatemala, Columbia, West Indies and Central America. Three ecological races *viz.*, Mexican, Guatemalan, and West Indian-are recognized, which are cultivated in tropical and subtropical parts of the world. Avocados are an ancient fruit as the undomesticated variety, known as a criollo, which is small, with dark black skin, and contains a large seed probably coevolved with extinct mega fauna. Over 14,000 years ago, mega fauna roaming Central and South America dined on avocados as a delicacy. The glyptodonts, or, massive armadillos, used to devour avocados whole and spread the seed throughout the region. The Aztecs (ethnic groups of Central Mexico) adored avocados as well. They considered that most of the available fruits are feminine in nature, while fruits of avocado were considered masculine. Avocado was known to them as *ahuacatl* or "testicles." This name might have arisen due to shape of fruits as well as the belief that they were aphrodisiac. The early Spanish explorers recorded its cultivation from Mexico to Peru but it was not in the West Indies at that time. It was introduced into Jamaica in 1650 and to Southern Spain in 1601. It was taken to the Philippines near the end of the 16th Century; to the Dutch East Indies (Indonesia) by 1750 and Mauritius in 1780; was first brought to Singapore between 1830 and 1840 but has never become common in Malaya. The plant was introduced to Brazil in 1809, South Africa and Australia in the late 1800s, and the Levant in 1908. It was reported in Zanzibar in 1892. It was planted in Hawaii in 1825 and was common throughout the islands by 1910; it was introduced into Florida from Mexico by Dr. Henry Perrine in 1833 and into California, also from Mexico, in 1871. Although avocados have existed for centuries, the Hass avocado, the bumpy skinned variant accounting for most of the avocados sold worldwide, has its origins in California during the early 1900s. Rudolph Hass convinced his children not to destroy a bizarre tree found in his neighborhood. By 1935, the small sapling grew magnificent fruits that he later patented them as the Hass avocado. The first trees were planted in Israel in 1908, but named cultivars ('Fuerte' and 'Dickinson') were not introduced until 1924.

Avocado entered in India by the way of travelers from Srilanka, to Waynad in Kerala, Coorg in Karnataka and Kodaikanal in Tamil Nadu in 19th century. It was introduced in Bangalore by an American missionary, residing in Bangalore between the years 1906 and 1914, who brought few seedlings of avocado from Royal Botanical Gardens of Sri Lanka and planted them in 'Lal Bagh Garden', Bangalore.

Avocado is not grown at large scale in India; however, certain Southern states grow them commercially. In South India, it is at present grown only experimentally in a few orchards at Bangalore, Shevroys, Nandi Hills, Lower Palnis, and at the Kallar and Burliar Fruit Stations in the lower foot hills of the Nilgiris. In western India, avocado was for the first time imported from Ceylon in the year 1941 and planted at the Ganeshkhind Fruit Experiment Station, Pune. It is also cultivated to some extent in Maharashtra and Kerala. Sikkim is the only eastern Himalayan state in the North, where avocado is grown successfully at elevations ranging between 800 to 1600 meters above mean sea level.

According to an estimate, 500-1000 MT of avocados is currently produced annually for the domestic market. At global level, Mexico tops the list of avocado producing countries distantly followed by Indonesia and Dominican Republic (Figure 4.1).

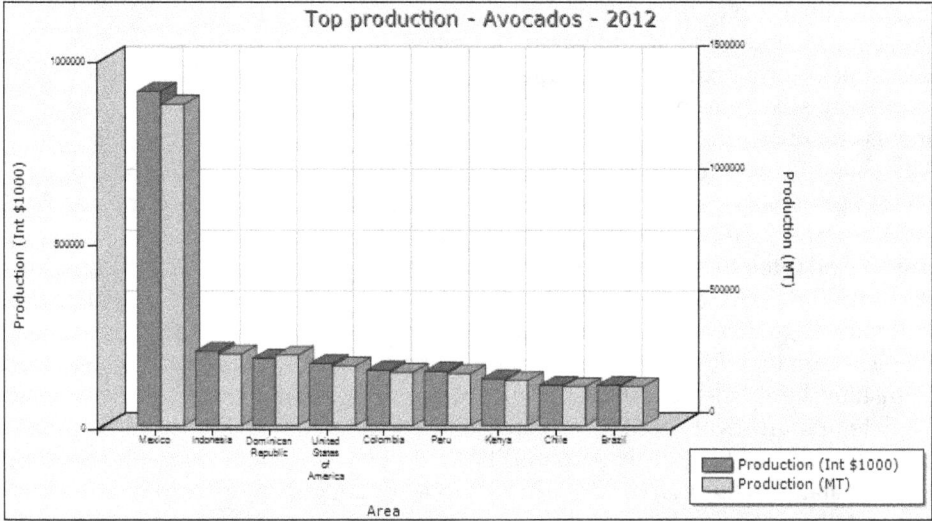

Figure 4.1: Top Avocado Producing Countries of the World (*Source*: FAO).

SOIL AND CLIMATIC REQUIREMENTS

Soil

Avocado trees can be grown in various soil types with good internal drainage. The tree will not perform well in poorly drained soils, nor will it tolerate flooding. Sites with underlying hardpan must be avoided. The water table should be at least 1m below the surface. The optimum pH ranges from 5 to 7. Before planting avocados in new site soil, profile aspects such as porosity and minimum depth for root penetration in soils of different types, soil bulk density, laboratory analyses, water-holding capacity and irrigation scheduling, and the influence of physical characteristics on root disease associated with *Phytophthora cinnamomi* infection should be taken into account. Avocados do best in soils with a clay content of between 20 and 40 per cent. If the clay content is below 20 per cent, the soil has a limited water-retention capacity and unless optimum irrigation is applied, the trees will sometimes suffer temporarily from drought. A too high clay percentage makes irrigation difficult because over irrigation and high rainfall lead to oversaturation of the soil. This means that water drains away relatively slowly, which promotes root rot. It does not tolerate salinity, excepting varieties of West Indian race. Certain cultivars like Fuchs-20 and Maoz have shown considerable salt-tolerance in Israel. Avocados grafted onto Fuch-20 rootstocks and irrigated with water containing 380 to 400 ppm Cl performed well in a commercial orchard.

Climate

Avocado is a tropical to subtropical tree that, with some exceptions, is best adapted to relatively frost-free areas with mild winter at an elevation of 600-1500 m and an annual rainfall of 125-180 cm. The 3 best-known avocado races each has specific climatic requirements as a result of adapting to their original environment. West Indian and some hybrid varieties are best adapted to continuous hot, humid conditions with a high summer rainfall. Like all avocado cultivars they are, however, extremely sensitive to drought and do not tolerate frost well. The cultivars in increasing order of sensitivity to cold temperatures are Edranol, Hass, Pinkerton, Fuerte and Ryan. The optimum temperature for growth is 25 to 28 °C. The humidity should preferably be above 60 per cent. High humidity during flowering and fruit set is necessary to secure a good crop. Mexican varieties are more cold tolerant and not well adapted to lowland tropical conditions. They can withstand temperatures of - 4 to -5 °C. The optimum temperature for growth is 20 to 24 °C. Guatemalan cultivars originated from the tropical highlands of Guatemala and require a cool, tropical climate without any extremes of temperature or humidity. The trees can withstand light frost, down to - 2 °C, but the flowers are very sensitive to frost. High temperatures of about 38 °C, especially if combined with low humidity, could cause flower and fruit drop. A humidity level of 65 per cent or higher is required. Guatemalan x Mexican hybrids such as Furete is generally more cold tolerant than West Indian x Guatemalan hybrid varieties. The most limiting factor to success with avocado trees is severe cold. For Fuerte, the daily mean temperature during flowering should preferably be above 18.5 °C, but definitely above 13 °C. Avocado is more sensitive than others to unfavourable weather conditions during flowering. Hot, dry conditions could result in low yields because of fruit and flower drop. A high humidity is desirable, because it decreases stress conditions (particularly high temperature), that play an important role during flowering and fruit set. Avocados should not be planted in windy locations as they tend to have brittle branches which are prone to damages caused by wind. In areas of strong winds, wind-breaks are necessary as wind reduces humidity, dehydrates the flowers and interferes with pollination, and also causes many fruits to fall prematurely. The majority of blemishes causing a downgrading of fruit most probably also result from wind damage.

IMPORTANT VARIETIES

Avocado varieties are classified in three groups, known as the West Indian, Guatemalan and Mexican "races", with distinguishing characteristics.

West Indian types are the least cold tolerant and somewhat watery in flavor, but they have the greatest tolerance to salinity and some diseases. West Indian avocados are useful primarily as rootstocks because of their high salt tolerance.

The Mexican race (cvs. Gott Fried, Duke, Pernod) is considered as most cold tolerant but the least salt tolerant. Its fruit ripens in the summer and is usually of good flavor. The fruit is rarely larger than 8 to 12 ounces, is green to purple or black, and has very thin skin. Because the skin is so thin, the fruit are very susceptible to disease. The crushed leaves of the Mexican race of avocados have a distinct odor of anise (licorice), which is absent in the other races.

The Guatemalan race (cvs. Taylor, Linda, Queen, Itsamma and Benik) of avocados is essentially intermediate between the other two, and its hybrids with the other two races include many of the more important varieties in commerce.

Early varieties are usually of West Indian (cvs. Pollock, Simmond, Black Prince, Fuchsia, Peterson and Waldin) and Mexican origin, whereas mid-season and late varieties are hybrids between the races and have intermediate characters.

Varietal Descriptions

☆ **Fuerte:** This cv. has good production potential. Tree growth habit is large and spreading. It tolerates temperatures as low as - 4 °C but alternate bearing, and sensitivity to microclimate for fruit set are limitations of its cultivation. Though, fruit set can be increased by making provision for pollinators. It is also susceptible to physiological disorders during storage.

☆ **Hass:** This variety has showed good production potential in cool areas. Fruit is smaller in warm areas. Trees are fairly upright and slow grower. It can tolerate temperatures as low as - 2 °C. Limitations to its cultivations are fruit becoming too small with age and in warm regions and susceptibility to environmental factors.

☆ **Pinkerton:** It is consistent heavy bearer. Trees are moderately spreading. It can also tolerate temperatures as low as - 1 to - 2 °C. Sometimes, fruit may develop internal disorders if picked when over mature.

☆ **Ryan:** It bears heavily and fairly consistently. It is also frost tolerant. It has shown its suitability for planting in drier inland areas.

☆ **TKD 1 or PA-2:** This variety was selection from germplasm pool at TNAU, T.N. It was developed in 1997. This variety registered the highest average fruit yield (266 kg/year, 214 per cent increase over control), fruit length (19.2 cm) and pulp thickness (2.17 cm) and recorded high values for number of fruits per tree (401.2) and single fruit weight (660.8 g). PA-2 also recorded the highest fat (23.8 per cent), vitamin A [retinol] (0.19 mg/100 g flesh), total soluble solids (8 °Brix) and protein (1.35 g) contents.

☆ **Lamb Hass:** Avocado [*Persea americana*] cv. Lamb Hass was derived from the cross of Guatemalan with a Mexican variety. Trees have upright growth habit, fruits have black, pebbly skin and good storage traits. Fruit weight is in the range 200-410 g and tree yields average 50 kg (20 t/ha). Lamb Hass has high fibre content and good yields.

☆ **Eden:** The avocado cultivar Eden was derived from a cross made between Pinkerton and N 151-2 in 1986 in Israel. Fruit weight of Eden ranges from 250 to 400 g and averages about 260 g, 30 per cent heavier than Hass. Flesh is yellow, buttery and largely free of fibres with a nut-like flavour. Fruits may be stored for at least 24 days at 5 °C and 95 per cent RH.

☆ **Adi:** This *Persea americana* variety (in flowering group A) from the cross between Hass and Horshim flowers about 2-3 weeks later than Hass and, unlike either parent, has short and compact inflorescence racemes. The fruit shape resembles that of Hass with fruits weighing about 230 g. The flesh is pale yellow with a green rim, buttery and of excellent, slightly less nut-like aroma than that of Hass. The harvest season is longer than that of Hass. Storage quality is good (7-9 days at 20°C).

☆ **Gil:** Originating from a cross between Tova and an unknown pollen donor, this *Persea americana* variety of the A-flowering group produces fruits with a somewhat pimpled skin and medium gloss, averaging 300 g and ranging from 250-400 g, compared with a range of 160-220 g for Hass. The fruit is black when ripe or when fully mature but can be harvested when still green. Moreover, it can be stored on the tree for 2.3 months after it turns black. The peel separates readily from the flesh and is of similar thickness as that of Hass or is slightly thicker. The seed represents 16-18 per cent of the fruit weight. The flesh is buttery firm, becoming somewhat doughy towards the end of the harvest season. Oil is recommended for areas with high summer temperatures, where fruit size might be reduced.

☆ **TKD-1:** TKD-1 avocado was released during 1996 by Horticultural Research Station, Thadiyankudisai, T.N. Fruits are dark green coloured, round shaped and medium in size with a fat content of 23.8 per cent. Yield is 26.4 t/ha.

POLLINATION

The avocado flower has both female and male organs, which means it is structurally "perfect", or "bisexual"; however, male and female organs within one flower do not function at the same time. The bisexual avocado flower opens twice, with an intermediate closing. The first flower opening is at a female stage and the second, usually on the following day, is at a pollen-releasing male stage. Both opening and closing of the female stage flowers, as well as of the male stage ones, occur simultaneously within the tree, and the cultivar, at a regular time daily. This unique flowering behavior, which is displayed by all avocados, termed 'diurnally synchronous protogyny dichogamous', with an intermediate closing. Therefore, each avocado flower is functionally unisexual. The avocado cultivars are divided into two complementary flowering groups A and B types, according to their daily flowering sequence. The A type flowers are those which show receptive female parts in the morning and receptive male parts in the afternoon of the following day. The B type flowers show receptive female parts in the afternoon and receptive male parts in the morning of the following day. With this mechanism the female parts of the A type flowers are thus receptive to pollen from the B type flowers in the mornings, whilst the female parts of the B type flowers are receptive to pollen from the A type flowers in the afternoon (Table 4.3). Some important examples of A types varieties are Hass, Gwen, Pinkerton, Reed, Anaheim, Lamb Hass and of B types are Fuerte, Zutano, Bacon, Whitsell, Sir Prize.

Table 4.3: Pollination Behavior of Avocado

Cultivar	First Day		Second Day	
	Morning	Afternoon	Morning	Afternoon
A type	Female (stigmata receptive)			Male (sheds pollen)
B type		Female	Male	

Avocado is not commonly self- pollinated, therefore it is important to encourage cross-pollination by having A and B type trees in the orchard or backyard. Therefore, avocado flowering behavior is a sophisticated mechanism, which prevents effective self-pollination (within a flower), enables close-pollination (between neighboring flowers within a tree) and encourages cross-pollination (between different cultivars). There are three methods to enhance cross pollination, *viz.*, planting trees of two varieties in the same hole, grafting of branches with pollinating varieties and alternating trees of different varieties. Self-pollination appears to be primarily caused by wind, whereas cross-pollination is caused by large flying insects such as bees and wasps.

The pollen must reach the stigma at the proper time in the flower and pollination of the same type, close or cross, must take place for first set. An adult avocado tree, in a good season, may bear about one million flowers, but only several hundred fruits. Namely, only 0.05 per cent of the flowers (one of 2,000) give fruits. This effect, which is very common for Sub-Tropical trees, was termed: "mass flowering". Varieties vary in the degree of self- or cross-pollination necessary for fruit set. Some varieties, such as 'Waldin', 'Lula' and 'Taylor' fruit well in solid plantings. Others, such as 'Pollock' and 'Booth S' (both B types) do not, and it is probably advantageous to plant them in rows alternating with other varieties (A types), which bloom simultaneously to facilitate adequate pollination. In most cases the cross-pollinated fruits were found to be not only stronger, but also larger than the self-pollinated ones. These phenomena are termed "metaxenic effects". Some avocado cultivars, like 'Ettinger', 'Bakon', 'Zutano' and 'Edranol', were found to produce a "positive metaxenic effects", namely, give an advantage to the out-cross pollen grains over the self grains, as well as to the out-cross fruits over the selves. These cultivars are termed "potent cultivars".

In parts of South India and Maharashtra, where avocado is successfully grown, 2 varieties namely, purple (West Indian race) and Green (Guatemalan race) are popular. Purple variety bears pear shaped fruits with a long neck weighing about 450 g. The fruits have smooth, moderately thick, leathery skin and the pulp is firm, deep yellow, fine in texture with a rich and nutty favour. The fruits of Green variety are oval to obovate, large (450-680 per cent) with rough, moderately thick, brittle skin. The flesh is soft, greenish-yellow with a mild nutty flavour. Single trees of avocado are not productive at times.

PLANT PROPAGATION

In India, avocado is commonly propagated through seeds. However, most avocado varieties do not come true from seed (*i.e.*, a seed will not render the same variety), so they must be propagated vegetatively. The viability of seeds of avocado is quite short (2 to 3 weeks); therefore, they should be sown soon after extraction from the fruit. Storing the seed in dry peat or sand at 5 °C can improve the seed viability. Treating the seeds with GA at 100 or 1000ppm or soaking of the seed in water for about 24 hr before sowing, or removing the seed coat and a thin slice at top and bottom may accelerate germination. Stratification of the seed hastened germination. Soaking the seeds in water at 30°C or 40°C for 8 hours improved the germination by 97 per cent. Seeds are planted in nursery at a spacing of 30cm x60 cm and then transplanted to polybags when they can be transplanted to their permanent location. Avocado plants are commonly grafted as this reduces the time taken for bearing, combines the best characteristics of varieties and improves resistance/tolerance to diseases. Different races/cultivars have also been recognized with specific objectives for using as rootstocks. For example, the West Indian stocks are preferred in warmer regions or where salinity, calcareous and alkaline soils are problems, while the Gautemalan race is more sensitive to cold and has also proved more susceptible to high pH chlorosis and to *Verticillium* wilt. Variable responses had been noted for poorly aerated soil but Mexican race is found to be more tolerant than West Indian. On a worldwide basis, Mexican rootstocks appear to dominate, including the clonal 'Duke 7' due to its large scale use in California for being more tolerant to cold and *Phytophthora*. Other *Phytophthora* tolerant rootstocks are 'Thomas', 'Spencer', 'Toro Canyon', 'Zentmyer', 'Barr Duke' G6, G22, G166, and G755/Martin Grande. West Indian seedlings, tolerant to saline soil, are mostly used in Israel. Clonal rootstocks like 'Dusa'/'Merensky 2', 'Duke 7' and 'Bounty' are very popular in South Africa, while rootstock 'Velvick' predominates Australian avocado industry. Likewise, seedlings of cv. Duke are resistant to root rot and cold and Pollock stock can overcome salinity problem. Cultivars Green and purple also do well as rootstocks. Green imparts more vigour to the scion than purple. Generally, young, vigorously growing seedlings, irrespective its source, are used for rootstocks, and terminals of leafy shoots are used for scion material. Wedge grafting is the type commonly used. The graft is wrapped securely and covered with clear plastic until it catches. This is seen if graft remains green and buds start to burst into leaf after 2-3 weeks. Established trees may be top-worked by cleft grafting scions of the desired varieties on stumps of cut-back trees or by veneer grafting new sprouts arising from stumped trees. During this entire process the plants should be protected from severe sunlight and receive adequate water until transplanting in field. Avocado plants are ready for planting out in the field approximately 6-9 months after grafting. In Nilgiri Hills of Tamil Nadu, layering as well as inarching gave up to 75 per cent success, while in West Bengal chip-budding is reported to be successful.

Etiolated shoots of avocado rootstocks also give response to various treatments including stooling, bark ringing and IBA application (5000 or 10000 mg/liter). Bark ringing combined with IBA at 10000 mg/liter is the best practice in terms of rooting and the highest numbers of primary and secondary roots.

PLANTING AND ORCHARD ESTABLISHMENT

Before establishing a plantation, the field should be well ploughed, harrowed and levelled, keeping in mind the possible intercropping, often with vegetables. In areas prone to excess water, they should be planted on mounds as avocados cannot withstand water logging.

Avocado is planted out to a distance of 6 to 12 metres depending on the choice of the variety and character of the soil. For varieties having a spreading type of growth, like Fuerte, a wider spacing should be given. Trees in deep soils with a high percentage of organic matter need more space, because they grow taller and larger under these conditions. Therefore, in light soil a spacing of 7.5x7.5 m may be sufficient, while in deep, rich soil, a spacing of 9-11 m may be necessary. In Sikkim, a planting distance of 10 x 10 metres on hills slopes (on half-moon terraces) is preferred. In general, final distances of less than 10 m will necessitate thinning before the orchard is 10 years old. High-density plantings can therefore be planted at less than half the "final" distance on the understanding that trees in the semi-permanent rows are removed timely.

By and large, high-density plantings have not been successful in most avocado growing areas. Under high density planting, eventually the trees start to crowd each other and production, therefore declines at later stages. Efforts to control shoot growth and maintain the trees at a manageable size have been generally unrewarding. Pruning often reduced productivity, at least for a few seasons. In the absence of any canopy management, large sections of the trees typically became shaded and unproductive. Large trees are also difficult to spray and harvest. Therefore, in the absence of dwarfing material, effective canopy management appears to be the largest barrier to success of high density orchards.

Pits of 60 cm x 60cm x 60cm are dug and left open to sun for about 10 days. These are then filled with top soil mixed with approximately 30 kg of well decomposed farmyard manure or leaf mould. Add 20 g of super phosphate at the base of the pit for good root growth. Planting can be carried out anytime during the year, but when adequate irrigation facilities are lacking, monsoon is the appropriate time for planting.

Avocado trees bought from a nursery should already have been hardened off. Trees should be planted as soon as possible; if kept too long they may become root-bound or suffer from nutrient deficiencies. They should not be placed in the sun because the containers will become hot and the roots could be burnt even before planting. Wherever feasible, support the young trees with sturdy props as soon as possible after planting. Make sure that the stems are whitewashed. Also remove the nursery tags and surplus graft strips after planting to prevent girdling. While planting grafts, it is important to keep the graft joint well above the ground.

Early Care

Once planting is done, regular watering is essential till the plants establish. Young trees must be irrigated to ensure a uniform stand. Over irrigation is just as harmful as too little water. Therefore, examine the soil moisture content of the subsoil regularly to prevent over irrigation. Avocados are sensitive to moisture stress. In the nursery the trees would have been accustomed to regular water applications and still have a limited root system as a result of the small bag. It is therefore essential that the water reaches the limited and shallow root system. A small basin around the tree will ensure that the roots get enough water.

Where hardening off has been inadequate (in the nursery) temporary shade should be provided. Frame covered with grass or shade netting over the trees can be erected to protect the leaves. The frame can be removed as soon as the leaves penetrate the grass because then they have become hardened off and need no further protection.

During the early years of an orchard a cover crop will protect and maintain the soil until the trees start providing shade. A cover crop must not, however, compete with the trees and must be restricted to the strips between the tree rows. Likewise, the drip area of the tree must be free of grass and other weeds and, if possible, this area should be covered with organic mulch.

INTERCULTURAL OPERATIONS

Pruning

Overcrowding poses a serious problem for orchard access and, more importantly, for adequate light interception needed for successful photosynthesis, flowering and fruit set. It is generally considered that canopy management should start early in the life of the orchard. Columnar cultivars require pinching at early age to form a rounded tree. Overall, regular light pruning is more effective than heavy pruning on a less regular basis. Upon heavy pruning, there are chances that bearing is reduced for several seasons. It is suggested that in warm subtropical coastal areas, hedging can be adopted for controlling tree growth and maintaining productivity, while in cool temperate areas where successive crops often overlap, selective limb removal may be followed. Commercial trees are pruned back quite severely to allow mechanical harvesting. Formative pruning during the first 2 years may be desirable to encourage lateral growth and multiple framework branching. Commercially, after several years of production, it is desirable to occasionally cut back the tops of the trees to 4.9 to 6.1 m. This reduces spraying and harvesting costs and possible storm damage. This operation should be done soon after harvest for early varieties, but after danger of frost has passed for late varieties. Severe topping and hedging (used to reduce canopy width) do not injure trees, but reduce production for one to several seasons. Planned tree removal is an option that should be seriously considered for commercial plantings before overcrowding and reduced yields begin. Preliminary studies to rejuvenate non-productive mature orchards with very tall (9.1 to 12.2 m) trees suggest that production on a per unit area basis can be improved when selected trees are removed and remaining trees are topped at 4.9 to 6.7 m. This is because the lower canopy of remaining trees is re-established and production per tree in the orchard exceeds yields of overcrowded trees. Further, avocado fruit is self-thinning.

Nutrition Management

In avocado flushes of shoot and root growth appears to be synchronizing and alternate on 30 to 60 day cycles. Shoot growth virtually ceases during late autumn and winter, but root growth, through it slows down during winter, continuous through out the year. Therefore, avocado requires heavy manuring.

Table 4.4: A General Recommendation of Manures and Fertilizers for an Avocado Tree

Tree Age (years)	N (g)	P (g)	K (g)	FYM (kg)
1-3	40	20	35	25
4-6	75	35	60	35
7-10	150	35	125	40
>10	200	45	165	50

Plants growing in calcareous soils should receive annual nutritional sprays of copper, zinc, manganese, and boron for the first 4 to 5 years. Thereafter, only zinc, manganese, and possibly boron are necessary. Avocado trees are susceptible to iron deficiency under alkaline conditions. Iron deficiency can be prevented or corrected by periodic soil applications of iron chelates formulated for alkaline conditions.

Water Management

Avocados are moderately drought tolerant, however, irrigation will be beneficial to plant growth and crop yields during prolonged dry periods. The specific water requirements for mature trees have not been determined. However, as with other tree crops, the period from bloom and through fruit development is important and drought stress should be avoided at this time with periodic irrigation. Once the rainy season arrives, irrigation frequency may be reduced.

An irrigation system should be established prior to planting the grove. A high volume system is essential. A high volume irrigation system should be a capable of applying a minimum of 0.25 inches of water per acre per hour. This high volume system may be used for cold protection and as a means of irrigation. A second low volume irrigation system (*e.g.*, micro-sprinkler) may also be established specifically for irrigation and application of selected fertilizers (called fertigation).

Irrigation management through the use of soil water content monitoring by tensiometers is recommended. Tensiometers are instruments that measure soil moisture tension and are valuable for monitoring soil moisture levels and scheduling irrigation. Properly installed, placed, and maintained tensiometers may save water, fuel, and fertilizer.

Mulching

Mulching is an effective tool to improve the soil environment, tree health, yield and fruit size, and ultimately profit as it simulated the rainforest floor from where avocado naturally occurs. The type of material that is used is important. It must have C:N ratio between 25:1 and 100:1, break down slowly, and not produce an anaerobic environment during decomposition, and should preferably be composted before application. Mulch should be applied in autumn, after the rains. Mulch should not be applied up against the tree trunk.

Use of Growth Regulators

GA$_3$ to Manipulate Flowering and Yield

Avocado trees (*Persea americana*) bearing a heavy crop produce a light off year flush of flowers the next spring. This results in a light crop and a subsequent intense flowering the year after. GA$_3$ (25 mg/litre) applied in November or January stimulated early development of the vegetative shoot of indeterminate inflorescences. GA$_3$ (25 mg/litre) applied in March at the start of an off year increased 2-fold the production of commercially valuable fruits (213-269 g size class).

Fruit Set and Development

Less than 1 per cent of the flowers ultimately produce fruit. Some varieties set a large number of fruits, most of which drop (absice) during early summer, while others set fewer fruits but retain most of them to maturity. Therefore, avocados can be divided into Type I cultivars that initially set a high number of fruit, most of which are subsequently shed; and Type II cultivars that initially set only a few fruit and these are mostly retained until maturity. Further, some avocado varieties, like Hass, Lamb Hass and Rinton, have been seen yielding typically alternate between heavy and light crops. This is also one of the constraints of avocado production.

Avocados follow a single sigmoid pattern of growth. Increase in fruit size results from cell division and expansion in the early period of growth, but cell division is the main component of growth in the latter stages of fruit development, apparently continuing throughout the time the fruit is on the tree.

Major Insect-Pests and their Management

Avocado Pests

Avocado producers must be familiar with the insects that occur in orchards as pests. Most of these are controlled by natural enemies. The injudicious use of agrochemicals on avocado trees could, however, allow minor pests to develop into major economic risks.

Fruit Flies

This pest has only recently gained economic importance in avocados. When the fruit is picked before it is ripe, the larvae never reach maturity. However, if the fruit remains on the tree for extended periods, as in the home garden, fruit flies may occasionally develop to maturity. The Natal fruitfly attacks both young and older fruit. It lays its eggs just under the skin surface. When the fruit is about golf ball size, a sting lesion appears as a slight puncture mark surrounded by white powdery exudates. As the fruit develops the lesion becomes dry and distinct star-shaped cracks in the skin surface occur. There are 2 methods of control, namely; (i) eradication of unwanted host plants as bug weed or bug tree, bramble and guavas and (ii) baiting.

Heart-Shaped Scale

This insect grows to about 3 mm and has a reddish-brown colour. A white, woolly edge can be seen at the rear end of the adult female. The female lays cream-coloured eggs which are kept underneath the body in the white, woolly secretion. The young scales, known as crawlers, eventually become permanently fixed in one spot. In this way the new leaves become infested. The scales occur on the back of avocado leaves where they suck the sap from the leaves. Fruit is never attacked, but the scales secrete considerable quantities of honeydew, landing on the leaves, branches and fruit on which sooty mould grows, causing a black discoloration of the plant and fruit and interfering with photosynthesis. Wasps, ladybirds, larvae of a lacewing and a fly species (Cecidomyidae) play an important role in the biological control of this scale. As a result of the many natural enemies, chemical control is usually not necessary.

Major Diseases and their Management

☆ *Cercospora* **spot (*Cercospora purpurea*):** Infection appears on fruits and leaves as small, angular, dark brown spots which coalesce to form irregular patches. These spots have a yellow halo. Fruit lesions are frequently the point of entry for other decay organisms, such as the anthracnose fungus. Infection usually occurs during the summer months. *Cercospora* spot can be controlled with two applications of benomyl (0.025 per cent a.i.).

☆ **Avocado scab (*Sphaceloma perseae*):** The scab fungus readily infects young, succulent tissues of leaves, twigs and fruit. These tissues become resistant as they mature. Lesions appear as small, dark spots visible on both sides of the leaves. Spots on leaf veins, petioles and twigs are slightly raised, and oval to elongate. Severe infections distort and stunt leaves. Spots on fruits are dark, oval and raised and eventually coalesce to form cracked and corky areas which impair the appearance but not the internal quality of the fruit. Begin a spray program for scab prevention when bloom buds begin to swell and continue spraying until harvest. The most susceptible commercial variety in Florida is 'Lula'.

☆ **Anthracnose (*Colletotrichum gloeosporoides*):** Anthracnose infection is important only on fruits. Infections occur through lesions caused by other organisms such as scab and *Cercospora* spot, or mechanical injuries. The fungus does not develop in actively growing fruits but causes a rot as the fruit ripens. Fruit lesions start as circular brown to black spots which enlarge and become sunken and crack.

☆ **Avocado root rot (*Phytophthora cinnamomi*):** Trees in areas with poorly drained soils and/or which are subject to flooding are likely to be affected by this fungus. This is the most serious disease in most avocado producing areas of the world. The disease appears to be serious only if trees are subjected to flooded conditions. Leaves of infected trees may be pale green, wilted, or dead, and terminal branches die back in advanced stages of the disease. Feeder roots become darkened and decayed, and severely affected trees usually die. Copper containing fungicides, ridomil or metalaxyl can effectively control this fungus. Further, scion may be grafted on tolerant rootstocks Duke 6, Duke 7 and G6.

☆ **Sun-blotch (Caused by a viroid):** Symptoms of infection include sunken yellow or whitish streaking or spotting and distortion of twigs, leaves, and fruit. It is transmitted through buds, seeds, and root-grafting of infected trees. There is no control for this disease, and infected trees should be destroyed. This disease is; however, rare in occurence.

☆ **Algal leaf spot (*Cephaleuros* sp.):** Symptoms appear first on upper leaf surfaces as green, yellowish-green, or rust-colored, roughly circular spots. This disease is most prevalent during summer and fall months.

☆ ***Diplodia* stem-end rot (*Diplodia* sp.):** This rot disease begins at the stem end of the fruit and develops as the fruit softens. It is usually only a problem with immature fruit after harvest and can be prevented by harvesting only mature fruit. Benlate or Thiabendazole can be used to suppress the infection.

Physiological Disorders and their Management

Chilling Injury

Due to their tropical and subtropical nature, avocados suffer chilling injury when exposed to low storage temperatures. In general, the Guatemalan and Mexican races are less susceptible to chilling injury than the West Indian race. The severity of this injury depends on the temperature, duration of exposure, cultivar, maturity at harvest and production area. Skin pitting, scalding, and blackening are the main external chilling injury symptoms on mature-green avocado kept at 0-2°C (32-36 °F) for more than 7 days before transfer to ripening temperatures. Avocados exposed to 3-5 °C (37-41 °F) for more than two weeks may exhibit internal flesh browning (grey pulp, pulp spot, vascular browning), failure to ripen, and increased susceptibility to pathogen attack.

Sunburn

Sunburn, sometimes called sunscald of fruits, typically occurs in the case of defoliation of trees, exposing fruit or previously shaded bark. Sunburning of immature and mature fruit on the tree may result in discolored peel and flesh damage below the affected peel. Newly planted trees that grew with the bark shaded in the nursery, and trees that are unable to take up enough water because of unhealthy roots or inappropriate irrigation, are highly susceptible to sunburn. Prevent sunburn by providing trees with good growing conditions and proper cultural care, especially suitable volume and frequency of irrigation.

Cukes

Cukes are seedless fruit. The cause of cuke formation is not known although there may be several reasons for its occurence. It is believed that either an environmental or hormonal stimulus induces the development of the cuke.

Double Fruit

This happen as sometimes flowers may have two ovaries and give rise to either a fruit with a two fully developed seeds or one normal seed with the other ovary expressed as a cuke.

Woody Fruits

Woody fruits actually are of wood structure; the normally soft parenchymatous ovary wall is displaced by a stem-like structure which may bear occasional small cauliflower-like appendages or excrescences suggestive of highly modified leaves.

Crick-side

In crick-side the upper half of the fruit on the high side fails to grow normally and as a result this part of the fruit becomes depressed. The part where deficiency of development occurs shows a crowding together of the prominences which make up the pebbling in the rough fruits. Affected flesh is denser and discolors more rapidly on exposure to the air, but otherwise appears about normal. Many crick-side fruits drop while small, and others are lost from a large black spot which develops in the depressed portion. Some affected fruits come to full maturity. It is believed that crick-side may arise due to water stress or high temperature during early fruit development or even may be due to insect feeding during early fruit development.

Ring-Neck

It is a blemish, usually on the fruit-stem or pedicel, consisting of irregular areas of superficial dried tissues which become more or less separated from the living tissue. Sometimes a complete ring of surface tissue dies, separates from the pedicel, and peels off, leaving a scar.

Carapace Spot

The name "carapace spot" was chosen because of the resemblance to a turtle's back. Fruits become corky, externally and usually cracked into somewhat regular, angular divisions. The flesh under carapace spot is undamaged but exterior appearance is such that the fruit is reduced in grade. Slight rubbing or bruising of tender young fruit on leaves or stems appears to cause this corky growth to start. Fruit on trees exposed to strong winds are more prone to develop the trouble.

Papacados

The occurrence of papacados is the extreme expression of avocado thrips damage. Many insects, however, can cause the fruit to be abnormal. Insect injury or stings during the early stages of fruit development probably are the cause of many types of irregular fruits. Injury of this nature is thought to cause some types of fasciation or flattened and irregular growth and woodiness of the fruit.

HARVESTING AND YIELD

Well cared grafted trees begin to bear fruit in their third year after planting, while those raised from seeds start bearing five to six years after planting. Avocados do not ripen as such on the tree but reach maturity when they are ready to pick.

It becomes soft and edible only after it has been picked. The edible part acquires a smooth, buttery texture and the peel shows no sign of shriveling. Mature fruit ripen evenly. Immature fruit, that is the fruit picked too early, will not ripen properly and the skin will eventually become shriveled. Dark skinned varieties reach their full colour when ripe *e.g.* Hass turn a purplish colour when mature. Mature fruits of purple varieties change their colour from purple to maroon, whereas fruits of green varieties become greenish-yellow. Some varieties show a slight thickening of the stem with a slight yellow tinge in the stem and skin, while in most of the varieties, the colour of seed coat within the fruit changes from yellowish white to dark brown when fruits are ready for harvest. These indicators must be within the known times for the fruit to mature.

As a guide pick one of the largest fruit and if it is mature, it should soften in 7 to 10 days without shrivelling. Fruits should never be pulled from the tree rather they should always be cut at the stem leaving a short piece on the fruits. If it is pulled this damages the fruit where the stem attaches and this will allow decaying organisms to enter the fruit. Picking poles with a net or cloth bag at the end can be used for harvesting the fruits. Commercially grown fruit are tested for moisture content to determine the mature fruit. The fruit is normally ready to be picked when it has a moisture content of about 80 per cent or less. In Tamil Nadu, July-August is the peak harvest time, while in Sikkim, fruits of Purple and Green varieties are harvested during July and September-October, respectively.

Storage

Harvested fruit should be cooled down soon after harvest to the recommended storage temperature of the cultivar. Precooling is done most effectively by forced-air or hydrocooling. To delay ripening, fruit must be stored at low temperatures. The lower the temperature, the longer the fruit will take to ripen. However, storage temperatures are that are too low will cause cold damage of fruit. A temperature of 5.5 °C is generally the best. Early maturing avocados may be kept at a slightly higher temperature while late season fruit may be kept at a temperature that is slightly lower.

5

Bael

INTRODUCTION

Aegle marmelos (L.) Correa is commonly known as *bael*, *bilva*, Bengal quince, Golden apple, Stone apple *etc.* It has been known from pre-historic times in India due to its greater mythological significance and its wide use in indigenous system of Indian medicine due to its wonderful medicinal properties. *Bael* is considered to be sacred tree by Hindu community in India and Nepal as it used in the worship of Lord Shiva. In the traditional Newari culture of Nepal, the *bael* tree is part of a fertility ritual for girls known as the *Bel baha*. Girls are 'married' to the *bael* fruit and as long as the fruit is kept safe and never cracks the girl can never become widowed, even if her human husband dies. However, all parts of this tree *i.e.*, root, bark, leaves, flowers, fruits and seed oil are used in various *Ayurvedic* medicines. Due to its great potential and wide prospects in terms of medicinal value and other nutritional purposes, this underutilized crop should be harnessed to its full extent by planting extensively in unproductive and waste lands. This will help in financial upliftment of poor and landless farmers. Also, a systematic and scientific research is necessary on this crop to harvest maximum benefits from this *Plant of Panacea* for human and environmental well-being.

COMPOSITION AND USES

Composition

Bael fruits are rich in carbohydrate, protein and riboflavin. In addition, it is rich in vitamin A (186 IU/100g pulp); volatile oils and marmelosines. Its food value is

88 calories/100g. Thus, it is richer than most of the reputed fruits like apple, guava and mango which have a calorific value of only 64, 59 and 36, respectively.

Table 5.1: Chemical Composition of Fresh *Bael* Fruit
(Food Value/100 g edible portions)

Characters	Composition	Characters	Composition
Water	61.5 per cent	Carotene	55 mg
Carbohydrate	31.8 g	Thiamine	0.13 mg
Protein	1.8 g	Riboflavin	1.19 mg
Fat	0.39 g	Niacin	1.1 mg
Minerals	1.7 g	Vitamin C	7-21 mg

Uses

Bael fruit is not much popular as a dessert fruit due to its hard shell and the mucilaginous pulp, but it is used for the preparation of products like squash, nectar, slab, toffee, powder, *etc.* The pulp from ripe fruits turns brown and develops off-flavour during extraction and processing. In India, green fruits are used for the preparation of *murabba* (preserve) whereas the ripe fruits are eaten fresh or used for the preparation of drinks, marmalade, *sherbets* and syrup. Green fruit slices are sun-dried and are stored for future use. The *bael* fruit also yields dried soluble gum, which is used to prepare adhesives and waterproof oil emulsion coating.

Bael tree has curative properties, which makes it one of the most useful medicinal plants. Several chemical constituents have been isolated and identified from various parts of the *bael* tree. These include alkaloids, coumarins and steroids sterol and aegelin. The unripe or half ripe fruit is regarded astringent, digestive and stomachic. It is beneficial in cases of diarrhea and dysentery. Marmelosin found in *bael* fruit has been found to be therapeutically active. It is prescribed in hepatitis and tuberculosis. Alcoholic extract of the fruit has shown anti-amoebic properties. The fruit is also rich in riboflavin and ascorbic acid. The leaf extract is considered anti-diabetic. Roots of the tree have been found to contain psoralin, xanthotoxin, scopoletin arid tembamide and hence used to treat intermittent fevers.

ORIGIN, HISTORY AND DISTRIBUTION

Bael is indigenous to Indian subcontinent and is grown mainly tropical and sub-tropical regions. It has its origin in Eastern Ghats and Central India. It is also found as a wild tree in lower ranges of Himalayas up to an elevation of 500 m. It is present throughout South-East Asia as a naturalized species, and is found growing in the foothills of Himalayas, Uttarakhand, Uttar Pradesh, Bihar, Rajasthan (Table 5.2), Madhya Pradesh and the Deccan Plateau and along the East coast of India and Pakistan, Sri Lanka, Myanmar, Thailand, Indonesia, the Philippines, Egypt, Vietnam and Bangladesh (Figure 5.1).

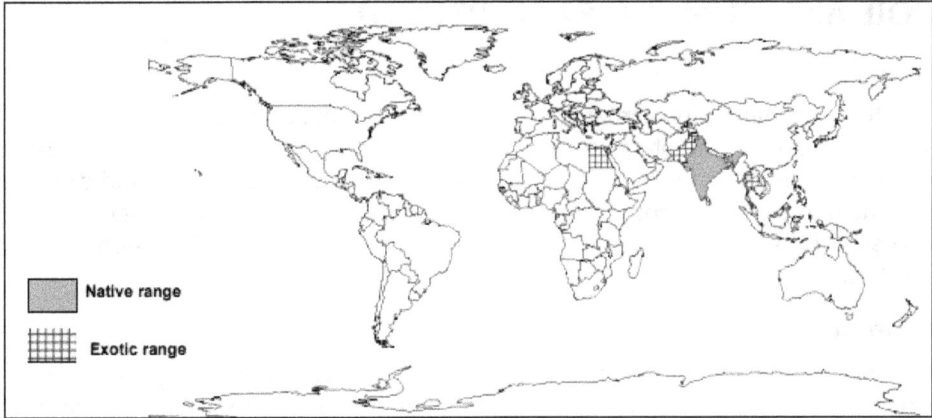

Figure 5.1: Distribution of *Bael* Across different Countries.
(*Source*: http://www.worldagroforestry.org)

Table 5.2: Major *Bael* Growing Belts in India

State	Districts
Bihar	Jamui, Banka, Bhagalpur, Nawada, Munger, Rohtas, Kaimur, Gaya, Aurangabad
Jharkhand	Deoghar
Rajasthan	Bikaner, Sriganganagar, Jhunjhunu, Ajmer, Jodhpur, Jalore, Nagaur, Jaipur, Bhilwara, Sikar, Alwar
Uttar Pradesh	Deoria, Gorakhpur, Basti, Lucknow, Kushinagar, Maharajganj, Ballia,Mirzapur, Rampur
Uttarakhand	Haridwar, Dehradun, Udham Singh Nagar

TAXONOMY AND BOTANICAL DESCRIPTIONS

Taxonomy

Aegle belongs to one of the three monotypic genera of orange sub-family Aurantioideae. The *bale* tree is a diploid species with a somatic chromosome number, $2n = 18$.

Botany

It is a deciduous, 6–8 meters tall, slow-growing tree which bears trifoliate aromatic leaves. The branches sometimes bear long straight spines. The bark is shallowly furrowed and corky. The bisexual flowers are sweet scented and greenish white. A clear, gummy sap, resembling gum arabic, exudes from wounded branches and hangs down in long strands, becoming gradually solid. It is sweet at first taste and then irritating to the throat. Fruits are oblong pyriform having gray or yellow colour; pulp sweet, thick yellow, orange to brown in color. Seeds are numerous and each enclosed in a sac of adhesive, transparent mucilage that solidifies on drying. Seeds have wooly hairs.

SOIL AND CLIMATIC REQUIREMENTS

Soil

It is very hardy and can grow in all type of soils including marginal lands like swamps, alkaline and stony soils. It can also grow well in poor dry soils where other trees fail. But it grows better in a humid climate on fairly rich and drained soils having pH range from 5.5 to 7.5. It can tolerate sodicity up to 30 per cent ESP and salinity of 9 dSm⁻¹ ECE. The tree can be used to control soil erosion and for sand dune fixation.

Climate

Bael is basically an sub-tropical tree but grows well both in tropical and sub-tropical climates up to an altitude of 1200 m and are not injured by temperatures as high as 49°C and as low as 7°C. The plants are drought hardy and moderately frost resistant.

IMPORTANT VARIETIES

Rich biodiversity of *bael* exists in India. Two distinct types of *bael*, the small fruiting type having acrid pulp, high seed and mucilage and the large fruiting type having thin skull, less seeds and mucilage and sweeter pulp. The former is used in medicinal preparations and the other is used for the preparation of *sharbat*, processed products and as a dessert fruit. Some popular types well known in different regions are Siwan in Bihar and Kagzi, Deoria Bada, Etawah, Gonda, Chakiya, Ayodhaya and Mirzapuri in Uttar Pradesh. From the existing genepool, some selections have been made which are NB 5, NB 9, NB 4, NB 7, and NB 9 at Faizabad, Pant Sujata, Pant Shiwani, Pant Aparna and Pant Suvarna at Pantnager and Dhara Road at Jodhpur. Three promising lines, Basti Collection-1, Basti Collection- 2 and Basti Collection-4, superior in physico-chemical characters, have also been identified. Superior cultivars for processing should be developed. Salient features of some of these cultivars and recommended areas for their cultivation are given in Table 5.3.

PLANT PROPAGATION

Bael is generally propagated by seeds. Seeds have no dormancy but get injured when allowed to dry. They can be stored at low temperatures of 4°C and germinate above15°C. Sowing is done in June or July. The development of seedlings is very slow. They require at least a year in the nursery to be fit for transplanting. They should be transplanted in rainy season, when the stem is ordinarily 5-7 cm tall with 3-5 leaves, and the taproot, 20-25 cm long. It is also propagated by root cuttings and stem cuttings treating with IBA (4000 ppm) using quick dip method. Shield and chip budding are considered the best. Rootstocks are raised in the nursery by sowing seeds during June. The seedlings are transplanted in the field after one year at 10 m spacing in square system. Budding on these rootstocks is done during June-July. Seedlings or budded plants are transplanted in the field at a spacing of 10-12 m. Budded plants start bearing fruits at the age of 4-5 years, whereas seedling trees require 7-8 years.

Table 5.3: List of Important Varieties

Name of Variety	Salient Features	Recommended Areas
Narendra Bael (NB)-4	Trees are spreading and oblong fruit in shape, fruit quality excellent and heavy bearer.	Uttar Pradesh
Narendra Bael (NB)-5	Prolific bearer and fruits are medium in size, round with thin skull, low fibre and seed content.	Uttar Pradesh, Rajasthan, Bihar, Jharkhand, Uttarakhand
Narendra Bael (NB)-7	Fruits are very large in size, flattened round, yellowish green in colour.	Uttar Pradesh
Narendra Bael (NB)-9	Prolific bearing, fuits are medium to large size with oblong shape, low fibre and seed content.	Uttar Pradesh, Rajasthan, Bihar, Jharkhand, Uttarakhand
Narendra Bael (NB)-16	Fruits are elliptical round, pulp yellow, average weight 1.3 kg, T.S.S. 31 per cent, and low fibre content.	Uttar Pradesh
Narendra Bael (NB)-17	Fruits are attractive, average weight 1.75 kg, fibre content low.	Uttar Pradesh
CISH B-1	It is a mid-season variety which matures during April-May. Trees are tall, vigorous with dense canopy, erect growth habit, precocious and heavy bearer. Fruit shape is oval to oblong. Average fruit weight 1.0 kg. Suitable for canning and slices preparation.	Uttar Pradesh, Uttarakhand, Bihar, Jharkhand, Rajasthan
CISH B-2	Trees are dwarf with medium spreading habit. Foliage is sparse and almost thornless, precocious with moderate bearing habit. Fruits are oblong to round in shape. Suitable for processing with pleasantly aromatic pulp.	Uttar Pradesh, Uttarakhand, Bihar, Jharkhand, Rajasthan
Goma Yashi	Good quality fruits with large in size. Ovate in shape, greenish yellow in colour. Flesh colour is straw.	Gujarat, Rajasthan
Thar Divya	An early maturing variety, which was developed based on selection from *bael* plants collected from Jaunpur district of Uttar Pradesh in 2006.	Gujarat
Pant Aparna	Its trees are dwarf with drooping foliage, almost thornless, precocious and heavy bearer. The leaves are large, dark green and pear shaped. Fruits are globose in shape with average weight 1.0 kg.	Uttar Pradesh, Uttarakhand
Pant Shivani	It is an early-mid-season variety. Trees are tall, vigorous, dense, upright growth, precocious and heavy bearer. Fruit weight range from 2 to 2.5 kg.	Uttar Pradesh, Uttarakhand
Pant Sujata	Trees are medium dwarf with drooping and spreading foliage, dense, precocious and heavy bearer. Fruit weight varied from 1 to 1.5 kg.	Uttar Pradesh, Uttarakhand
Pant Urvashi	It is mid-season variety. Trees are tall, vigorous, dense, upright growth, precocious and heavy bearer. Fruits are ovoid, oblong. The fruit weight range from 1.5 to 2.5 kg.	Uttar Pradesh, Uttarakhand

Grafting is also found successful in *Bael*. It could be grafted on a number of closely related plants like *Afraegle gabonensis, Afraegle paniculata, Swingla glutinosa* and *Aeglopsis chevalieri*. Successful grafting was also reported on *Afraegle gabonensis* and *Aeglopsis chevalieri* in extreme South-Eastern Florida.

In arid regions, *in situ* budding is considered desirable. This is done by sowing 2-3 seeds directly in the field or by planting polythene raised seedlings. After one year, budding is done in the field. Wild sprouts emerging from the rootstock portion should be removed from time to time.

Root cuttings and layers are also successful in *bael*. Separation and planting of root suckers is done during monsoon. To ensure establishment, suckers are planted in nursery beds for about two years after uprooting before shifting to the field. Top working of old and uneconomic trees can also be done by heading them back during March and budding on the new shoots during June-August.

PLANTING AND ORCHARD ESTABLISHMENT

Bael is planted in square (6 x 6 m) or rectangular (5 x 7 m) system. Pits of 1 x 1 x 1 m size are dug during May-June which are kept open for 15 days. By the end of June, each pit is filled with a mixture containing top soil, 10-15 kg FYM, 100 g diammonium phosphate (DAP) and 50 g chloropyriphos dust. In sodic soils, while filling, 5-10 kg gypsum, depending on the level of sodicity is added to each pit. The best time for planting is July. However, planting can also be done during February-March. Light irrigations must be done until the newly planted saplings establish. Later the *bael* can grow under rainfed conditions. Timely weeding, nutrient application, plant protection, *etc.* also ensure their establishment. Shelter belts and windbreaks should be planted around the orchard to protect from hot and desiccating winds during summer and cool waves during the winter. Fast growing, trees of *gonda, boradi, neem, etc.*, planted in 2-3 staggered rows are considered good.

INTERCULTURAL OPERATIONS

Training and Pruning

Young plants are trained with the help of stakes so that they can grow straight. The plants are trained to a straight central stem on which branches are not allowed up to a height of 60-90 cm. Then scaffold branches are permitted in all directions. Later *bael* does not require any pruning. Dried, diseased and cris-crossing branches are removed when the plants are leafless during winter.

Nutrition Management

In sandy loam soils, besides 10-15 kg FYM, an annual dose of 100 g nitrogen, 50 g phosphorus and 75-100 g potash per plant should be given. The doses should be increased each year by the same quantity up to 7 years of age and then stabilized. July-August is the best time for the application of manures and fertilizers. In case of a dry spell, irrigation should be given after fertilizer application. Deficiencies of micronutrients adversely affect plant growth. These can be corrected by 3 sprays of

0.6 per cent solution containing zinc sulphate, borax and ferrous sulphate in equal proportion during July, October and December.

Water Management

After the plants have established after planting, these can successfully grow under rainfed conditions if 300-500 mm annual rainfall is well distributed. Basin system providing more uniform distribution of water should be used for irrigation of young plants. Irrigation to young plantation should be given just after manuring and fertilization. Productivity and quality of fruits can be increased in arid region if irrigation is done during summer months. Provision of mulching increases productivity of tree by conserving moisture and reducing run-off losses.

Intercropping

During the early age of orchard, interspace between tree rows can be utilized by growing crops of clusterbean and *moth* bean in rainfed areas. Under Irrigated conditions, however, high value crops like vegetables can be grown.

After Care

Bael is susceptible to water logging and hence care should be taken to avoid such conditions in the field. Suckers on the rootstock below the bud union should be removed periodically and keep the stem free of suckers all times. Keep the field weed-free.

FLOWERING, POLLINATION AND FRUIT SET

Vegetatively propagated *bael* trees start bearing 4-5 years after planting while the seedling plants take 7-10 years to commence fruiting. Commercial yield in budded plants comes after about 10 years of planting. In North India, the tree flowers during March-April and the fruits are harvested during February-March. It is a cross-pollinated fruit tree and honeybees are the chief pollinating agents. After successful pollination and fertilization takes place. After set, fruit development is slow initially for about a month, which is followed by rapid development for next four months and thereafter the growth remains more or less static until the fruits are harvested.

PLANT PROTECTION

Major Insect-Pests and their Management

No pest causing economic losses is reported. However the chrysomellids, *Clitea picta* Baly and *C.indica* J. are specific on *bael* which cause noticeable damage to the leaves and shoots sometimes. The coccid, *L. canium viridae* Green, at times, appears in large numbers and suck the plant sap.

Major Diseases and their Management

☆ **Bacterial canker:** *Xanthomonas campestris* pv. *bilvae* causes canker and bacterial shot holes in the leaves, twigs, thorns and fruits. The symptoms appear first on leaves in the form of round, water soaked spots, surrounded by a clear halo which later turn into brown lesions. The pathogen also infects the twigs and thorns and causes canker. Affected portions should be removed along with antibacterial sprays. The disease can be managed by 2-3 sprays of 500 ppm streptomycin or 1 per cent Bordeaux mixture at 15 days interval.

Physiological Disorders and their Management

Fruit drop and cracking before ripening are main pre-harvest physiological disorders in *bael*.

Fruit Drop

Fruit drop in *bael* may occurr due to embryo abortion, physiological imbalances of nutrients such as boron, calcium and zinc, fruit borer attack, fruit rotting and fruit cracking. Growth regulators, 2, 4-D @ 20-30 ppm, NAA @ 20-30 ppm and GA_3 @ 50 ppm if sprayed twice (first after seven days of initiation of growth and second spray after an interval of 15 days) and nutrient like borax @ 0.1 per cent (twice at full bloom and after fruit set) may check fruit drop to a reasonable extent.

Fruit Cracking

Fruit cracking is the physiological disorder in some genotypes of *bael*, which occurs just before ripening. In general, cracking occurs as a result of excessive water absorption by fruits by way of root absorption, while at the same time the ripening and/or other factors are reducing the strength and elasticity of the epicarp/peel. The environmental factors which make epicarp rigid are low humidity in environment or surrounding fruits. In a study at ICAR-CIAH, Bikaner, *bael* fruits were wrapped with cling film, firmly. The wrapping of fruits with cling film resulted in reduced fruit cracking. Further, cracking can be minimized by maintaining optimum moisture regime in soil upto full growth or maturity of fruit, provisioning wind breaks around the orchard to reduce the impact of desiccating winds and by spraying borax @ 0.1 per cent twice at full bloom and after fruit set.

Temperature Related Injuries during Storage

Bael fruits should not be stored below 9°C as brown spots appear on the surface, which gradually increase in number, size, and colour intensity indicating chilling injury. Prolonged storage at low temperatures may lead to internal fruit break down. Likewise, at high temperature (above 14°C) brown-to-black patches appear on the surface of the fruit together with heavy fungal growth inside the fruit indicating fungal spoilage.

MATURITY, HARVESTING AND POSTHARVEST MANAGEMENT

Maturity and Harvesting

Bael is a climacteric fruit. Under North Indian conditions, the fruit takes 8-10 months to mature and 10-12 months for ripening after fruit set. Under such conditions, maturity takes place in December-January and ripening in March-April, continuing upto June. In South India, the fruits start ripening in February and the harvest extends up to April. Stray fruits can be seen on the tree during rest of the period. Fruit should be harvested when they change colour from green to yellow during April-May. Fully mature fruit easily slips from the peduncle indicating correct stage of harvesting. Mature fruits alongwith about 5 cm stalk are harvested with the help of a fruit picker. Fruit skull becomes thick and stony on ripening. After harvesting, its fruit is kept for 8 days while it loses its green tint. Then the stem readily separates from the fruit. The quality of fruits is greatly associated with the weight and size of the seed-sacs. The larger and heavier the seed sacs, the greater is the amount of mucilage and poorer the quality. The fruits can be ripened artificially in 18 to 24 days by treatment with 1,000 to 1,500 ppm ethrel (2-chloroethane phosphonic acid) and storage at 30° C. Care is needed in harvesting and handling to avoid causing cracks in the rind. A tree may yield as many as 800 fruits in a season but an average crop is 150 to 200, or, in the better cultivars, up to 400. *Bael* fruits should be stored at 9° C.

Value Addition

Bael fruits may be cut in half, or the soft types broken open, and the pulp, dressed with palm sugar, eaten for breakfast, as is a common practice in Indonesia. The pulp is often processed as nectar or 'squash' (diluted nectar). A popular drink (called '*sherbet*' in India) is made by beating the seeded pulp together with milk and sugar. A beverage is also made by combining *bael* fruit pulp with that of tamarind. These drinks are consumed perhaps less as food or refreshment than for their medicinal effects.

Mature but still unripe fruits are made into jam, with the addition of citric acid. The pulp is also converted into marmalade or syrup, likewise for both food and therapeutic use, the marmalade being eaten at breakfast by those convalescing from diarrhea and dysentery. A firm jelly is made from the pulp alone, or, better still, combined with guava to modify the astringent flavor. The pulp is also pickled.

Bael pulp is steeped in water, strained, preserved with 350 ppm SO_2, blended with 30 per cent sugar, then dehydrated for 15 hrs at 89° C and pulverized. The powder is enriched with 66 mg per 100 g ascorbic acid and can be stored for 3 months for use in making cold drinks ("squashes"). A confection, *bael* fruit toffee, is prepared by combining the pulp with sugar, glucose, skim milk powder and hydrogenated fat. Indian food technologists view the prospects for expanded *bael* fruit processing as highly promising.

6

Carambola

INTRODUCTION

Carambola (*Averrhoa carambola*) is a multipurpose, drought resistant, subropical tree commonly known as "*kamrakh*", which belongs to family Oxalidaceae. The word 'carambola' is originally a Malayalam language's word, which has been derived from the Sanskrit word *karmaranga* meaning 'food appetizer and when the Portuguese introduced it to Africa and South America they retained the original name. It is also known as five corner/five edge fruit or star fruit, which has been derived from the shape of the fruit when cut crosswise. It is known for its heavy citrus fragrance when ripe. The fruit color can range from a pale white to a golden yellow. The skin has a waxy look to it with heavy ribbing at the edges. Further, the tree makes an ideal specimen plant due to its attractive foliage and branches laden with fruits during the season. Carambola trees are excellent for home landscaping. The foliage is dark green, attractive and flowers and fruit are beautiful. The fruit is valued for its appearance and unusual shape. A close relative of carambola is *bilimbi* (*Averrhoa bilimbi*), which has a more rounded shape and greenish color and produces more acidic fruits and; thus, rarely eaten raw, but usually used for preparing pickles, jellies, and curries or cooked with fish.

COMPOSITION AND USES

Composition

Apart from reducing and non-reducing sugars, edible portion of the carambola fruit is a good source of minerals, volatile favours, natural antioxidants like L-ascorbic acid, (-) epicatechin and gallic acid in gallotannin forms, tannins, dietary fibers, pectin, cellulose, hemicelluloses, iron, calcium, phosphorus and carotenoid compositions.

Carambola fruit are a good source of vitamins and minerals as depicted in Table 6.1.

Table 6.1: Nutritive Value per 100 g of Edible Portion of Carambola Fruits

Constituent	Content	Constituent	Content
Calories	35.7	Citric acid	1.32mg
Carbohydrates	9.38g	Phosphorus	15.5-21.0mg
Protein	0.38g	Carotene	0.003-0.552mg
Fat	0.8g	Thiamin	0.03-0.038mg
Dietary Fiber	0.80-0.90g	Riboflavin	0.019-0.03mg
Calcium	4.4-6.0mg	Niacin	0.294-0.38mg
Potassium	2.35mg	Ascorbic acid	26.0-53.1 mg
Iron	0.32- 1.65mg	α-ketoglutaric acid	2.2mg
Oxalic acid	9.6mg	Methionine	2mg
Tartaric acid	4.37mg	Lysine	26 mg

Uses

The unique star shape and rich golden color makes it attractive garnishes for salads, drinks, sweet and savory dishes particularly seafood dishes after cutting into cross sections. In Western countries, the fruit is generally eaten at a ripe stage when it is yellow. On the other hand, in some Asian countries, the green mature fruit is relished and consumed as fresh and in pickle preparations. A relish can be made from unripe fruits by combining horseradish, celery, vinegar, seasonings and spices.

In India, the ripe fruits or its juice are recommended as appetite stimulant, anti-pyretic, laxative, astringent and antiscorbutic. In Brazil, the fruit is used as diuretic for kidney and bladder related problems. In Ayurveda system of medicines, the ripe fruit is considered as digestive, tonic and causes biliousness. Similarly, in Chinese Materia Medica, it is used to quench thirst and to increase the secretion of saliva. Moreover, the fruits are also used to treat throat inflammation, mouth ulcer, toothache, cough, asthma, hiccups, indigestion, food poisoning, colic, diarrhea, jaundice, malarial splenomegaly, hemorrhoids, skin rashes, pruritis, sunstroke and some eye related problems. Recent studies have shown that the fiber-rich fractions of the fruit possess hypoglyceamic properties, and thus help the body help control blood glucose levels. Besides, carambola's extracts have anti-ulcer properties on account of their ability to protect beneficial gastric mucosa. In certain medicines practices, fruits are used as aphrodisiac for both men and women. In women, the fruits can be used to increase lactation and in large doses to stimulate menstruation. The juice is also applied to remove rust stains. Likewise, the juice can be used to polish brass because of its acidity. Unripe fruits are used in place of a conventional mordant in dyeing, sometimes. The roots of carambola are used to treat arthralgia, chronic headache, epitaxis and spermatorrhea. Furthermore, the roots with sugar are considered as an antidote for poison. The wood of carambola tree is made into furniture.

ORIGIN, HISTORY AND DISTRIBUTION

Carambola is considered to have originated in Southeast Asia (Sri Lanka and the Moluccas) but has been in cultivation in South-East Asia and Malaysia for hundreds of years. Carambola trees were introduced into southern Florida before 1887 and were viewed mainly as a curiosity until recent years. Fruit from the first introductions into Florida were tart. More recently, seeds and vegetative material from Thailand, Taiwan, and Malaysia have been introduced and sweet cultivars have been selected. Several cultivars from Taiwan are being grown at the United States Department of Agriculture's Subtropical Horticulture Research Unit in Miami, including 'Mih Tao' (Plant Inventory No. 272065) introduced in 1963, also 'Dah Pon' and 'Tean Ma' and others identified only by numbers, and Fwang Tung' brought from Thailand by Dr. R J. Knight in 1973. There are certain 'lines' of carambola, such as 'Newcomb', 'Thayer' and 'Arkin' being grown commercially in Southern Florida. Carambolas arrived to Europe in 1598 by way of the Dutch traveler, Linschoten, who amusingly described the fruit as 'a sour apple with ribs', while the British settlers referred to star fruit as 'Coromandel gooseberries'. Despite its prehistoric status, carambola is a relatively new fruit in many parts of the world as short seed viability is the single biggest factor for its slow spread across the continents. The Americans were not growing carambola until the end of the 18th century; Africa the 19th century, while Israel's experiments began in 1930s. Several trees have been growing since 1935 at the Rehovoth Research Station in Israel. In the same year, seeds from Hawaii were planted at the University of Florida's Agricultural Research and Education Center in Homestead. A selection from the resulting seedlings was vegetatively propagated during the 1940's and 1950's and, in late 1965, was officially released under the name 'Golden Star' and distributed to growers.

In many areas, it is grown more as an ornamental than for its fruits. There are some specimens of the tree in special collections in the Caribbean islands, Central America, tropical South America, and also in West Tropical Africa and Zanzibar. It is commonly grown in the provinces of Fukien, Kuangtung and Kuangsi in Southern China. In Florida, carambolas are grown commercially in Dade, Lee, Broward, and Palm Beach counties. It is commonly grown in Malaysia, Taiwan, Thailand, Israel, Florida, Brazil, Indonesia, in the warmer parts of India, Bangladesh and other isoclimatic parts of the world. It is rather popular in the Philippines and Queensland, Australia, and moderately so in some of the South Pacific islands, particularly Tahiti, New Caledonia and Netherlands New Guinea and in Guam and Hawaii. The native and exotic range of cultivation of carambola is presented in Figure 6.1. In India, carambola grows primarily in the Southern states and along the West coast, extending from Kerala up to West Bengal.

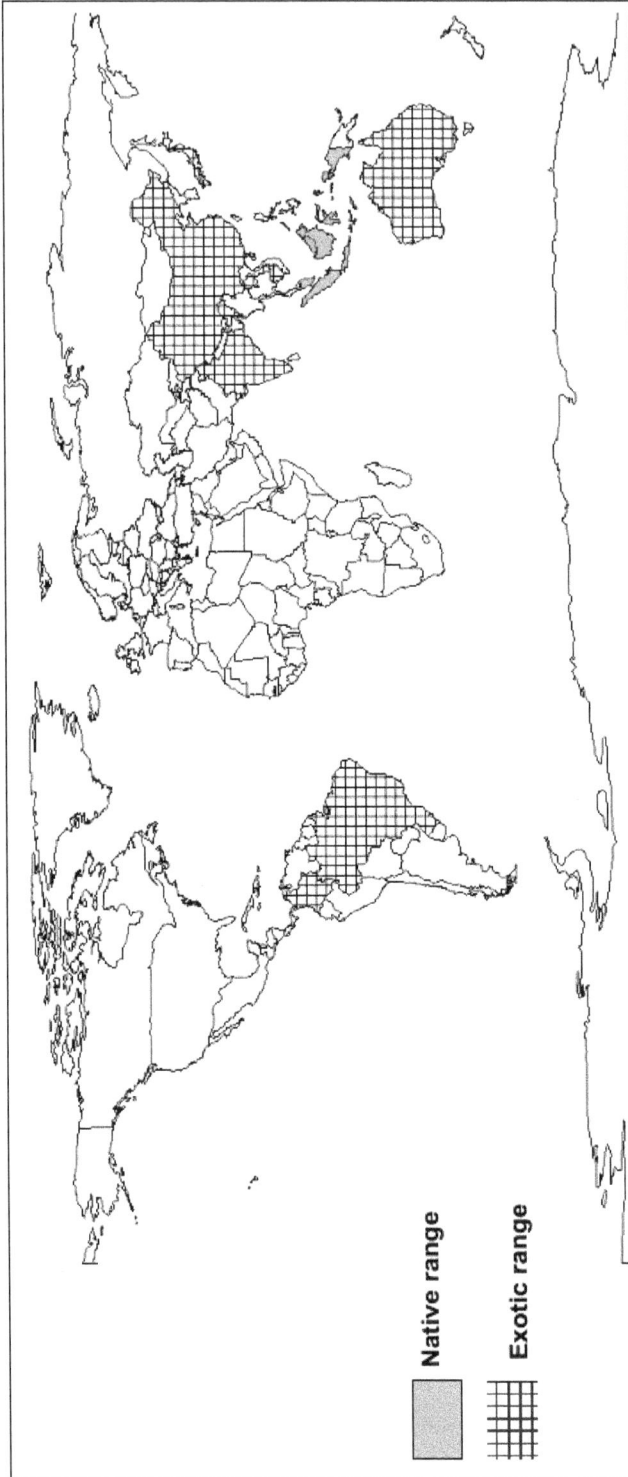

Figure 6.1: Distribution of Carambola.
(*Source:* **http://worldagroforestry.org).**

TAXONOMY AND BOTANICAL DESCRIPTIONS

Taxonomy

Averrhoa carambola is the most important member of the family Oxalidaceae of sub-order Geraniineae and order Geraniales with diploid number of chromosomes; $2n = 2x = 22$ or 24.

Botany

It is a small, attractive, multi-stemmed, slow growing evergreen tree with a short trunk or a shrub, 5-7m of height or rarely, 10m high, spreading 7-8m in diameter. It has a bushy shape with many branches producing a broad, rounded crown. At the base, the trunk reaches a diameter of 15cm. Leaves are 15-25cm long, alternate, spirally arranged, ovate to ovate-oblong in shape, imparipinnate, shortly petiolate with 5-11 green pedant leaflets of 2-9cm long and 1-4.5cm wide. The compound leaves are soft, pubescent, medium-green, smooth on the upper surface and whitish on the underside. The leaflets are reactive to light and tend to fold together at night; they are also sensitive to abrupt shock. Purple to bright purple colored flowers are produced in the axils of the leaves. The flowers are arranged in small clusters and each cluster is attached to the tree with red stalks. The flowers are small, about 6mm wide, pedicellate with 5 petals (having curve ends) and sepals. Depending upon the cultivar, carambola flowers have either long or short styles (*i.e.* heterostyly), a factor which affects self- and cross-pollination. The fruits are green when small and unripe but turn to yellow or orange when matured and ripe. The fruits are pulpy with an oblong shape, longitudinally 5-6 angled, 5-15 cm long and up to 9cm wide. The fruits are crunchy, having a crisp texture and when cut in cross-section are star shaped, hence its name. The odor of the fruits resembles oxalic acid and their taste varies from very sour to mildly sweetish or sweetish. The pulp is light yellow to yellow, translucent and very juicy without fiber. There may be up to 12 flat, thin, 5 mm long seeds or none at all. The brown colored seeds are enclosed by a gelatinous aril and lose viability in a few days after removal from the fruit. Star fruit is easily propagated from fully developed seeds.

SOIL AND CLIMATIC REQUIREMENTS

Soil

Carambola is not too particular as to soil; thus, it can be grown in a wider range of soils *viz.*, right from sand to heavy clay loam. Though, the best suited soil is rich loam with a pH of 5.2 to 6.2. It is often chlorotic on limestone as trees generally develop iron, magnesium, and manganese deficiencies when grown in soils with a pH above 7. Symptoms of iron deficiency are interveinal chlorosis (green veins with yellowing in between), reduced leaflet size, and, with severe deficiency, leaflets may become almost white. Symptoms of magnesium deficiency include a mottling of green and yellow areas. Symptoms of manganese deficiency include reduced leaf size and yellowing. Likewise, carambola trees cannot tolerate saline soil and water. Symptoms of injury include browning of leaflet margins, leaf drop, stem and limb dieback, reduced fruit size, and in severe conditions, tree death.

Carambola needs a good drainage. Carambola trees are moderately tolerant of excessively wet or flooded soil conditions for about 2 to 10 days depending upon tree health, air temperatures (less time when it is hot), and presence of root diseases. However, during flooding time the tree stops growing, and if wet conditions persist, symptoms of flooding develop. Symptoms of excessively wet soil conditions include leaf wilting, yellowing and browning of leaves, leaf and fruit drop, stem and limb dieback, and, if wet conditions persist, tree death. In addition, root rotting fungi may attack the root system, causing trees to decline, die back, or die. Similarly, carambola trees have only limited tolerance to drought. Symptoms of excessively dry soil conditions (drought) include leaf folding, leaf wilting, yellowing and browning of leaves, leaf drop, reduced flowering and fruit size, stem and limb dieback, and in severe drought, tree death.

Climate

Since carambola is a tropical and sub-tropical fruit in nature; therefore, it thrives well in warm, humid condition with long hours of sunshine. It requires an annual rainfall of 1800 mm. The optimum temperature for the growth is 20-35 °C. It is best grown in frost free subtropical climate. It can withstand light frost when established. Its cultivation can be taken up to an elevation of 1200 m in India. Low air temperatures of -1.1° to 0°C may kill immature leaves, while young trees, mature twigs and leaves may be killed at -2.8° to -1.7°C. Small branches on mature trees may be damaged after exposure to -3.9° to -1.7°C, whereas temperatures of -6.7° to -4.4°C may kill large branches and mature trees. Carambola cultivars may vary in their susceptibility to wind damage. Symptoms of wind damage include defoliation, desiccation, twig dieback, stunted growth, and fruit damage (wind scar). Cultivars such as 'Golden Star' and 'Newcomb' can withstand windy conditions better than 'Arkin' and 'Fwang Tung.' Trees protected by wind breaks are generally more vigorous and productive than wind-exposed trees.

IMPORTANT VARIETIES

Carambola varieties are grouped into two main types *viz.*, sour and sweet. The sour types are smaller, more sour (contains as much as 1 per cent acid), richly flavored, while the sweet types are mild-flavored, rather bland, with less oxalic acid (0.4 per cent) with 5 per cent sugars. However, tart cultivars, such as 'Golden Star' attain a sweet flavor if they are allowed to ripen on the tree (*i.e.*, become golden yellow). Sweet types are recommended for fresh fruit while both sweet and tart types are useful for processing and home recipes. Some Chinese types like Fuang Tung are very sweet and Brazilian ones are rich in vitamin C. Some superior types of carambola are available at Columbia (Icambola), Taiwan (Tean Ma, Min Tao) and Hawaii (Golden Star).

Brief description of some important cultivars, commercially grown in other countries, is given below:

☆ **Golden Star:** Developed at Florida. The wing edges are slightly rounded. Fruits are golden yellow, crisp each weighting 90m- 200 g.

☆ **B1:** Developed at Malaysia having lemon yellow fruit which are crisp each weighing 100-300 g. The edges are rounded.

☆ **B6:** Another variety developed at Malaysia with orange fruits. The edges are slightly rounded.

☆ **Sri Kembanqan (Kembangan):** Originated in Thailand. Elongated pointed fruit, 5-6 inches long. Bright yellow-orange peel and pulp. Juicy, firm pulp with few seeds. Flavor rich and sweet with excellent dessert quality.

☆ **Wheeler:** Medium to large, elongated fruit. Orange skin and pulp. Mildly sweet flavor. Tree a heavy bearer.

☆ **Maha:** Commonly grown in Florida. Fruits pale lemon yellow each weighing 100-200g.

FLOWERING AND POLLINATION

Flowering

Flowering in carambola appears on small branches from axillary buds in short panicles or cymes. These are also formed on axils of leaves. Flowering and fruiting take place also on large and thick branches and even on trunks. Fruits are borne in cluster, year round under Indian conditions, but peak flowering occurs in three flushes *i.e.* April-May, July-August and September-October and fruits are harvested in July-September, November-December and January-February, respectively. Grafted trees start bearing 1-2 years after planting, while seedling varieties may take longer *i.e.* from 4-8 years.

Pollination

All the flowers on a given carambola variety have either long or short styles; this condition is called heterostyly. Heterostyly is defined as the occurrence in a species of two or more floral morphs that exhibit reciprocal herkogamy, which is the spatial separation of pollen presentation and pollen receipt within or between blossoms of an individual plant. The heterostyly in carambola is of distylous nature, where two morphs or forms of flowers are found *i.e.* pin (long styled flowers) and thrum (short styled flowers). Self-incompatibility is also reported in carambola. In distylous plants the supergene determining floral morphology also controls a diallelic sporophytic self-incompatibility system, so that only pollinations between morphs are compatible (Figure 6.2). Tristyly has been reported in bilimbi.

Some carambola cultivars may require cross pollination (short-styled by long-styled cultivar or *vice versa*) for

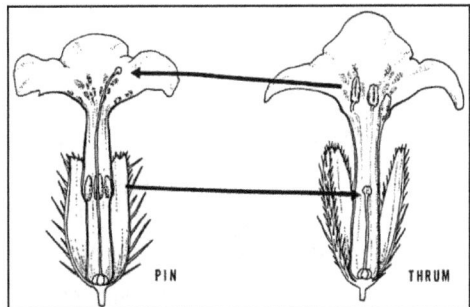

Figure 6.2. Reciprocal Herkogamy *i.e.* Anther and Stigma Positions in the Pin and Thrum Forms of a Distylous Plant. The arrows indicate the directions of compatible pollinations.

good fruit set and yields. However, varieties such as 'Fwang Tung', 'Golden Star' and 'Arkin' produce abundant crops when planted in solid blocks, indicating that the need for cross pollination by opposing stylar types is not always necessary. Other varieties such as 'B-10' and 'B-17' produce more fruit when cross pollinated with another variety. Carambola is insect pollinated, the pollinators being honeybees and Diptera species. Studies have shown that the fruit set takes place on the fourth day of opening of flower in both 'sour' and 'sweet' types. Natural open pollination promotes fruit set, while no set was noted under 'natural self-pollination'. The fruit set takes place on the fourth day of opening of flower ('sour' and 'sweet'). Initial fruit set under natural condition per panicle was recorded to be 58.9 per cent in 'sour' and 74.5 per cent in 'sweet' variety, respectively. The number of mature fruit obtained per panicle is approximately two in sour and four in sweet type. Out of the growing fruits, 84 per cent drops off from the tree and the remaining only 16 per cent reach final maturity. About half of the total drop occurrs during the later half period of fruit growth. Fruits mature at the end of 14th or 15th week after set and the total period required from opening of flower till maturity is in between 101 to 108 days. The growth of fruit continues till maturity after which it remains hardly for a week on the tree. During this period, growth of the fruit remains completely ceased.

PLANT PROPAGATION

Carambolas raised through seeds are unlikely to produce fruit which closely resembles the mother tree owing to the fact that it is genetically heterozygous. Seeds are used for generating variability or raising rootstocks. When it has to be raised through seeds, it must be done quickly because the seeds have viability only for a few days. Only plump, fully developed seeds should be selected. When seedlings are to be utilized as rootstock, they should be at least one year old, on which budding (shield) or grafting can be taken up. Asexual propagation by grafting is recommended as it is relatively simple. The different techniques of grafting in common use are wedge, splice (whip and tongue), side veneer and approach grafts. Approach grafting is, however, very successful. Forcing of bud wood by removing leaves of the scions before 3 to 7 days of budding is suggested. Use of M-18960 seedling rootstock and cv. Golden Star as scion are recommended for calcareous soils in Florida. Air layering of one year old terminal shoots, treated with IBA 10,000 ppm at the time of layering, is also one of the methods of vegetative propagation recommended for carambola production.

PLANTING AND ORCHARD ESTABLISHMENT

Proper planting is one of the most important steps in successfully establishing and growing a strong, productive tree. The plants can be planted in pits of size 1m x 1m x 1m dug at the spacing of 8m x 6m or 8m x 8m.

For better orchard establishment, operations like regular irrigation of young plants, staking, maintenance of weed-free basin and protection from frost are important and should be carried out as per need.

INTERCULTURAL OPERATIONS

Training and Pruning

For commercial orchard, tree size control is important for retaining fruit production in the lower tree canopies and facilitating foliar spraying and harvesting. Since carambola is a fast-growing tree, it requires pruning and thinning of excess branches at an early stage. In young trees 4-6 wide crotch angled branches are allowed. Young trees should be pruned by tipping shoots in excess of 2 to 3 ft to increase branching. Low hanging, criss-cross, diseased or infested branches should be pruned one year after planting. If desired, trees may be trained to a modified central leader or open center configuration. Selectively removing a few upper limbs back to their origins (crotches) each year will help prevent the loss of the lower tree canopy due to shading by the upper canopy. Researchers at University of Florida, U.S.A. had shown that removal of upright limbs during late winter followed by removal of selected new re-growth (shoots) and heading back remaining new shoots to one-half their length in early fall, can maintain mature (7 to 9 year old) trees at a height of 2.7 to 4.0 m without significantly affecting fruit production. On the other hand, a healthy tree in the home garden requires no pruning except for occasional removal of dead branches. Removal of upright limbs will reduce tree height and maintain fruit production in the lower canopy if tree size or crowding becomes a problem.

Nutrition Management

It is a good practice to manure the trees once in a year (40-50kg of farmyard manure/tree) and also apply some nitrogenous and phosphatic fertilizers, the doses depending upon the age and size of the tree and the fertility of the soil. One year old tree is applied 40 g N, 10g P and 70 g K in 4 equal splits during April, June, September and January. Every year it can be increased in the same ratio. However, an 8 years or older tree may be supplied with 600 g N, 120 g P and 1000 g K/tree, annually. Occasionally, it is prone to minor mineral deficiencies, which can be corrected by application of micronutrients. In acid to neutral pH soils, micronutrients such as manganese, zinc, and iron may be applied in dry applications to the soil or in a liquid form and sprayed onto the leaves. Three to 6 applications should be made per year. Trees growing in calcareous soils should receive 4 to 8 foliar applications per year of zinc and manganese. Iron deficiency may be corrected by 3 to 6 yearly soil drench applications of chelated iron specifically formulated for calcareous soils.

Water Management

Irrigation is recommended for commercial carambola orchards as it is very sensitive to water stress. It demands very high quantity of water (2000 l/week/ tree/in maximum demanding months). Young trees should be irrigated regularly to facilitate tree establishment and growth. Once trees begin to bear (1 to 2 years after planting), trees should be irrigated regularly from flowering through harvest. Mulching the tree basin can help prevent evaporative loss; thereby, conserving available moisture. Mulch the tree basin with a 5-15 cm thick layer of locally available materials such as grass clippings, paddy husk and leaf moulds *etc.* or black polythene mulch. Besides conserving moisture, mulches can help smothering weeds being grown under tree. Keep mulch 20-30 cm away from the trunk. High volume under or over tree-sprinkler irrigation has been observed to adequately protect carambola trees during frost incidence.

Weed Management

A weed-free area (2 to 5 or more feet away from trunk of the tree) should be maintained. Utmost care should be taken to avoid any mechanical damage to trunk during cultivation. Mechanical damage to the trunk of the tree will weaken the tree and, if severe enough, can cause dieback or kill the tree.

Intercropping

During pre-bearing phase of orchard, the vacant spaces in between plants and rows may be utilized for cultivation of short duration fruit crops and vegetable crops for generating the extra income from the orchards.

PLANT PROTECTION

Major Insect-Pests and their Management

Carambola is relatively pest-free; however, some pests like fruit borer and fruit sucking moth do attack trees. In other countries, fruit flies are menace of carambola cultivation.

Fruit borer can be managed by spraying chlorpyriphos @ 1 ml/l, while for the management of fruit sucking moth, enclosing the fruits with mesh less than 10 mm is recommended.

Major Diseases and their Management

No serious diseases are known to be of sufficient importance to require control measures; however, carambola is sometimes affected by the occurrence of *Cercospora* leaf spot, which can be managed by application of Indofil M. 45@ 2g/lt.

Physiological Disorders and their Management

Physical Injury

The major problem is physical injury, especially on the rib edges, that leads to browning. Injury due to abrasion and impact can be avoided by careful handling. Browning due to mechanical injury can intensify with water loss. Fruit that have lost about 5 per cent of their weight due to water loss show visible symptoms of dehydration.

Chilling Injury

Carambola is not especially sensitive to chilling. However, during low-temperature storage at 0°C or 5°C for 2 and 6 weeks, respectively, some small surface pitting and rib edge browning can occur. The severity of injury increases with storage time.

HARVESTING AND YIELD

Carambola fruits are usually harvested by hand from the tree or shaken down manually. Yield varies according to the age, variety and plant health. The yield ranges from 200-300 kg/tree from 7-8 years onwards. Harvesting is based on physiological and horticultural maturity as indicated by skin color change from green to yellowish-green, then to full yellow or yellowish-orange. Color represents an accurate means to determine the fruit ripeness. Optimum sugars are achieved at the full yellow color; however, ripe fruit are more fragile and easily damaged; therefore, commercially, carambola fruit are picked at color break. Fruit that are 50 to 75 per cent yellow are firmer than full-color fruit and are therefore regarded as commercially mature. Fruit continue to develop color after harvest, although there is little other change in quality. The capability of the fruit for storage at longer periods without damaging its quality and its ability to ripen normally at 23°C, can be programmed to supply fruits of required maturity such as green mature, half ripe or full ripe, to the consumer market depending upon the preference of consumers as well as to the fruit processing industries. Carambolas do not increase in sugar content after picking; therefore, those interested in fruit with optimum sweetness should pick fruit when all traces of green disappear from the fruit surface (yellow to golden yellow color). Fruit picked at color break will develop yellow color, at the temperature of 21-23 °C, after cold storage. Fruit can be stored at cold temperatures (5 to 7 °C) in the refrigerator for three weeks. Even at 21 °C, carambola fruits can be stored for two weeks. Before sending the fruits to the market, they can be graded into 3 sizes *viz.*, small (130-160 g), medium (160-190 g) and large (190 g).

VALUE ADDITION

Carambolas are cooked in puddings, tarts, stews and curries. Slightly under-ripe fruits are salted, pickled or made into jam or other preserves. The juice makes a delicious iced drink alone or in combination with other beverages. In China and Taiwan, carambolas are sliced lengthwise and canned in syrup for export. Sour fruits can be pricked to allow absorption of sugar and cooked in syrup, at first 33 °Brix and later 72 °Brix, to make acceptable candied product. The ripe fruits are sometimes dried in Jamaica. In Hawaii, the juice of sour fruits is mixed with gelatin, sugar, lemon juice and boiling water to make beverages.

7

Cherry

INTRODUCTION

Cherry occupies an important place among temperate stone fruits world over. The cherry is the fruit of many plants of the genus *Prunus*, and is a fleshy drupe (stone fruit). The cherry fruits of commerce are usually obtained from a limited number of species such as cultivars of the sweet cherry (*Prunus avium* L.), sour cherry (*Prunus cerasus* L.) and duke cherries (*Prunus avium x Prunus cerasus*). Cherry is delicious fruit, and has more calorific value than apple. Due to higher return, cherry is gaining popularity in temperate regions of our country.

COMPOSITION AND USES

Composition

Sweet as well as sour cherries are attractive and delicious fruits and contain some amount of proteins, dietary fibre, sugars and have higher calorific value. Among minerals, K, Ca, Mg, Fe and Zn are found in measurable amounts (Table 7.1). Compared to sweet cherries, raw sour cherries contain higher content per 100 g of vitamin C (Table 7.1).

Uses

Cherry fruits are very rich in stable anti-oxidant 'melatonin'. Melatonin can cross the blood-brain barrier easily and produces soothing effects on the brain neurons, calming down nervous system irritability, which helps relieve neurosis, insomnia and headache conditions. Methyl anthranilate and methyl salicylate are major flavouring agents.

Table 7.1: Nutritional Value of Sweet and Sour Cherries per 100 g of Edible Portion

Attribute	Contents	
	Sweet Cherry	Sour Cherry
Energy	63 kcal	50 kcal
Carbohydrates	16 g	12.2 g
Sugars	12.8 g	8.5 g
Dietary fiber	2.1 g	1.6 g
Protein	1.1 g	1.0 g
β-carotene	38 µg	770 µg
Vitamin C	7 mg	10 mg
Calcium	12 mg	16 mg
Phosphorus	21 mg	9 mg
Magnesium	11 mg	15mg
Potassium	222 mg	173 mg

Sweet cherries are mainly used for table purposes, although some value added products are also made.

ORIGIN, HISTORY AND DISTRIBUTION

The cherries are considered native of South-East Europe, Western Asia and Asia Minor. The different forms of cherries are cross compatible to form hybrids easily as well as graft compatible. The cherries were domesticated in Greece around 300 BC and subsequently spread to England, Germany, Belgium, Portugal and other parts of Europe. Later, cherries eventually spread to other parts of the world and were then domesticated in North America and Asia.

Table 7.2: Top-10 Cherry Producing Countries of the World

Sweet Cherry		Sour Cherry	
Country	Production (MT)	Country	Poduction (MT)
Turkey	480,748	Turkey	187,941
United States	384,646	Russia	183,300
Iran	200,000	Poland	175,391
Italy	104,766	Ukraine	172,800
Spain	98,400	Iran	105,000
Chile	90,000	Serbia	74,656
Uzbekistan	84,000	Hungary	53,425
Syria	82,341	United States	38,601
Ukraine	72,600	Uzbekistan	34,000
Russia	72,000	Romania	70,542

Source: FAO-2012.

The major cherry producing countries are Italy, USA. Germany, France, Russia, Hungary, Spain, UK, Denmark, Poland, Ugoslavia, Spain, Turkey, Romania, Bulgaria, Italy, Greece and Japan. In India, the sweet cherries are cultivated on a commercial scale in Jammu and Kashmir and to a limited extent in Himachal Pradesh.

TAXONOMY AND BOTANICAL DESCRIPTIONS

Taxonomy

Cherry belongs to the family Rosaceae, sub-family Prunoideae, genus Prunus, subgenus Cerasus and section Eucerasus. The cultivated cherries are divided into three groups *viz.*, Sweet cherries (*Prunus avium* L.), sour cherries/tart cherry (*Prunus cerasus* L.) and Duke cherries (*Prunus avium x Prunus cerasus*). In addition to 3 named species of cherry, several other species have horticultural importance. For instance, *P. besseyi, P. pumila, P. tomentosa, P. padus, P. nipponica, P. sargentii* and *P. mahaleb* are mainly used for ornamental and rootstock purposes.

Several other species of cherry such as *Prunus apetala* ((clove cherry), *Prunus campanulata* (Taiwan cherry or bell-flowered cherry), *Prunus cornuta* (Himalayan bird cherry), *Prunus cyclamina* (cyclamen cherry or Chinese flowering cherry), *Prunus fruticosa* (European dwarf cherry, dwarf cherry or steppe cherry), *Prunus grayana* (Japanese bird cherry), *Prunus humilis* (Chinese plum-cherry), *Prunus incisa* (Fuji cherry), *Prunus japonica* (Korean cherry), *Prunus myrtifolia* (West Indian cherry), *Prunus nipponica* (Alpine cherry), *Prunus pumila* (Sand cherry), *Prunus rufa* (Himalayan cherry), *Prunus salicifolia* (Singapore cherry) and *Prunus tomentosa* (Nanking cherry or Chinese bush cherry) are also grown to some extent.

Most of its varieties are diploid (2n = 16). The sour cherries and duke cherries are mainly used for processing purposes and not popular commercially, and most of their varieties are tetraploids (2n = 32), and only a few cultivars are diploids.

Botany

Sweet cherry is a deciduous tree growing to 15–32 m tall. Young trees show strong apical dominance with a straight trunk and symmetrical conical crown, becoming rounded to irregular on old trees. The bark is smooth purplish-brown with prominent horizontal grey-brown lenticels on young trees, becoming thick dark blackish-brown and fissured on old trees. The leaves are alternate, having a serrated margin and an acuminate tip and strongly veined, with a green or reddish petiole bearing two to five small red glands. In autumn, the leaves turn orange, pink or red before falling. The white flowers are produced on short spurs in early spring at the same time as the new leaves, borne in corymbs of two to six together. Spurs are long-lived, producing for 10-12 years. The fruit is a drupe, bright red to dark purple when mature in mid-summer. All parts of the plant except for the ripe fruit are slightly toxic, containing cyanogenic glycosides.

Sour cherry is a medium sized deciduous tree with a round, and more spreading canopy than sweet cherry. Leaves are elliptic with acute tips, mildly serrate margins, smaller than sweet cherry, with long petioles. Individual flowers are the same as for sweet cherry. Sour cherry inflorescence buds usually produce 2-4 flowers, with long pedicels, as in sweet cherry. However, many are borne laterally on 1-yr wood, not exclusively on spurs as in sweet cherry. Spurs are shorter-lived than sweet cherry, gradually declining in productivity over 3-5 years. Sour cherries are the latest blooming among the stone fruits. Its fruit is drupe as in sweet cherry but have lower sugars and higher organic acid contents than sweet cherries, giving them distinct flavour. Fruits are generally bright red in colour and exhibit less colour variations than sweet cherry.

SOIL AND CLIMATIC REQUIREMENTS

Soil

Cherries grow in different types of soils but for profitable production, but fertile, well drained chalky or deep sandy loam soils with pH 6.5-7.0, and can hold, moisture during summer are most suitable for its commercial cultivation. Soil should not be waterlogged as cherries can't tolerate wet feet.

Climate

Sweet cherries are more exacting in climatic requirements than other stone fruits. The cherries are well adapted to cooler climates and require 1000-1500 chilling hours below 7 °C during winters. Thus, cherries are grown successfully in areas between 2,000 and 2,700m in India. In general sweet and sour cherries differ in climatic requirements. Sour cherry varieties can tolerate more cold than sweet cherry varieties. Sweet cherry is also more susceptible to spring frost. Sour cherry does not survive in warm climate while sweet cherry is adverse to both severe cold and warm climate.

Cherry blossoms are badly damaged by spring frosts; therefore frost-free sites of hill slope and valley areas with and outlet for the cold air are preferred. Since Southern and South-Western aspects are warmer, they should not be selected at lower elevations. Generally, North-Eastern aspects are most suitable for cherry cultivation. Cherries require about 100-120cm of fairly well distributed rainfall throughout the year. However, rains during flowering and fruit ripening may results in to blossom wilt and fruit cracking, respectively. Hence, weather should be dry at the time of fruit ripening. In India, climate of Jammu and Kashmir, high hills of Himachal Pradesh and Uttar Pradesh is highly ideal for its commercial cultivation.

IMPORTANT VARIETIES

Sweet Cherries

The varieties of sweet cherries are divided into two groups, *viz.,* Heart group and Bigarreau group. Varieties of Heart group have soft and tender flesh and ripen early. The Lyons, Early Winkler, Black Heart, Red Heart, Black Tartarian, Early Purple, Coe, Ida and Elton. The varieties of Bigarreau group are usually round in shape, with firm flesh and late ripening. Important varieties of this group are Lambert, Stella, Bing, Van Windsor, Schmidt, Napoleon, Emperor Francis, Ranier, Yellow Spanish and Lambert. Nearly all sweet cherry cultivars appear to be self-unfruitful. They produce viable pollen, but not all cultivar combinations are fruitful.

There are many cross-incompatible groups of sweet cherries. Cultivars within a group should not be planted together without a suitable pollinizer. Some of the most common cross-incompatible groupings are as under:

  I. Bing, Lambert, Napolean, Emperor Francis, Somerset, and Vernon

 II. Windsor and Abundance

III. Black Tartarian, Somerset, Black Eagle, Knight's Early Black, Bedford Prolific, and Early Rivers

IV. Centennial and Napolean

 V. Advance and Rockport

VI. Elton, Governor Wood, Stark's Gold, and Hartland

VII. Early Purple, Royalton, and Rockport

VIII. Black Tartarian, Early Rivers, and V29023

IX. Sodus, Van, Venus, and Windsor

 X. Velvet, Victor, Gold, Merton Heart, Viva, and Vogue

XI. Hedelfingen, Vic, and Ulster

XII. Hudson, Giant, Schmidt, Ursala, Chinook, Ranier, and Viscount

Due to the problem of cross-incompatibility in sweet cherries, fruit breeders have developed self-fertile sweet cherry cultivars, which can easily be planted in solid blocks. Some self-fertile varieties are Sonata, Stella, Lapins, Skeena, White Gold, Black Gold, Symphony *etc.*

Sour Cherries

Tart or sour cherries are well adapted to various soil and climatic conditions than sweet cherries. Tart cherries are self-fruitful and produce commercial crops when planted in solid blocks. They do not have severe cracking problem like sweet cherries. The important varieties of this group are Montomorency, Early Richmond, English Morellos, Ferracida, Hebros, Gabrovskea, Leto, Raxershen and Okinawa. However, sour cherry varieties are neither not popular not grown commercially in our country.

Duke Cherries

This group is also not popular in Indian conditions. Some important varieties are May Duke, Royal Duke, Late Duke, Brassington and Reine Hortense.

Table 7.3: Recommended Varieties of Sweet Cherry for different States of India

State	Varieties
Jammu and Kashmir	Black Heart, Early Purple Black Heart, Guigne Noir Gross Lucenta, Guigne Noir Hative, Guigne Pour ova Precece, Bigarreau Napoleon and Bigarreau Noir Gross
Himachal Pradesh	Black Tartarian, Bing, Napoleon White, Sam, Sue, Stella, Van, Lambert, Black Republican, Pink Early, White Heart and Early Rivers
Uttar Pradesh	Bedford Prolific, Black Heart and Governor's Wood

The chief characteristics of some important varieties of cherries are briefly described hereunder.

- ☆ **Lambert:** Trees are medium in vigour, mid flowering, fruits are large, round in shape, skin deep ruby red blackish in colour, susceptible to rain cracking and mid to late season maturity.

- ☆ **Bing:** Trees are upright to spreading and vigourous, mid flowering, very large sized fruit with roundish to slightly heart shape, skin maroon or deep red in colour.

- ☆ **Stella:** The first self-fertile sweet cherry. Its fruits are dark red in colour, medium-sized, firm, and of good quality, containing small stone. The tree is medium spreading in vigour, precocious, and moderately hardy. It is a good pollinizer for other cultivars and produces fruits without pollenizer.

- ☆ **Sam:** Its fruits are dark mahogany, large, firm, and with good fruit quality. The tree is vigorous and hardy.

- ☆ **Black Heart:** Vigourous trees, mid-season flowering, fruits are medium in size and heart shaped with compressed apex, skin colour deep shiny blackish purple and early to mid-season maturity.

- ☆ **Napoleon (Royal Ann):** This is a widely grown mid-season cultivar. Its fruits are large and yellow with pink blush. It is firm but tends to crack badly. The tree is hardy and productive, but susceptible to freeze damage. Sometimes it becomes over-productive, resulting in small fruits and biennial bearing.

- ☆ **Rainier:** Its fruits are yellow with considerable pink blush, firm, and of high quality. The tree is vigorous and as hardy like Van. Rainier is a good pollinizer for Van and Sam.

- ☆ **Schmidt:** Its fruits are large, very dark mahogany in colour, firm and of high quality. It is relatively fruit cracking resistant. The tree is hardy, but blossom buds are relatively susceptible to spring frost. It is poor producer but a good pollinator for Napoleon.

- ☆ **Lapins:** Its fruits are large, firm, and very dark red. Flowers are self-fertile and it is a good pollenizer for other varieties.

- ☆ **Van:** Its fruits are dark mahogany in colour, sweet and firm, with excellent quality. The tree tends hardy. It is a good pollinizer for Napoleon and other cultivars.

- ☆ **Windsor:** The standard late ripening cultivar. Its fruits are dark mahogany in colour, medium sized, tender, and of fair to good quality. The tree is hardy and productive, with good bud hardiness.

- ☆ **Gold:** Gold ripens later than Emperor Francis. Its fruits are lemon yellow coloured with no red blush; resistant to cracking, and of good quality. The tree is vigorous and very winter hardy, with hardy fruit buds.

☆ **Bigarreau Napolian (Double):** Trees are medium to vigorous, early season flowering, fruits are longish large, medium to large and heart shaped, skin colour creamy red yellow and early to mid-season maturity.

☆ **Bigarreau Noir Grossa (Mishri):** Trees are medium to large in vigour, fruits are with obtuse heart shape, skin dark red colour which finally changes to blackish purple and mid to late season maturity.

☆ **CITH Cherry-01:** Trees are semi spreading, precocious, regular and prolific bearing cultivar selected from Bigarreu Napolion (Double Gilass) cherry orchard. Fruits are large, ovoid in shape, attractive, dark red colored with long pedicels. Fruits have good acid/sugar balance and high in TSS.

☆ **CITH Cherry-02:** Trees are upright, precocious, prolific and regular bearer selected from Local Mishri. Fruits are large with attractive red and high in TSS as compared to Mishri and mature 10 days earlier than 'Mishri.

FLOWERING, POLLINATION AND FRUIT SET

Cherry starts flowering after 3rd year of planting on dwarf rootstocks and 5th year of planting on seedling rootstocks. Flowering commences in spring season. The flowers appear laterally on one-year-old spurs and shoots. Honeybees are the chief pollinating agents. Most of the sweet cherry varieties appear to be self unfruitful as well as cross incompatible despite having viable pollens. So the cherry varieties must be planted in proper combinations. Stella, Vista, Vega and Seneca are universal donors for cross-pollination. Some of the cross compatible groups of sweet cherries are:

a) Bing, Lambert, Napoleon, Emperor Francis, Ohio Beauty

b) Windsor and Abundance

c) Black Tartarian, Black Eagle, Bedford Prolific, Early Rivers, Knight's Early Black

d) Bing, Emperor Francis, Lambert, Napoleon and Vernon.

Most of the sour cherry varieties are self-fruitful and can be planted in solid blocks. Sour cherry varieties cannot be used as pollinizers for sweet cherries as their flowering period does not overlap each other. Duke cherries are also not reliable cross pollinizers for sweet cherries, however, sweet cherries cultivars act as good pollinizers for Duke cherries. Most adequate fruit set for commercial crop in cherry ranges from 7-50 per cent, which depends on the cultivar, orchard and tree management and climatic conditions. However, fruit setting can be increased by spraying NAA or 2,4,5-T (10 ppm) during petal fall stage. On the basis of compatibility, cherry varieties have classified in to 10 groups (Table 7.4).

For maximum pollination, planting should be done in such a manner that plants of one variety face the other in the planting arrangement (Table 7.5).

Table 7.4: Incompatibility Groups of some Important Varieties of Sweet Cherry

Group	Variety	Sterility Allele
I	Bedford Prolific, Black Downton, Black Eagle, Black Tartarian and Early Rivers	S1S2
II	Bigarreau Schrecken, Black Elton, Black Heart, Clauster Black Victoria Black Waterloo, Merton Bigarreau and Van	S1S3
III	Bigarreau Napoleon, Emperor Francis, Ving, Lambert and Star	S3S4
IV	Kentish Bigarreau, White Bigarreau, Sue, Victor, Merton Heart and Velvet	S2S3
V	Bohemian Black, Late Black and Turkey Heart	S3S5
VI	Elton Heart, Governor Wood and Early Amber	S3S6
VII	Hedelfingen Monstreuse Mezel, Black Republican and Vic	S4S5
VIII	Noir de Schmidt, Peggy Rivers, Schmidt and Giant	S2S5
IX	Red Turk, Black Giant and Ursula Rivers	S1S4
O	Stella, Vista, Vega and Seneca	Universal donors

Table 7.5: Different Planting Plans for an Effective Pollination

(a) Planting plan for fully compatible cultivars of groups VI, VII and IX where A is an important common cultivar

A	A	A	A	A	A	A	A	A	A	Age of plants
B	C	B	C	B	C	B	C	B	C	(per cent)
A	A	A	A	A	A	A	A	A	A	A=60
A	A	A	A	A	A	A	A	A	A	B=20
B	C	B	C	B	C	B	C	B	C	C=20

(b) Where a culture is of less importance

A	B	A	B	A	B	A	B	A	B	Age of plants
B	C	B	C	B	C	B	C	B	C	(per cent)
A	B	A	B	A	B	A	B	A	B	A=40
C	A	C	A	C	A	C	A	C	A	B=40
A	B	A	B	A	B	A	B	A	B	

(c) For partially compatible cultivars of groups I, II and III

A	B	C	A	B	A	B	C	A	Age of plants
B	C	A	B	C	B	C	A	B	(per cent)
C	A	B	C	A	C	A	B	C	A=34
A	B	C	A	B	A	B	C	A	B=34
B	C	A	B	C	B	C	A	B	C=21

In addition, 4-5 honeybee hives/ha should also be kept in a cherry orchard for better pollination and fruit set.

ROOTSTOCKS AND PROPAGATION

The cherries are usually propagated by stooling, cutting, budding and grafting. Budding and grafting is done either on seedling of clonal roostocks. The most common seedling rootstocks are Mahaleb (*P. mahaleb*) and Mazzard (*P. avium*), Paja (*P. cerasoides*) and Bird cherry (*P. padus*). F12/1 is a good clonal rootstock of cherry which is resistant to bacterial canker and vigorous in growth. Colt is easy-to-root, semi-dwarfing rootstock suitable for high density plantings but is susceptible to bacterial crown gall. Other important clonal rootstock series for cherries are CAB, W GM, M x M and OCR clones. The Gisela series of rootstocks was developed in Germany in the 1960s at Justus Leibig University in Giessen by Drs. Werner Gruppe and Hanna Schmidt. The commercially available rootstocks are: Gisela 5, Gisela 6, Gisela 7, and Gisela 12. Size control ranges from 45 to 80 per cent of the size of similar cultivars on Mazzard.

Montmorency, Northstar and Kansar Sweet can be used as interstocks for sweet cherry varieties on Mahaleb and Mazzard rootstocks to reduce the tree size and precocity in fruiting.

Seedling Rootstock

Seedling of *paja* (*Prunus cerasoides*), bird cherry (*Prunus paddum*), mahaleb and mazzard are used for raising sweet cherry plants in India. Seeds of *paja* do not require chilling treatment to break dormancy but seeds of mahaleb and mazzard require stratification before sowing. Seeds are extracted from fully ripe fruits. They are dried and stored in a cool place. Seeds are soaked in 500 ppm GA_3 for about 24h, then they are stratified by placing between the layers of sand in a cool place at 2-4°C for 80-120 days for mahaleb and 120-50 days for mazzard to break seed dormancy. During stratification, the medium is kept moist. As the embryonic root comes out from seed coat, these are transplanted 6cm deep and 10-15 cm apart in rows spaced at 20-25 cm in nursery beds. The nursery beds are mulched with 10-15 cm thick hay and irrigated lightly. Mulch material is removed when seedlings attain 5-6 cm height. The nursery should be watered twice a week and weed-free.

Clonal Rootstock

The clonal rootstocks such as Colt, and Mazzard F 12/1 have recommended for cherry in India. Colt is semi-dwarf, compatible with almost all varieties of sweet cherry, has good anchorage, and is resistant to gummosis, crown-rot, moderately resistant to stem-pitting virus and bacterial canker but susceptible to oak-root fungus. Mazzard F 12/1 is semi-vigorous and difficult-to root. The clonal rootstocks are usually propagated by mound layering or hardwood cuttings.

The scion varieties are propagated by 'Tongue' grafting or 'T' budding on seedling or clonal rootstocks. T budding is usually done in the last week of April or first week of May. Among different methods of grafting, cleft or tongue grafting gives good success, which is done in towards the end of dormant season.

PLANTING AND ORCHARD ESTABLISHMENT

The best time for planting cherries is winters when the plants are dormant. Pits of 1 x 1 x1 m are dug and filled with the mixture of soil and FYM about one month prior to planting. In flat areas, the square or hexagonal system of planting should be followed, however, in hills the planting is done in terraces or contours. For standard plants on seedling rootstock the plant-to-plant spacing should be 6 m but plants on dwarfing rootstocks or for the compact cultivars like Compact Stella or Compact Lambert, the planting distance can be reduced to 3-4 m either way.

INTERCULTURAL OPERATIONS

Training and Pruning

Cherry plants grow more or less upright and scaffolds develop in whorls of 3 to 5. The plants are usually trained on modified leader system. It is important to prevent the leader and upper branches from being choked by opposite closely placed scaffold limbs. At the time of planting, the plants are headed back to about 60-80 cm at the time of planting. The central leader is retained and 3-5 wide-angled branches, 20-25 cm apart spirally around the tree are selected in first dormant pruning. The lowest branch should be 40-60 cm above the ground level. The selected scaffold branches are headed back to minimum and only one-fourth of the growth is pruned off. In second dormant pruning, 3-4 well-spaced main branches are selected whose one-fourth growth is pruned off and on each main scaffold well-spaced 3-4 secondary branches are selected. After 3-4 years, central leader is headed back at 5-6 m height and lateral branches are allowed to grow, resulting in the development of strong and moderately spreading tree. If not trained properly, cherry plants attain huge growth.

The cherry plants bear fruits mainly on spurs and thus require light pruning. Usually corrective pruning is desirable rather than heading back or thinning out of branches. Heading back of more than 2.5 cm thick branches should be avoided. When pruning of thick branches is absolutely necessary, thinning out is preferred to heading back. The pruning cuts should be treated with fungicidal paste. The pruning should be done in the late dormant season but well before the new growth starts.

Nutrition Management

Like other stone fruits, cherries respond adequately to nutrient application. However, it is very responsive to nitrogen application and clean cultivation. Ten-year-old full-grown tree requires 60 kg FYM, 2.0 kg calcium ammonium nitrate (CAN), 1.6 kg single super phosphate and 1.0 kg muriate of potash annually (Table 7.6). Farmyard manure should be applied in December along with a full dose of super phosphate and muriate of potash. Half dose of N is applied in spring before flowering and the other half dose of N is applied in spring before flowering and the other half one month later. Fertilizers are broadcast in tree basin about 30 cm away from the tree trunk.

Table 7.6: Manure and Fertilizers Schedule for Sweet Cherry

Age of Tree	Farmyard Manure (kg)	Calcium Ammonium Nitrate (g)	Super Phosphate (g)	Muriate of Potash (g)
1	10	200	160	100
2	15	400	320	200
3	20	600	480	300
4	25	800	640	400
5	30	1,000	800	500
6	35	1,200	960	600
7	40	1,400	1,120	700
8	45	1,600	1,280	800
9	50	1,800	1,440	900
10 and above	60	2,000	1,600	1,000

Water Management

Due to sloppy lands and non-availability of irrigation water, cherry is grown under rain fed conditions in our country. The distribution of rainfall throughout the year is uneven and owing to less rainfall during April-May, its plants should be watered frequently. However, irrigating its trees at weekly intervals during fruit growth and development is recommended for better fruit size and quality.

Weed Management

Cherry orchards are maintained under permanent sod with a clean basin management. The basins are kept clean by hand-weeding or using weedicides. Application of diuron 4kg/ha as pre-emergence and paraquat (0.5 per cent) as post-emergence are recommended to suppress the growth of weeds for 4-5 months. Mulching tree basin in April with 10-15 cm thick layer of hay after spring rains also helps in weed control soil moisture conservation, and improvement in soil structure and fertility. Green manuring crops-bean, pea, red clover and white clover-can also are grown in tree basins to improve soil texture and fertility.

PLANT PROTECTION

Major Insect-Pests and their Management

There are several insects, which attack on cherry crop and wreak havoc.

☆ **Black cherry aphid (*Myzus cerasi*):** Black cherry aphid is the most serious pest of cherries. When in large numbers, they cause severe damage by sucking the sap from developing buds, young leaves, flowers, shoots and developing fruits. Young shoots and leaves are smothered with masses of young and adult aphids. The leaves curl, clustered and drop; and the affected shoots may die. Honeydew produced by insects falls on fruits and makes it unfit for consumption. Spray metasystox or rogor (1 ppm) before or at bud swell stage to control this pest.

☆ **Cherry fruit fly (*Rhagoletis cerasi*):** It is also a serious pest of cherry. Its small maggots tunnel into the fruits and feed on fruit tissues and may devastate the entire crop. It is very difficult to control this pest. However, if systematic insecticides are sprayed just before fruit start maturing, it is likely to control it to some extent. In addition, use of poison baits in the orchards is helpful in killing the adult flies.

Major Diseases and their Management

Most important diseases of cherry are bacterial canker, leaf spot, brown rot and crown and root galls, which have been described briefly hereunder:

☆ **Bacterial canker:** It is the most serious disease of sweet cherry, which is caused by a bacterium, *Pseudomonas syringae*. The affected portions like bark, outer sap wood and fruits develop circular to purplish spots surrounded by a ring of small green tissues. Thse spots become irregular in shape and turn brown. It causes extensive killing of leaves, flower buds and spurs as the bacterium destroys the conducting tissues. For its management, spray streptocycline @ 10g/100 L water before the onset of rainy season or alternatively spray copper oxychloride/Bordeaux mixture @ 0.3 per cent after leaf fall. In addition, cut and burn the affected plant parts immediately to restrict the further spread of the disease.

☆ **Bacterial spot:** Like other stone fruits, this is also a serious disease of cherry, which is caused by a bacterium, *Xanthomonas campestris* pv. *pruni*. Its symptoms appear on leaves as small, circular, or irregularly shaped, pale green lesions. During early development, lesions almost always are concentrated near the leaf tip. In advanced stages, the inner portion of the lesion falls out, giving the leaf a 'ragged' or 'shot hole' appearance. Leaves heavily infected with bacterial spot turn yellow and fall. Symptoms first appear on fruit as small, olive brown, circular spots. Spots become slightly darker and depressed as the bacteria develops. Leaf infection is more common than fruit infections. Copper containing fungicides should be applied just as the leaves begin to shed. Resistant varieties should be grown.

☆ **Brown rot:** This disease is caused by *Monilinia fructicola*, and appears in severe form during hot and humid conditions at harvest time. Sweet and Duke types are more affected by this disease. Small circular brown spots appear on the fruit and the lesions spread over the whole fruit. Powdery, light brown or grey masses of spores appear on the rotten areas. The fungus lives over winter in mummied fruits. Spray captan 50 WP (300g/L water) before harvest.

☆ ***Rhizopus* rot:** It is major postharvest disease of cherries. It is caused by a fungus, *Rhizopus stolonifer* that is found in fruit exposed to temperatures of 5°C or greater. Proper temperature management (rapid cooling to optimum storage temperature) can completely control *Rhizopus* rot.

✰ **Leaf spot:** It is one of the most dreaded diseases of cherries particularly in the sour types. In this case, the leaves develop purplish spots, which coalesce to form brown spots surrounded by a reddish brown zone. When the dead tissues fall out, holes are formed on the leaf surface, chlorosis develops and ultimately the leaves drop prematurely. Spray captan 75 WP (300 g/L) after petal fall. The subsequent spray should be done after 20 days and at harvest.

✰ **Crown and root gall:** This disease is caused by a bacterium, *Agrobacterium tumefaciens*. There is development of wart like tumors or swellings at the crown or the roots. Initially these galls are light in colour and soft in texture but late may turn as black and become quite hard. The incidence of this disease reduced by dipping the plants in copper sulphate (0.4 per cent) solution before planting. In the field, its incidence can be reduced by applying Chaubattia paste to the lesions.

Physiological Disorders and their Management

Fruit Cracking

It is one of the most serious problems in cherry. Fruits of some cultivars like Bing, Stella crack badly before attaining full maturity. In some areas the losses may be even up to 50 per cent. Cracked fruits are susceptible to decay and attack of insect-pest during storage, and are very difficult to sell in the market. The problem takes bad shape in rain-fed areas where arrangements for regular supply of irrigation are not made. Cracking of fruits takes place when sudden rains occur after long dry spells of orchards are irrigated at regular intervals. The difference in cracking of fruits among different cherry cultivars is primarily due to the difference in the rate of absorption of water and capacity of expansion of peel. For example, cultivars with a rapid rate of absorption and a low capacity for expansion crack readily (*e.g.* Bing, Lambert) as compared to those which have a slow rate of absorption and a high capacity of peel expansion (Early Rivers, Sam, Victor). Fruit cracking can be reduced or checked by spraying calcium chloride (2-3 per cent), GA_3 (2,000 ppm) and by regular irrigation in the orchard.

Surface Pitting and Bruising

This is a very common problem for sweet cherries. Symptoms are primarily caused by a mechanical impact or compression. Bruising appears to be associated with injury to cells much below the epidermis. Bruising leads to increased fruit respiration, ethylene production, and susceptibility to decay. Fruit factors such as high soluble solids concentration, warm temperature, pre-harvest use of gibberellic acid and high fruit weight have been shown to reduce fruit susceptibility to bruising.

MATURITY, HARVESTING, AND YIELD

Maturity

Cherry fruits are very delicate and must be harvested at appropriate maturity. If harvesting is done too early, fruits may be smaller in size with poor colour and flavour. Similarly, if harvested too late, the fruits may be soft with increased susceptibility to decay, shriveling, stem browning and pitting. Hence, cherries should be harvested at appropriate maturity to get maximum benefits in the market. There is no recognized maturity standard for picking cherries. However, the colour development is considered as the best guide, although it varies with the cultivar and the purpose for which it has to be used.

Harvesting

Cherries are picked manually, which requires lot of labour. Harvest cherries early in the day when temperature is quite low. Harvest cherries by their stems, lifting the cluster to remove from the spur. Avoid impact injury during harvest operations and keep fruits in the shade and transport to cooler or packing units as soon as possible. Cherries to be used for Sweets intended for processing are also hand harvested, but without pedicels. Sour cherries intended for processing are shaken from trees when ripe. Ethephon, an ethylene releasing compound, is applied about 2 weeks prior to harvest to reduce fruit removal force, and increase percentage fruit harvested.

Yield

Cherry tree starts bearing after the age of 6-8 years. The yield of cherries is highly variable from year-to-year but a full bearing tree may produce 30-40 kg fruits.

Packing

After harvesting, the fruits should be pre-cooled to reduce losses from rotting. The fruits are graded according to size and packed in corrugated cartons or baskets of 2 to 5 kg.

POSTHARVEST HANDLING

Cherries do not stand long storage but can be held for 10 to 14 days at –1 to 0 °C and 85-90 per cent relative humidity. Cherries can also be stored for about 24 days in CA storage maintained with 5 per cent CO_2 and 3 per cent O_2 at 2°C. However, the storage life of cherries can be extended by spraying the fruits with GA_3 (30-40 ppm), vapour guard and calcium nitrate in the orchard and by the use of edible coating and captan dips after harvesting.

8

Custard Apple

INTRODUCTION

Custard apple or sweet sop or *sitaphal* (*Annona squamosa* L.) is a drought resistant tree or shrub, which is widely distributed throughout the tropics and does well in hot and relatively dry climates such as those of the low-lying interior plains of many tropical countries. Fruits are dark greening brown in colour and marked with depressions giving it a quilted appearance; its pulp is reddish yellow, sweetish, and very soft; the kernels of the seeds are said to be poisonous. Apart from delicious fruits, plants parts are used for curing different diseases. Therefore, it is considered as the 'New Super Fruit of the 21st Century'. The genus 'Annona' contains approximately 2300 known species in the world. These include numerous fruit-trees, especially of the genera *Annona* and *Rollinia*; the majority of *Annona* species and all the *Rollinia* species originate from the New World. Other Annonaceous fruits of the New World include species of *Asimina, Duguetia, Fusaea* and *Porcelia*. Some related species such as *A. reticulata* (*Ramphal*, Bulock's heart or Bull's heart): larger fruits, heart shaped, smooth, less seeded and inferior quality pulp; *A. cherimola* (*Laxmanphal*, Cherimoyer or Cherimola or Cherimoya): most delicious fruits among *Annona* species, pulp slightly acidic, sweet with buttery consistency having few seeds; *A. atemoya* (*Hanumanphal*): F_1 hybrid between *A. squamosa* x *A. cherimola* with a better quality fruits like that of *A. cherimola* and adaptability to high temperature as that of *A. squamosa*; *A. muricata* (Sour sop): highly acidic with many soft spines; *A. diversifolia* or the ilama is a small tree adaptable to subtropical climate and is a popular fruit in Mexico and Central America; and *A. glabra* (pond apple, alligator apple, swamp apple, corkwood, bobwood, and monkey apple), which grows in swamps and is tolerant of saltwater, are also important. The name custard apple is an umbrella term for any edible *Annona* species. Therefore in different countries, the name custard apple is given to separate species of genus *Annona*. For instances, in India, the custard apple is *Annona squamosa*, in the USA, *A. reticulata* or *A. glabra* and in Australia *A. atemoya*.

COMPOSITION AND USES

Composition

The fruit of custard apple is very sweet and delicious. The edible portion or pulp is creamy or custard-like, granular with a good blend of sweetness and acidity which vary with the species. The pleasant flavour and mild aroma have a universal liking. Custard apple fruits are rich in carbohydrates, protein, calcium, iron and mineral matter. Despite its high sugar content, the glycemic index of custard apple is low (54) and the glycemic load moderate (10.2). Custard apples contain anti-oxidants like Vitamin C, which helps to fight free radicals in our body. These are also rich in potassium and magnesium that protect us from cardiac diseases. The food value of custard apple fruit pulp is presented in the following Table 8.1.

Table 8.1: Nutritive Value per 100 g of Edible Portion of Sugar Apple Fruits

Constituent	Content	Constituent	Content
Calories	80-101	Phosphorus	14.7-32.1 mg
Moisture	68.3-80.1 g	Iron	0.42-1.14 mg
Protein	1.17-2.47 g	Carotene	0.007-0.018 mg
Fat	0.5-0.6 g	Thiamine	0.075-0.119 mg
Carbohydrates	20-25.2 g	Riboflavin	0.086-0.175 mg
Crude Fiber	0.9-6.6 g	Niacin	0.528-1.190 m
Ash	0.5-1.11 g	Ascorbic Acid	15.0-44.4 mg
Calcium	17.6-27mg	Nicotinic Acid	0.5mg

Uses

Fruits are eaten fresh or utilized as flavouring for ice- cream and milk beverages. Delicious products such as jam and squash can also be made from the pulp. The sweet sop starch has higher swelling power, solubility, paste clarity and freeze–thaw stability and lower pasting temperature. The functionality of sweet sop starch is much comparable to those of the waxy corn and *Amaranthus hypochondriacus* starches, thus making it a candidate for the usage in the instant or as the frozen foods. The seed yield about 30 per cent oil. Non-edible seed oil is used as insect repellant as it contains acetogenins that are toxic to insects. Pesticides derived from plants like sweet sop can play a major role in pest management in sustainable agriculture. Seed cake can be used as manure. Fiber extracted from the bark has been employed for cordage. The tree serves as host for lac-excreting insects also. This plant is reputed to contain several medicinal properties. For centuries, Ayurvedic practitioners in India have extensively used various parts of the sugar apple (*Annona squamosa*) tree for the management of diabetes. In Aligarh district of U.P., villagers used to consume a mixture of 4–5 newly grown young leaves along with black pepper, earlier in the morning in the treatment of diabetes. Similarly, Inca tribes in the Peruvian Andes have used cherimoya (*Annona cherimola*) as a medicinal plant. Cherimola is often considered as one of the 'lost crops of Inca'. Folkloric record reported the use of *Annona squamosa* as an anti-tumor agent, anti-diabetic, antioxidant, anti-lipidimic and anti-inflammatory agent, which has been characterized due to the presence of the cyclic peptides. Diterpenes which was isolated from the *Annona squamosa* possess the anti-HIV principle and the anti-platelet aggregation activity. The partially purified flavonoids were reported from the same source as the responsible agent for the anti-microbial and other pesticidal activities. Currently the Annonaceae remain a "hot' family for the discovery of new anti-cancer drugs. Custard apple contains a class of chemicals called acetogenins, which are very long chain fatty acids, and only found in Annonaceous species. *In vivo* testing has shown that bullatacin, an acetogenin found in custard apple, to be 300 times more potent than Taxol (paclitaxel), a standard anti-cancer drug. In India the crushed leaves are sniffed to overcome hysteria and fainting spells; they are also applied on ulcers and wounds and a leaf decoction is taken in cases of dysentery. The seeds are also ground and applied to hair to get rid of lice. Seeds, fruits and leaves were found to be effective as an insecticide, fish poison, and as a powerful irritant of the conjunctiva.

ORIGIN, HISTORY AND DISTRIBUTION

It is a native semi-deciduous tree in tropical America and the West Indies. It was carried in early times through Central America to Southern Mexico. It has long been cultivated and naturalized as far South as Peru and Brazil. It was planted in Puerto Rico as fruit trees in 1626, spreading from cultivated areas to roadside and valleys. Apparently, it was introduced into tropical Africa early in the 17th century and is grown in South Africa as a dooryard fruit tree. The Spanish probably carried seeds from the new world to the Philippines. The British recoded their widespread assimilation on India's soils in 1835, but no Sanskrit word appears for any *Annona* fruit. The generally accepted theory is that the Portuguese brought the fruits from the New World *circa* 1590, around the same time they introduced other fruits like the cashew apples, papayas and peppers. It is commonly grown in the Bahamas, and occasionally in Bermuda and Southern Florida. It has become fairly common on the East coast of Malaya, and more or less throughout South-East Asia and the Philippines. Eighty years ago it was reported as thoroughly naturalised in Guam, in the Pacific, though it is not known in Hawaii. The sugar apple is one of the most important fruits in the interior of Brazil and is conspicuous in the markets of Bahia. Today, *Annona* fruits are cultivated in Australia, Brazil, Chile, Egypt, Israel, Burma, the Philippines, Spain, Sri Lanka, South Africa, the West Indies, and parts of the US.

Annona has adapted so well in India that a considerable variability is found in Aravali hills in North India and in Andhra Pradesh in South India. Custard apple plants can also be seen wild in UP, Rajasthan, Karnataka, MP, Maharashtra and Tamil Nadu. Approximately 55,000 hectares are dedicated to custard apple cultivation in India. Pune's Purandar Tehsil district is a very well known sugar apple growing area. Some important custard apple growing regions in India are presented as under;

Table 8.2: Major Custard Apple Growing Belts in India

State	Districts
Chhattisgarh	Mahasamund, Rajnandgaon, Kabirdham, Jagdalpur, Kondagaon, Kanker, Dantewada, Bilaspur, Mungeli, Korba, Raigarh, Jashpur, Koriya, Narayanpur
Gujarat	Vadodara, Bhavnagar
Rajasthan	Chittorgarh, Udaipur, Pratapgarh
Madhya Pradesh	Sheopur, Gwalior, Shivpuri, Jhabua
Maharashtra	Nanded, Beed, Aurangabad, Ahmednagar, Pune, Solapur
Uttar Pradesh	Jhansi, Lalitpur, Jalaun, Hamirpur, Mahoba, Agra

The state of Gujarat has lion's share in custard apple production followed by A.P. and Chhattishgarh (Figure 8.1).

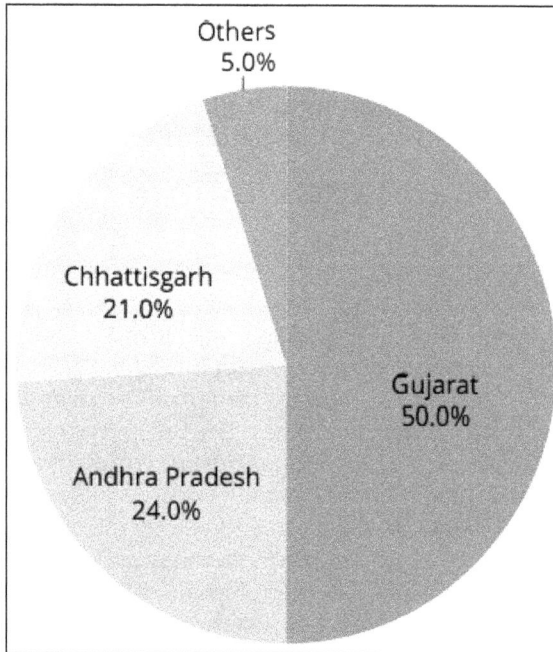

Figure 8.1: Leading Custard Apple Producing States (Per cent share in production) [*Source*: Checklist of Commercial Varieties of Fruits, 2012, Department of Agriculture and Co operation, Ministry of Agriculture, GoI].

TAXONOMY AND BOTANICAL DESCRIPTIONS

Taxonomy

The genus name *'Annona'* is from the Latin word *'anon'* meaning *'yearly produce'*, (referring to the production of fruits of the various species in this genus). Species name *'squamosa'* refers to the knobbly appearance of the fruit. Genus *Annona* belongs to Sub-class Magnoliidae, Order Magnoliales, family Annonaceae and Subfamily Maloideae. The *Annona squamosa* is a diploid species with 2n=2x=14. All the important cultivated *Annona* species have the haploid chromosome number as seven except the species *A. glabra* which has fourteen.

Botany

The sugar apple tree ranges from 3-6 m in height with open crown of irregular branches, and some-what zigzag twigs. Deciduous leaves, alternately arranged on short, hairy petioles, are lanceolate or oblong, blunt tipped, 5-15 cm long and 2-5 cm wide; dull-green on the upper side, pale, with a bloom, below: slightly hairy when young; aromatic when crushed. Along the branch tips, opposite the leaves, the fragrant flowers are borne singly or in groups of 2 to 4. They are oblong, 2.5-3.8 cm long, never fully open; with 2.5 cm long, drooping stalks, and 3 fleshy outer petals, yellow-green on the outside and pale-yellow inside with a purple or dark-red spot at the base. The 3 inner petals are merely tiny scales. Despite being morphologically perfect, neither self or cross-pollination takes place effectively in *Annona* spp. Self-pollination is hindered by the presence of protogyny, whereas cross-pollination by insects is limited because the flowers are not brightly colored and lack fragrance and nectar. The aggregate fruit (etario of berries) is nearly round, ovoid, or conical; 6-10 cm long; its thick rind composed of knobby segments, pale-green, gray-green, bluish-green, or, in one form, dull, deep-pink externally (nearly always with a bloom); separating when the fruit is ripe and revealing the mass of conically segmented, creamy-white, glistening, delightfully fragrant, juicy, sweet, delicious flesh. Many of the segments enclose a single oblong-cylindric, black or dark-brown seed about 1/2 in (1.25 cm) long. There may be a total of 20 to 38, or perhaps more, seeds in the average fruit. Some trees, however, bear seedless fruits.

SOIL AND CLIMATIC REQUIREMENTS

Soil

The sugar apple is not particular as to soil and can perform well on sand, limestone and heavy loam with good drainage. Well drained loam soils having normal pH value *i.e.* 6.0-6.5 is considered the best for custard apple. Light soils having gravels and small pebbles are also satisfactory. The tree is shallow-rooted and doesn't need deep soil. Heavy soil having high clay content is not suitable for its cultivation. Sugar apples are not tolerant of excessively wet or flooded soil conditions. Flooding for as little as 7 to 10 days may result in plant death. Symptoms of flooding stress include leaf chlorosis (yellowing), stunted leaf and shoot growth, leaf wilting and browning, leaf drop, stem dieback, and tree death. Likewise, it is not tolerant to saline soil and water conditions. Symptoms of salt stress include marginal and tip necrosis (death) of leaves, leaf browning and drop, stem dieback, and tree death.

Climate

The sugar apple tree requires a tropical or near-tropical climate. However, it thrives well in hot and dry climate of subtropics as well. It grows well with more than 700 mm of rainfall per year. Temperature is a limiting factor. Frost causes severe damage to young plants but older trees show some tolerance. Seedlings have high photosynthesis activity at 30°C and show vigorous shoot growth. Being deciduous, it sheds leaves during winter and new growth flush comes during the spring along with flowers. During the blooming season, drought interferes with pollination; therefore, there should be high atmospheric humidity but no rain during flowering. Poor pollination is a frequent problem under high temperatures (>30°C) and low humidity (<60 per cent relative humidity, RH), even with hand pollination. Lower temperature (25°C) and higher humidity (70 per cent or higher RH) greatly improves pollination. High temperature (40 °C) at the time of flowering causes flower drop and rainfall obstructs the pollination process. In severe droughts, the tree sheds its leaves and the fruit rind hardens and will split with the advent of rain. Further, high wind velocity may also impact tree's growth. Constant winds may distort the tree canopy, making tree training and pruning more difficult. Strong winds along with heavy crop loads may result in limb breakage.

IMPORTANT VARIETIES

Indian horticulturists have studied the diverse wild and cultivated sugar apples of that country and recognized some promising different types: 'Red' (*A. squamosa* var. *sangareddyiz*)-red-tinted foliage and flowers, deep-pink rind, mostly non-reducing sugars, insipid, with small, blackish-pink seeds; poor quality; comes true from seed. 'Red-speckled'-having red spots on green rind. 'Crimson'-conspicuous red-toned foliage and flowers, deep-pink rind, pink flesh. 'Yellow'; 'White-stemmed'; 'Mammoth' (*A. squamosa* var. *mammoth*)-pale yellow petals, smooth, broad, thick, round rind segments that are light russet green; fruits lopsided, pulp soft, white, very sweet; comes true from seed. 'Kakarlapahad'-very high yielding. 'Washington'-acute tuberculate rind segments, orange-yellow margins; high yielding; late in season, 20 days after others. 'Barbados' and 'British Guiana'-having green rind, orange-yellow margins; high-yielding; late. A spontaneous seedless type known 'Brazilian seedless' had been developed in Brazil. This mutant represents a particular case of stenospermocarpy, resulting in a failure in seed development as the ovules or embryos get aborted.

Characters of some of the promising cultivars in India are given below:

- ☆ **Balanagar**: Fruits are comparatively smaller. The variety tolerates stress conditions Island Gem: The fruit has nearly 600 g weight and contains 93 areoles and less number (31) of seeds.

- ☆ **Pink's Mammoth**: Average fruit weight is 595 g with more (113) areoles and less number (27) of seeds. The pulp content is high.

- ☆ **Arka Sahan**: It is an interspecific hybrid, between Island Gem (atemoya) and Mammoth (custard apple), developed at IIHR, Bangalore. Fruits are big (210g), skin is light green in colour with waxy bloom, moderately thick with large flat eyes. Fruits have improved shelf life *i.e.*, take 7 days to ripe, 4 days more than 'Mammoth'. The creamy white flesh is juicy with mild pleasant aroma and tender with sparse seeds (9/100g of fruit weight). The fruit of this variety is also characterized by large segments or flakes and many of which are seedless. Flesh is very sweet (30° brix) compared to 24° brix in Mammoth. Average yield is 12 tonnes/ha.

☆ **APK (Ca)-1:** It is a clonal selection from a high yielding type in State Horticultural Farm, Courtallam of Tirunelveli District of Tamil Nadu and developed at Regional Research Station, Aruppukkottai. It is a high yielder in rainfed vertisol (Black soil) 14.90 kg/tree, 30.7 per cent more than Balanagar. Each fruit weighs 207g. Average number of fruits would be 72/tree. TSS 24.5° Brix, acidity 0.2 per cent. Adapted to semiarid plains.

Besides, these some local selections are also popular among farmers such as Saharanpur Local, Raidurg Local, Mehboobnagar, Kakarlapahad *etc.*

Mr. N.M. Kaspate, a progressive farmer from Solapur, Maharashtra has developed some promising varieties through selection. Of which, Annona 2 and NMK 1 are suitable for all types of soils for plantation with a distance of 10 x 15 feet.

☆ **Annona-2:** Attractive in colour, have less number of seeds per fruit and berry is quite big. It requires the heading back practice at the height of 2 to 2.5 ft., which helps expediting growth of side branches and alternatively more number of fruits.

☆ **NMK-1:** Big fruit size, attractive golden yellow colour fruits with tinge of pink colour between two berries, suitable for export because of its uniform size, on-tree storage up to 20-25 days on plant itself.

Another progressive farmer from Maharashtra, Mr. Suresh Patil, also developed a variety 'Saraswati Saat'. The fruit of Saraswati Saat is quite big (500-1100 g) with almost 80 per cent pulp, fewer seeds, thin skin, less prominent eyes and lesser sweetness. It is ideal for use in processing of ice-cream as well as for making juice.

Among the varieties mentioned above, cv. Arka Sahan and Balanagar are grown most extensively in different parts of India (Table 8.3).

Table 8.3: Recommended Areas for Growing Custard Apple Varieties

Variety	Recommended Areas
Arka Sahan	Rajasthan, Chhattisgarh, Madhya Pradesh, Maharashtra
Balangar	Maharashtra, Rajasthan, Chhattisgarh, Gujarat, Madhya Pradesh, Uttar Pradesh

PLANT PROPAGATION

Custard apple is commonly propagated by seeds in India which has resulted in considerable variability. The seeds have a relatively long life and can be kept well for 3 to 4 years. To raise true-to-type plants in the orchard, plants are multiplied by shield budding, veneer grafting, softwood grafting and side grafting. Seeds are sown in polythene tubes filled with mixture of FYM sand and clay in 1: 1: 1 proportion during June-July. Seedlings become ready after one year for planting in the field or for budding/grafting. For softwood grafting, one year old rootstocks are grafted with suitable scion stick during March-April to obtain about 90 per cent success. Veneer grafting is also done on one year old rootstocks which gives nearly 40 per cent success. Side-grafting can be done only from December to May, which requires much skill with the success rate of 58.33 per cent. Shield-budding gives 75 per cent success and is the only commercially feasible method. Seedlings of Laxmanphal, Sitaphal, Ramphal, Mamphal and pond apple can be used as rootstock. The selected clones grafted on *A. reticulata* seedlings have flowered within 4 months and fruited in 8 months after planting out, compared with 2 to 4 years in seedlings. The grafted trees are vigorous, the fruits less seedy and more uniform in size. *A. glabra* is suitable but less hardy. The sugar apple itself ranks next after *A. reticulata* as a rootstock. Cherimoya was noted to be a vigorous rootstock for Pink's Mammoth, while atemoya was not compatible with *A. glabra*, *A. montana*, *A. muricata* and *A. reticulata* as rootstocks. Custard apple can also be propagated by stool layering during September-October using 20,000 ppm IBA. It gives more than 90 per cent rooting and 80 per cent field survival. Sugar apple can also be propagated successfully through micropropagation using Murashige and Skoog's (MS) medium. Different types of explants such as anther, endosperm, hypocotyls, leaf explants, nodal segments *etc.* have given varying results.

PLANTING AND ORCHARD ESTABLISHMENT

Planting is done at 5 x 5 m spacing. Pits of 60 x 60 x 60 cm size should be prepared and filled with FYM + top soil + fertilizer (50: 50: 25 g NPK) mixture alongwith chlorpyriphos dust (50 g/pit) before the onset of monsoon. During July-August, after 1 or 2 rains, the saplings are planted in the centre of each pit. Small basin is prepared around each plant.

CROP REGULATION

Leaf shedding in winter season, and production of floral buds on new shoots in sugar apple indicate towards scope of crop regulation. The completion of leaf fall is essential for the initiation of new growth. Under moderate climatic conditions, custard apple trees remain dormant and put forth new growth during April. Since, early initiation of new growth and extent of fruit set determine the yield in custard apples; therefore, by early induction of new growth, the span of fruit harvest can be prolonged to increase fruit yield. This can be done by defoliation of plants by two sprays of 0.5 per cent potassium iodide at weekly interval during December- January. This has resulted in not only early and prolonged harvest but also in realization of higher yield. Alternatively, for defoliation and bringing the plants under uniform rest Ethrel @ 1000 ppm is sprayed one month after the harvest of the fruits.

Furthermore, the problem of poor fruit set in custard apple can be overcome considerably by application of GA_3 @ 50ppm. Dipping of freshly opened flowers in GA_3 ensures better fruit-set, better fruit retention, improved fruit size and weight with less seed. Likewise, spray of 10 to 20 ppm NAA just prior to flowering period reduces the flower and fruit drop. During fruit development, foliar spray of 10 ppm GA_3 + 5 ppm CPPU, improves the fruit size and luster of the fruits.

At Queensland, Australia, experiments have shown that shoot tipping and leaf stripping are useful techniques to increase late flowering and fruit set in custard apple. Tipping comprised removal of about 10 per cent of the shoot length and leaf stripping involved removal of about 70 per cent of the leaves on a shoot. This technique, if done at the appropriate time during early summer, releases the sub-petiolar buds (buds below the leaf stalks), increasing and in some cases doubling, the number of new season laterals and flowers. Leaf stripping involveed removing 5-10 terminal leaves, including the tip (about the last third depending on shoot length).

HAND POLLINATION

The custard apple flower is hermaphroditic (male and female parts in the same flower) and exhibits protogynous dichogamy (stigmas = female parts are receptive before the pollen is shed by the anthers). The separation of the female and the male stages appears to be the main factor limiting the level of self pollination. Fruit set can occur as a result of self pollination, insect-assisted pollination (by *Apis mellifera*, nitidulid beetles (*Carpophilus* and *Uroporus* spp.) and *Grills*) or wind pollination, which may vary (near zero to about 3 per cent) and consequently fruit set may be low. The low fruit set in custard apples due to poor pollination is attributed to both the external and internal factors, such as very high and low humidity prevailing at the time of flowering, soil moisture stress, competition between vegetative and floral growth, hypogyny, dichogamy, poor pollen germination and lack of insect pollinators. Differences between varieties has also been noticed *e.g.* variety Balanagar sets satisfactory fruits without any assisted pollination; however, variety Arka Sahan needs hand pollination for a good yield. In each tree of Arka Sahan, about 150 flowers should be hand pollinated to achieve an expected yield of 25 tonnes/ha.

Hand pollination should be effected in the morning hours before 9.30 am by first collecting pollen of custard apple (*e.g.* Balanagar) from the previous day opened flowers. Flowers from which pollen is to be collected should be picked from trees during mid to late afternoon. The petals should be nearly fully opened and the pollen sacs should have turned creamy-grey in colour and less tightly held together. Immature pollen sacs are white and tightly packed. On hot, low humidity days, the flowers will release their pollen sacs readily. These can be collected by shaking the flowers over a piece of paper. The pollen sacs should be placed in a small container and stored at room temperature (<20°C) overnight for use the following morning. Alternatively if the pollen sacs do not readily separate from the flowers, the flowers should be spread thinly over shallow layered trays and left overnight. Pollen sacs should be then separated on the following morning and pollens may be collected in a cup. Flowers should not be placed in a closed container as a buildup of ethylene and moisture will cause the pollen sacs to turn brown and pollen germination will be severely reduced. Studies suggest that custard apple pollen can only be stored for very short periods of time (2-3 days) at ambient conditions. Collected pollens can be applied on flowers by dipping a clean dry camel hair brush or ordinary paintbrush into the cup. The pollen coated brush can be used to evenly smear the pollen on the stigma of flowers.

INTERCULTURAL OPERATIONS

Training and Pruning

Custard apple tree is of medium stature with upright growth habit. Strong tree frame raised by training to single stem which is kept clean up to 50-70 cm height.

After that scaffold branches are allowed in all directions at proper spacing between them. Once tree has developed a desirable framework, pruning is confined mostly to removal of cris-crossed and overlapping branches. Further, custard apple bears on new growth but occasionally fruiting also occurs on old branches. Therefore, a combination of young and old branches is maintained by judicious pruning. At Hyderabad, light pruning (removing 25 per cent top of branches) during March-April is found to give good results.

Nutrition Management

Application of an annual dose of 50: 50: 25 g NPK alongwith 10 kg FYM is found to be sufficient in light sandy and gravelly soils. The doses are increased up to 5 years of age of trees and thereafter a fixed dose is given every year. Whole amount of FYM, phosphorus, potash and half dose of nitrogen should be applied during monsoon and remaining nitrogen dose should be given during spring. Application of organic manures such as castor cake and bone meal are found beneficial. Green manuring with *sanai, moong, urd* and *dhaincha* is also found to be good.

Water Management

Sugar apple trees are considered tolerant to drought conditions; therefore, it is grown usually as a rainfed crop, and no irrigation is given. However, fruit set and fruit size may be impacted and defoliation may occur due to drought stress. During the initial stage of orchard establishment, frequent irrigations at 15-20 days interval should be given particularly during the summer months. After the plants have established, only life saving irrigations may be needed during summer and sometimes during winter. However, for early and satisfactory harvest of the crop, irrigation during flowering *i.e.* from May should be given till regular monsoon starts. In light soils, water harvesting by giving 5 per cent slope to the basin around the tree concentrated sufficient water. Besides, mulching of the basins with black polythene or locally available organic waste materials is helpful to conserve soil moisture. A thin layer of mulch, 2 to 4 inches thick, applied from the tree drip-line to within 6 inches of the trunk is recommended. Keep mulch away from the trunk of the tree to prevent the trunk area from becoming too moist, which can, lead to bark disease problems. Improper nutrition and water management causes the trees to enter dormancy even when they are still laden with fruits. Consequently, the fruits get mummified *i.e.* become hard and brown.

Weed Management

Sugar apple is a shallow rooted tree, which limits the possibility of deep cultivation under the tree. The young trees should be protected from weed competition by manual weeding, mulching or use of contact herbicides. Perennial weeds that survived this treatment were eliminated by glyphosate at 2 or 3 applications per year. The treatment can be completed by manual weeding thrice a year.

Intercropping

During the initial years after planting, economic return can be obtained from the unutilized spaces between the plants by cultivating intercrops. Inter-row spaces in custard apple orchard can be utilized to raise intercrops such as *moong, urd,* cluster bean, moth bean and gram under rainfed conditions. Vegetables such as pea, chilies, okra, brinjal *etc.* can also be grown if irrigation facilities are available. Timely hoeing and weeding results in good growth and production.

HARVESTING AND YIELD

Flowering in custard apple starts during spring and the fruits mature for harvest by August-September in humid areas and by October-November in dry areas. Grafted plants start fruiting 3-4 years after planting but the commercial yield is obtained only after 5-6 years. The fruits are to be harvested at correct stage of maturity. Light green fruit colour, yellowish white colour between the carpels and initiation of cracking of the skin between the carpels may be taken as maturity indices. The fruits are usually hand-picked. A well managed grown up custard apple tree produces 80-100 fruits each weighing about 200 g.

The custard apples ripen within a few days after harvest. The mature fruits can be stored at 15° to 20°C with RH of 85-90 per cent and low oxygen and ethylene but with 10 per cent CO_2. Under such storage conditions, the fruits can be kept intact for 12-18 days. Additives like Saccharified starch (1:1), high voltage treatment and packing under nitrogen gas cover, sugar (1:2) were quite effective in extending the storage of pulp at 4°C (45 days) and –18°C (90 days) temperature.

POSTHARVEST MANAGEMENT

Custard apple is a climacteric fruit and ripens after harvest. Therefore, the fruits must be harvested at proper maturity when they are still hard. Maturity of fruits is indicated by yellow colour of skin and interspaces. Since the fruit has very poor shelf life, comparatively hard ripe fruits should be harvested for transport to long distances. Immature fruits do not ripe well and remain less sweet. Custard apple pulp can be processed to prepare products like pastries, chocolates, ice creams, mixed fruit jam, *etc.*

Extraction of pulp is a major constraint in processing of custard apple fruits. Development of enzymatic browning within an hour of pulp extraction, bitterness, unpleasant repulsive off-flavour in the pulp on heating beyond 65°C and presence of gritty cells are problems encountered during processing of fruits. At Maharana Pratap University of Agriculture and Technology, Udaipur, a technology as well as a machine for Browning free mechanized pulp extraction from the custard apple has been developed. The machine has two parts. First part scoops out the pulp with seed from fruit leaving behind the peel (*i.e.* pulp extraction). The second part separates the seed from the pulp which can be stored for up-to one year. The technology is reported to be very useful as it reduces the cost of pulp extraction and enhances of the quality of the pulp.

Plant Protection

Custard apple is generally free from pests and diseases. However, sometimes insect-pests and diseases becomes problem if not taken care well at initial stages.

Major Insect-Pests and their Management

Mealy Bugs (*Ferrisia virgata, Maconellicoccus hirstus*)

Mealy bugs infect the fruits reducing their market value. Tree branches also get infected. The pest can be controlled by 2-3 sprays of 0.1 per cent monocrotophos at 15 days interval after the appearance of larvae. Mealy bugs infect the fruits reducing their market value. Tree branches also get infected. For its management through holistic approach, collect and destroy the mealy bug infested leaves, twigs and fruits. Flooding of orchard with water in the month of October is recommended to kill the eggs of pest. Ploughing of orchard in November also help reduce pest population. Further, wrapping polythene sheets around the tree trunk prevents crawling up of the pest on the trunk from the ground. For biological control, release *Cryptolaemous montrouzieri* beetles @ 10/tree or @ 30 larvae/plant twice at 15 days interval. Besides, coccinellid *Scymnus coccivora* (@ 10 beetles/tree or @ 30 larvae/plant) is a good predator of both nymphs and adults of mealy bug.

Scale (*Parasaissetia nigra*)

Scale sucks saps from the tender parts of the plant, especially, leaves. It can be managed culturally by collecting and destroying damaged leaves. Further, application of well rotten sheep manure @ 4 t/acre in two splits or poultry manure in 2 splits also help mitigating the pest. Control of ants and dust, which can give the scale a competitive advantage, is also suggested. For the biological control, field release of ladybird beetle, spray dormant oil in late winter before spring and spray horticultural oil, are suggested.

Major Diseases and their Management

Anthracnose (*Glomerella cingulata*)

Infection begins at blossom-end of the fruit and later spreads on entire fruit surface, affected fruits shrivel and they may cling to the tree or fall down. For its management, spray of Indofil M.45 (0.02 per cent) at 15 days interval is suggested.

Alternaria Leaf Spot (*Alternaria* sp.)

Alternaria leaf spot causes some damage to custard apple. It appears in the month of November at the fag end of harvesting causing considerable loss in production. The affected leaves drop down prematurely. It can be controlled by 3 sprays of 0.5 per cent benzimidazol or carbendazim or 0.2 per cent mancozeb at 15 days interval starting from the initiation of the symptoms.

Physiological Disorders and their Management

Several internal disorders have been recorded in custard apple; especially common is 'woodiness' which is characterized by the presence of woody seed pockets and gritty lumps in the flesh and 'brown pulp', which is a discolouration of the pulp and is thought to be associated with calcium and boron deficiency. Excessive vegetative growth has been shown to reduce fruit set and significantly increase the severity of the internal fruit disorder called 'woodiness'. Experiments have shown that shoot growth greater than 60cm should be considered excessive. The growth retardant, uniconazole (Sunny®) has been shown to effectively control vegetative growth in custard apple in Queensland, Australia. Two applications of this growth retardant during the first vegetative flush reduced shoot extension growth by about 20 per cent and increased flowering by about 30 per cent. Another disorder, 'translucent pulp' is characterized by the development of translucent areas around the seeds and sometimes in a ring around the seed cavity. The etiology of translucent pulp remains unidentified. Cold russet on fruits develops as a result of exposure of developing fruits to either temperature below 13°C or cold. The other disorders are fruit cracking (owing to soil moisture and temperature fluctuation during fruit development), sun burn (due to exposure of fruits to sun during dry weathers), suture line darkening (due to rapid changes in water content of fruit during development) and blackening of fruits at blossom end (owing to calcium deficiency).

9

Fig

INTRODUCTION

Fig is one of the first few plants that were popular worldwide for its dry and fresh consumption. This is an important fruit due to its high economic and nutritional values and also an important part of the biodiversity in the rainforest ecosystem. It is also a good source of food for fruit-eating animals in tropical areas.

COMPOSITION AND USES

Composition

The dried fruits are an important source of vitamins, minerals, carbohydrates, sugars, organic acids, and phenolic compounds including gallic acid, chlorogenic acid, syringic acid, catechin, epicatechin and rutin (Table 9.1). The fresh and dried figs also contain high amounts of fiber and polyphenols. Figs are an excellent source of phenolic compounds, such as pro-anthocyanidins, whereas red wine and tea, which are two good sources of phenolic compounds, contain phenols lower than those in fig.

Table 9.1: Nutritional Value of Dried Uncooked Figs per 100g

Component	Quantity	Component	Quantity	Component	Quantity
Energy	249 kcal	Protein	3.3 g	Vitamin K	15.6 µg
Carbohydrates	63.87 g	Niacin (B_3)	0.619 mg	Calcium	162 mg
Sugars	47.92 g	Pantothenic acid (B_5)	0.434 mg	Phosphorus	67 mg
Dietary fiber	9.8 g	Vitamin (B_6)	0.106 mg	Potassium	680 mg
Fat	0.93 g	Vitamin C	1.2 mg		

Uses

Its common edible part is the fruit which is fleshy, hollow, and receptacle. Fruits can be eaten raw, dried, canned, or in other preserved forms. Its fruit, root, and leaves are used in traditional medicine to treat various ailments such as gastrointestinal (colic, indigestion, loss of appetite, and diarrhea), respiratory (sore throats, coughs, and bronchial problems), and cardiovascular disorders and as anti-inflammatory and antispasmodic remedy. Fig leaves are used for fodder in India. They are plucked after the fruit harvest. The latex is collected at its peak of activity in early morning, dried and powdered for use in coagulating milk to make cheese and junket. From it can be isolated the protein-digesting enzyme 'ficin', which is used for tenderizing meat, rendering fat, and clarifying beverages. Dried seeds contain 30 per cent of a fixed oil, which is an edible oil and can be used as a lubricant.

ORIGIN, HISTORY AND DISTRIBUTION

Common fig has possibly originated from the Middle East, where it is one of the earliest cultivated fruit species and currently is an important crop worldwide. Nowadays, the common fig still grows wild in the Mediterranean basin. It is also an imperative world crop today. Turkey, Egypt, Morocco, Spain, Greece, California, Italy, Brazil, and other places with typically mild winters and hot dry summers are the major producers of edible figs. Habitual fig cultivation areas have significantly decreased, and genetic variability was reduced due to disappearance of many cultivars selected in the past. Actually almost all grown cultivars are the result of old selections and are maintained by cutting as a way of vegetative propagation. Its annual estimated global production of one million tons of fruit of which about 30 per cent is produced by Turkey. The other major producers are Egypt, Morocco, Greece, California, Italy, Algeria, Syria and Tunisia. In India, fig is considered to be a minor fruit crop and the commercial cultivation of common (edible) fig is mostly confined to Western Parts of Maharashtra, Gujarat, Uttar Pradesh, Karnataka and Tamil Nadu (Table 9.2). Fegra fig (*Ficus palmata*), a small fruited fig of excellent taste, grows wild in the mid-hill region of the Western Himalayas.

Table 9.2: Major Fig Growing Belts in India

State	Districts
Gujarat	Rajkot, Junagarh, Bhavnagar, Porbander, Amreli, Surendranagar, Jamnagar
Himachal Pradesh	Una, Hamirpur, Kangra, Bilaspur
Karnataka	Dharwad, Belgaum, Davangeri, Hubli, Bellary, Raichur
Maharashtra	Aurangabad, Pune, Kolhapur, Solapur, Karad (Sangli)
Tamil Nadu	Coimbatore, Dindigul, Virudhunagar (Hindupur)
Uttar Pradesh	Lucknow, Saharanpur

TAXONOMICAL AND BOTANICAL DESCRIPTIONS

Taxonomy

Common fig (*Ficus carica* L.) belongs to the genus *Ficus* of Moraceae family with

Fig | 115

chromosome number of 2n = 26. The genus *Ficus* comprises one of the largest genera of angiosperms with more than 800 species of trees, shrubs, hemiepiphytes, climbers, and creepers in the tropics and subtropics worldwide, some of them produce aerial and creeping root systems. A number of them are functionally female and produce only a seed-bearing fruit, whereas others are functionally male and produce only pollen and pollen-carrying wasp progeny.

Botany

Fig is a gynodioecious (functionally dioecious), deciduous tree or large shrub, growing to a height of 7–10 metres, with smooth white bark. It bears fragrant leaves which have three or five lobes. The complex inflorescence consists of a hollow fleshy structure called the syconium, which is lined with numerous unisexual flowers. The flower itself is not visible outwardly, as it blooms inside the infructescence. Although commonly referred to as a fruit, the fig is actually the infructescence or scion of the tree, known as a false fruit or multiple fruit, in which the flowers and seeds are borne. It is a hollow-ended stem containing many flowers. The small orifice (ostiole) visible on the middle of the fruit is a narrow passage, which allows the specialized fig wasp, *Blastophaga psenes*, to enter the fruit and pollinate the flower, whereafter the fruit grows seeds. The edible fruit consists of the mature syconium containing numerous one-seeded fruits (druplets). The peel of the fig fruit is thin and tender, usually green, and turns purple or brown after ripening. *Ficus carica* has milky sap (laticifer). The sap of the fig's green parts is an irritant to human skin. The edible seeds are generally hollow, unless pollinated. Pollinated seeds provide the characteristic nutty taste of dried figs.

SOIL AND CLIMATIC REQUIREMENTS

Soil

The fig can be grown on a wide range of soils; light sand, rich loam, heavy clay or limestone, provided there is sufficient depth and drainage. Sandy soil that is medium-dry and contains a good deal of lime is preferred when the crop is intended for drying. Highly acid soils are unsuitable. The pH should be between 6.0 and 6.5. The tree is fairly tolerant of moderate salinity.

Climate

Fig being a deciduous and sub tropical tree, prefers areas having arid or semiarid environment, high summer temperature, plenty of sunshine and moderate water. Although the plants can survive temperature as high 45°C, the fruit quality deteriorates beyond 39°C. Though the mature tree can withstand low temperature up to 40°C, it makes good growth when the temperature is above 15 - 20°C. The size, shape, colour of the skin and pulp quality are markedly affected by climate. But quality figs are produced in the region with dry climate especially at the time of fruit development and maturity. High humidity coupled with low temperature usually results in fruit splitting and low fruit quality.

TYPES OF FIG

Figs have been grouped into four types depending on the sex of the flower and the method of pollination or 'caprification' in order to set a crop. These are:

☆ **Capri fig or wild fig**: It has male and female flowers enclosed in the synconium and is generally considered the 'male' fig. All caprifigs are placed in this class without regard to whether the synconia persist or not. Capri fig are not edible but grown because they harbour fig wasp (*Blastophaga psenes*), which is necessary for pollination and setting fruits. The most common cultivars of caprifigs grown are Brawley, Croisic, Roeding #3 and Stanford. Several cultivated varieties of caprifigs are sweet and palatable, including the Cordelia, Brawley, Enderud and Saleeb.

☆ **Smyrna fig**: It has only female flowers and needs cross pollination by Caprifigs in order to develop normally. This crop sets virtually no breba crop. Only one cultivar 'Sari Lop' (Calimyrna) is cultivated extensively. Other cultivars include 'Marabout' and 'Zidi'.

☆ **San Pedro fig**: It is an intermediate type where the first crop (known as Breba) is parthenocarpic, while the 2nd crop (main) requires pollination like Smyrna type. Some important cultivars of this kind are King, Lampeira, and San Pedro.

☆ **Common Fig or Edible Fig**: All the flowers are female and need no pollination to produce fruit (parthenocarpic fruit set). Some cultivars in this class set no breba crop, some set a moderate crop and some set a good breba crop. Most popular varieties under this group are Poona, Conardia, Mission Kadota, Brown Turkey, Celestial, Conadria, Excel, Flanders, Italian Everbearing, Osborn, Panachee, Peter's Honey, Tena.

IMPORTANT VARIETIES

There are about 20 popular varieties of fig that are being grown in different parts of the world. Some famous varieties of common fig grown in different countries are 'White Adriatic' 'Black Mission', 'Kodota' and 'Conadira' in California; 'Kalamon' in Greece and Sultani in Egypt. The 'Smyrna types' includes popular Turkish cultivar known as 'Saricop' in Turkey and 'Calimyrna' in United States. In India 'Poona' is the most popular cultivar grown for consumption as fresh fruit. Most of the fig grown in Mangalore, Bellary, Coimbatore, Daulatabad, Ganjam, Lucknow and Saharanpur resembles in plant and fruit morphology to that of Poona Fig. Recently, a variety 'Dinkar', an improvement over 'Daulatabad' variety for yield and fruit quality is gaining commercial importance. Some hybrids from California have reportedly performed better over Poona Fig under Mangalore conditions.

☆ **YCD.1 TIMLA fig**: It is an introduction and released from Horticultural Research Station, Yercaud. The trees are well adapted to the rainfed situations of Shevroys hills and to the poor shallow and rocky soils. The trees are suited to growing in home gardens and parks. The trees showed high drought tolerance besides its exceptionally hardy nature and free

Fig |117

from any pest or disease including the common fig rust. The plants are spherical in canopy and are elegant with dense dark green leaves often growing to a height of 7.0 m with a spread of 12.0 m. Fruits are in attractive reddish purple color and are large in size measuring 7.0 cm in diameter, each weighing 100-200g. Higher harvests are made from each tree, the maximum being 4000 fruits. The bearing is throughout the year excepting winter months. Fruits are a rich source of vitamin C (500 mg/100g) besides β-carotene and lycopene.

PLANT PROPAGATION

Figs plants can be easily propagated by several methods. Seed propagation is not preferred generally because quicker and more reliable vegetative methods exist which do not yield the inedible caprifigs. However, those desiring can plant seeds of dried figs with moist sphagnum moss or other media in a zip lock bag and expect germination in a few weeks to several months. The tiny plants can be transplanted out little by little once the leaves open, and despite the tiny initial size can grow to 30 cm or more within one year from planting seeds.

Main Vegetative Propagation, or Spring Propagation

Before the tree starts growth, cut 15–25 cm (6–10 inch) shoots that have healthy buds at their ends, and set into a moist mix of soil and peat-moss located in shade in first time, buried 3/4 of their length. Larger diameter stems are better – intermediate cuttings on branches can be done too (up to diam. 3/4") – but in this case the upper side must be cut inclined, thus marking the upper part, to avoid planting upside-down. Grow one year in a nursery, in a pot or in-ground spaced one foot apart, till winter. Before the plant starts growth, plant it in the desired final location.

For propagation in the mid-summer months, air layer new growth in August (mid-summer) or insert hardened off 15–25 cm (6-10 inches) shoots into moist perlite or a sandy soil mix, keeping the cuttings shaded until new growth begins; then gradually move them into full sun. For spring propagation, before the tree starts growth, cut 15–25 cm (6-10 inches) shoots that have healthy buds at their ends, and set into a moist perlite and/or sandy soil mix located in the shade. Once the cuttings start to produce leaves, bury them up to the bottom leaf to give the plant a good start in the desired location.

An alternative propagation method is bending over a taller branch, scratching the bark to reveal the green inner bark, then pinning the scratched area tightly to the ground. Within a few weeks, roots will develop and the branch can be clipped from the mother plant and transplanted where desired.

PLANTING AND ORCHARD ESTABLISHMENT

Fig trees are recommended to be planted at a spacing of 8 x 8 m or even 15 x 15 m. Closer spacing have resulted in increased yields per tree and per unit area along with early ripening of fruits. Manures can be applied once/twice per annum *i.e.* in July-September or/and January-February. In general, cattle manure alone is applied @ 2-10 baskets/tree depending on its age. Under south Indian conditions, heavy manuring atleast twice a year is essential for getting good crop.

INTERCULTURAL OPERATIONS

Training and Pruning

Fig trees are usually fan-trained. The tree should not be allowed to become tall, as it is easy to harvest fruits by hand picking from low headed trees. Pruning is also done to encourage new growth. In Pune, notching of buds is done in July for inducing fruit bearing shoots by giving slant cuts over dormant buds to remove a small slice of bark with wood and two buds are notched in each shoot.

Nutrition Management

Nutrient requirements may vary according to need of the variety and type of soil. A general recommendation for manurial and fertilizer requirement of fig is presented in Table 9.3. Fertilizers can be applied with the onset of monsoon in juvenile plants, while just after pruning in bearing trees. The annual application of manures and fertilizers can be split in to two, half after pruning and remaining 2 months later when the syconia are developing. Nitrogen is essential for rapid growth of foliage and development of syconia, fruit color and maturation and K for yield and quality. Better fruit quality can be achieved if N and K are applied in the form of ammonium sulphate and sulphate of potash, respectively.

Table 9.3: Recommended Dose of Manures and Fertilizers for Fig

Age of Plant (year)	Organic Manure (kg)		Inorganic Manure (g)		
	Farmyard Manure	Oil Cake*	N	P	K
1-2	15.0	0.5	75	50	50
3-5	25.0	1.0-1.5	150	100	100
Above 5	40.0	2.0	300	200	200

* *Neem, Pongamia* or castor.

Some fig-growing soils may be deficient in micronutrients. This can be corrected by either soil or foliar application of deficient micronutrients. Foliar spray is applied when the plants are flushing or putting forth new growth.

Table 9.4: Micronutrients to be Applied for Correcting Micronutrient Deficiencies

Micronutrients	Soil Application	Foliar Application*
Zinc	30kg $ZnSO_4$	3-4 sprays of 0.25 per cent $ZnSO_4$ (unneutralized) at 10 days interval
Iron	–	3-4 sprays of 05 per cent $FeSO_4$ at 10 days interval
Boron	12kg Borax/ha	–
Magnesium	50 kg $MgSO_4$/ha	2-3 sprays of 0.5 per cent Mg SO_4 at 10 days interval

Fig | 119

Water Management

Fig can sustain heat and drought. However, for commercial production timely irrigation is necessary. Flood irrigation at an interval of 10-12 days during summer is ideal. However, if drip irrigation is adopted 15-20 litres of water/day/plant needs to be provided.

FLOWERING, POLLINATION AND FRUIT SET

The common fig (*Ficus carica*) is a gynodioecious species with bisexual trees (functional male caprifigs) and unisexual female trees. The infructescence is pollinated by a symbiosis with a kind of fig wasp (*Blastophaga psenes*). The fertilized female wasp enters the fig through the scion, which is a tiny hole in the crown (the ostiole). She crawls on the inflorescence inside the fig and pollinates some of the female flowers. She lays her eggs inside some of the flowers and dies. After weeks of development in their galls, the male wasps emerge before females through holes they produce by chewing the galls. The male wasps then fertilize the females by depositing semen in the hole in the gall. The males later return to the females and enlarge the holes to enable the females to emerge. Then some males enlarge holes in the scion, which enables females to disperse after collecting pollen from the developed male flowers. Females have a short time (<48 hours) to find another fig tree with receptive scions to spread the pollen, assist the tree in reproduction, and lay their own eggs to start a new cycle.

PLANT PROTECTION

Although fig trees are generally considered to be hardy, yet they are susceptible to a number of pests and diseases.

Major Insect-Pests and their Management

☆ **Dried fruit beetle** (*Carpophilus* spp.): This small beetle enters the eye of the fruit, creating a favourable situation for the feeding of adults and larvae. Feeding adults carry yeasts into the fruit, causing souring. Carpophilus is more likely to attack developing figs rather than ones that are starting to dry. Carpophilus attack is likely if figs are grown near stone fruit and citrus trees, as the beetles over-winter in fallen fruits. All unwanted fruit should be destroyed after harvest.

☆ **Queensland fruit fly** (*Dacus tryoni*): Female fruit flies lay their eggs in ripening fruit. The larvae hatch and feed inside the fruit, which is then spoiled. The flies pupate in the soil. It can be monitored by using pheromone traps. These traps attract and kill male flies. The presence of flies in the traps indicates that a registered chemical should be applied according to label directions. Destroy fallen fruit by burning or boiling before disposal. This will kill pupae and help break the breeding cycle.

☆ **Fig blister mite** (*Aceria ficus*): They are found in green fruit but they can also damage ripe fruit. They enter the eye and feed near the opening causing rust colored and dry patches on the florets. The damage appears when the fruit is cut open and the exterior of the fruit shows no symptoms.

The affected fruits have reduced eating quality and the blister mites are known to spread fig mosaic virus. No chemicals were registered so far to control this pest and as a precautionary measure, the early fruit can be discarded if the pest is detected in the samples.

☆ **Root knot nematode** (*Meloidogyne* spp.): These nematodes attack a wide range of plants and will carry over from an infected crop. These nematodes damage the roots of the tree and cause significant damage on newly planted trees. Trees on light sandy soils are more susceptible to nematode attack. The females live inside the roots, causing galls to form. The 'knotted' roots have a reduced capacity to take up water and nutrients. The tree becomes stunted and yellow, and produces less fruit. Heavily infested roots may die and decay.

In the past, fumigants were commonly used, but many of these chemicals are being withdrawn from sale. Planting a green manure crop of mustard can act as a 'bio-fumigant before fig trees are planted. Resistant varieties such as Zidi can be grafted on to *Ficus cocculifolia*.

☆ **Birds:** Birds attack and eat developing figs, especially as they start to ripen. Fruit damage ranges from small claw marks and pecks to the whole fruit being eaten. Birds can be a serious pest in some areas, and permanent bird-netting is necessary. Once the birds have located your fig planting they are likely to continue returning to this good source of food, making them a persistent problem. Bird netting is most commonly used. Scare devices vary in effectiveness.

Major Diseases and their Management

☆ **Anthracnose (*Glomerella cingulata*):** It is a fungal diseases in which the fungus infects leaves and fruit under warm, humid conditions. The disease starts off as small brown to black spots, which grow out to form a larger 'target spot' of infected tissue. On leaves, the tissue surrounding the spots turns yellow. Infected fruit and leaves will eventually fall off if the disease is not controlled. Copper-based fungicides are normally used to control the disease.

☆ **Fig rust (*Physopella fici*):** Young leaves are usually attacked by this disease. The fungus produces powdery yellow spots on the leaves. If the disease is not controlled the leaf tissue will turn yellow and the leaves will fall off. Trees can be seriously defoliated. Copper-based fungicides are normally used to control the disease. Consult your local horticulturist for registered products.

☆ **Fig Mosaic:** Fig mosaic is caused by a virus that affects leaf pigment and causes a mottled pattern on the leaf. It is spread mechanically by grafting and taking cuttings from infected trees. Photosynthesis is affected, as the leaf pigment is damaged. Trees become stunted and fruit production is reduced. There is no cure for fig mosaic virus. Affected plants should be removed so the virus does not spread. Virus-infected plants should not

Fig | 121

be used for propagation. Orchard hygiene is important. Work on infected trees last so that the virus is not spread via sap on pruning tools. Clean tools thoroughly in a bleach solution.

Physiological Disorders and their Management

Fig is susceptible to sun burn, fruit splitting and fruit drop. Sun born is noticed mostly in young plants and those subjected to excessive pruning. The trunk and shoots that are exposed to direct sun are prone to sun burn. The affected parts crack and the bark peels off, providing easy access for fungi and other infection. Developing a good canopy by proper pruning and coating the exposed limbs with lime protect the plants from sunburn. Fruit splitting is attributed to sudden change in atmospheric humidity during ripening. This makes the fruit unfit fro consumption as the pulp is exposed to insect and microbial infection. Fruit drop may result from excessive drought and heat, cold nights or light frost. Lack of pollination also causes fruit drop in figs.

MATURITY, HARVESTING AND POSTHARVEST MANAGEMENT

Maturity and Harvesting

Though fig starts bearing fruits from the second year, commercial harvesting is done from the third year. The yield increases with increase in canopy size of the tree and stabilizes during seventh-eighth year. The economic life of the plant is about 35 years. The harvesting season starts in February - March and is over by May - June. The fruits are harvested in 2-3 day intervals manually. Figs destined for the fresh fruit market or canning should be picked when they become fully coloured and still firm. They are harvested by hand with a twisting and pulling motion. Pickers should wear gloves, as the latex from the tree can cause skin irritation. The fruit is then placed into buckets or shallow flats for transport to the packing shed. Fruit should be carefully packed to avoid latex drops staining the skin of harvested fruit. The fruit should be cooled to 0°C as soon as possible.

Storage

Optimum cold room temperature is between –1°C and 0°C, with 90 to 95 per cent relative humidity. At 4.4°C to 6.1°C and 75 per cent relative humidity, figs will keep for eight days. The shelf life out of the cool store is only 1 or 2 days.

Processing

The most common method of processing in fig is drying. Smyrna and Calimyrna figs are the most suitable for drying, as the seeds contribute to the final flavour. In California, drying figs is still a viable industry. Fruit is allowed to ripen and fall naturally before it is collected from the ground. This practice requires mechanized collection and a bare orchard floor. Figs can be picked off the tree, but care must be taken to pick with the stem intact.

After harvest, the figs are immersed in a boiling brine solution (100 g salt per 5 L water). This removes soil and cracks the skin to assist with drying. The figs are rinsed in clean water, then dried further by sun drying or dehydration. The final moisture content is aimed at 17 per cent.

Dried fruits are graded to remove damaged, sunburnt, split, diseased and defective fruit. They are then sent to processors in bulk bins.

10

Gorgon Nut

INTRODUCTION

Gorgon nut (*Euryale ferox*) is also known as fox nut, and *'thangjing'* in Manipuri, *makhana, nikori* in Assamese, *Onibas* in Japanese or gorgon plant/Gorgan nut/ Prickly water lily in English and *makhana* in Hindi. Cultivation of *Makhana* is highly cumbersome, labour intensive and involve human drudgery while sweeping bottom of the water body for seed collection. It is followed by processing of raw seeds, which is equally painstaking activity. Fishermen community belonging to the weaker sections of the society is mainly involved in *makhana* sector.

ORIGIN, HISTORY AND DISTRIBUTION

It is native of South-East Asia and China. Its distribution is limited to tropical and sub-tropical regions of South-East and East Asia. It has been cultivated in China for 3000 years. It occurs in wild form in Japan, Korea, Bangladesh, China and Russia *etc.* and in India, it is grown as a natural crop commonly in fresh water tanks and lakes of Northern, Central and Western India. It is mainly cultivated in Motihari, Purnia, Darbhanga, Madhubani, Saharsa, Katihar and Sitamarhi districts of North Bihar and in some scattered areas of Jammu and Kashmir, West Bengal, Assam, Tripura, Manipur, Rajasthan and Uttar Pradesh. It is cultivated as a cash crop in Bihar and parts of West Bengal and Assam. Its plant requires shallow water depth having thick layer of mucky bottom that is rich in organic nutrients. Bihar accounts for more than 85 per cent of the *makhana* produced in the country. Northern part of Bihar, constituting districts of Madhubani, Darbhanga, Sitamarhi, Saharsha, Katihar, Purnia, Supaul, Kishanganj and Araria, is agro climatically suitable for *makhana* cultivation. As per the estimates of the ICAR–RCER Research Center for Makhana, Darbhanga (Bihar), total area under *makhana* cultivation in India is estimated to be 15000 Ha.

COMPOSITION AND USES

Composition

The plant produces starchy white seeds, and the seeds are edible. The plant is mainly cultivated for its seeds, which are often roasted or fried, which causes them to pop like popcorn. These are then eaten, often with a sprinkling of oil and spices. *Makhana* pop is considered to be nutritious and healthy food with a protein content of 10 -12 per cent. 100 g of raw and popped *makhana* gives a calorific value of 362 kilo cal and 382 kilo cal respectively. It is a good source of carbohydrates and proteins (Table 10.1).

Table 10.1: Nutritional Content of Raw and Popped Makhana

Parameters	Raw Makhana Content	Makhana Pop Content
Carbohydrate	76.9 per cent	84.9 per cent
Protein	9.7 per cent	9.5 per cent
Fat	0.1 per cent	0.5 per cent
Moisture	12.8 per cent	4 per cent

Source: National Research Centre for Makhana, Darbhanga.

Uses

Fruit pulp, stems and rhizome are edible. The seeds are easily digestible, tonic, astringent and are sold as farinaceous food. Fried *makhana* with salt or sugar is widely used as snack food. The seeds are also dried and made into flour. The flour is used as a substitute for arrowroot. It is also used in *kheer* (rice pudding), snacks and curry. The thorny shape of leaves makes it a good aquatic ornamental. *Makhana* is also an auspicious ingredient in offerings to the Goddesses during festivals in India especially Bihar. *Makhana* bran can also be used as poultry (broiler) feed ingredient *i.e,* up to 6 per cent level.

TAXONOMICAL AND BOTANICAL DESCRIPTIONS

Taxonomy

Gorgon nut (*Euryale ferox*) is the only species in the genus *Euryale* which is actually an aquatic herb grown in large numbers in stagnant freshwater pools. It is a flowering plant classified in the water lily family, Nymphaeaceae, although it is occasionally regarded as a distinct family Euryalaceae. Unlike other water lilies, the pollen grains of *Euryale* have three nuclei. Its somatic chromosome number, 2n = 58. In India, *Euryale* normally grows in ponds, wetlands *etc*.

Botany

Makhana is a perennial and an aquatic and floating leaf emergent macrophyte. It does not bear stem but the rootstalks are short, thick and fibrous comprising 3 to 5 clusters, each consisting of about 15 rootlets. The plant roots make their way into the fine clay bottom soil while the plant shows very fast vegetative growth. The leaves are large and round, and often, are than 3 feet across, producing bright purple flowers. The leaves are large and round, often more than a meter (3 feet) across, with a leaf stalk attached in the center of the lower surface. The underside of the leaf is purplish, while the upper surface is green. The leaves have a quilted texture, although the stems, flowers, and leaves which float on the surface are covered in sharp prickles. The colouration and puckering of the leaves makes it attractive. Petioles are prickly and deep green or pink. The flowers are about 5-6 m in diameter and are violet-blue or dark pink in colour, prickly outside and open during the day. Flowers are short lived and are barely held above water level. The calyx is reddish inside and 20-30 petals are violet blue or purple and are shorter than the calyx lobes. Stamens are numerous. Each plant produces 15-20 fruits, which are small, round to oval, spongy and prickly outside. Each fruit consists of 20-25 seeds, which are small (0.75 cm in diameter), black and encrusted with a thick sheath around the white edible part which can be eaten either raw or after roasting.

SOIL AND CLIMATIC REQUIREMENTS

Soil

It can grow in any type of soil except sandy and stony soils. Clay and clay loam with plenty of humus are ideal for its growth. Makhana grows in shallow fresh water bodies. It can also be cultivated successfully in neat and clean aquatic weed free ponds. If aquatic weeds such as *Ipomea* and water hyacinth are present, they should be thoroughly removed before sowing.

Climate

Gorgan nut requires subtropical, tropical and sub-temperate climate with full sunshine and 30-49°C temperature with high humidity (60-70 per cent) for better growth and performance.

PLANT PROPAGATION

Makhana plants germinate from the left over seeds of the previous season. When *makhana* is grown for the first time in a new pond, the rate of seed for sowing is about 80 kg/ha. However when sowing is done annually, 35 kg of seed is required for 1 ha of water spread. It is appropriate to get the seeds sprouted before sowing. The sprouting is encouraged by sprinkling water frequently over the gunny bags in which the seeds are stored under shade. Seed sowing is done in ponds with the help of a boat. After broadcasting, the seeds are pressed slightly into the mud for better germination, anchorage and growth of plants. Sprouting takes place by December-January and the early leaves appear on the pond surface during January- February. During April-May, the entire water surface gets covered with huge, sprawling and thorny leaves, which float on the surface of water. Flowering begins in the month of April when the temperature is around 30° C and maximum flowering occurs in the month of May. *Makhana* flowers stay afloat for two days and then submerge inside water. Fruiting begins by mid of May.

INTERCULTURAL OPERATIONS

Cultivation is done in shallow ponds or tanks. It can also be cultivated in stagnant areas or where the water flow is slow. The water level in the pond must be at least 1.5-2 m high during October-November and 0.6-1 m deep during May-June. This affects the shape, size and colour of the seeds. Before sowing, the pond should be thoroughly cleaned of all weeds and water hyacinth. The seeds are sown at one meter spacing in November-December.

Usually, large seeds are selected for sowing. For better germination, the seeds are kept in a sack under shade before sowing and sprinkled with water for a few days. If germination is poor, seedlings are transplanted during April-May. The seedlings are uprooted with soil without disturbing the roots and central growing points. For distant transportation, these are uprooted without soil and tied in bundles. The water level in the pond is maintained at about 1 m particularly during summer and the rainy season. Since *makhana* is grown in the same pond year after year, organic matter content of the pond soil also increases significantly. Application of 5 q/ha *neem* or castor cake and urea or calcium ammonium nitrate @ 50-100 kg/ha after transplanting during May gives better production. The doses are varied depending on the general vigour of the plants and the previous yield record. Recently the ICAR–RCER Research Center for Makhana has found out a technique for the field cultivation of *Euryale*. By their technology, *makhana* can be cultivated only at 0.3 m water depth. Under this system, *makhana* is transplanted in April and subsequently harvested in August. Seed yield of *makhana* was recorded to be 21.6 q/ha. In another system, cultivation was taken up in low-lying field with clay soil in which direct sowing of *makhana* seeds was done in a chosen formation (as against transplantation in natural ponds) after raising bunds along the periphery. Good agronomic practices that were observed, resulting in a yield of 22 to 30 q/ha as against the 12 to 15 q/ha in natural ponds.

Integration of *makhana* with water chestnut and fish can also be taken up for improving the return as *makhana* grown as a sole crop utilizes water bodies for only seven months (February to August). With the intervention, the water bodies are utilized throughout the year. Further, a refuge area covering 10 per cent of net water bodies is used as vacant space for integration of Indian major carps. Under this system, water chestnut is taken as a tertiary crop.

FLOWERING, POLLINATION AND FRUIT SET

Each *makhana* plant produces nearly 20-25 branches, and each branch bears nearly 10-15 fruits. Flowering continues from May-to-July. It is a self–pollinated crop, and hence no agent has been reported to pollinate it. Fruits are formed within 10-15 days after flower opening. About 8-12 follicles are formed on one plant and a follicle may contain 100-200 g raw *makhana*. The follicle is reddish, similar to a small banana inflorescence.

PLANT PROTECTION

Several insect pests like *makhana* beetle, aphids and snails attack gorgon nut. When cultivated in deep and stagnant water ponds along with fishes, it is not advisable to use any insecticides to control these pests. In the ponds free from fishes, the pests can be controlled by spray of 0.14 per cent monocrotophos. The main diseases that infect gorgan nut are stem rot, root rot, leaf rot and yellowing of leaf. Cleaning of pond water and seed treatment should be done. If there is no fish in the pond, 250 ppm streptocycline or agrimycin and 0.25 per cent blitox-50 can be sprayed along with 10 ml of linseed oil per 10 litres of spray solution.

MATURITY, HARVESTING AND POSTHARVEST MANAGEMENT

Maturity and Harvesting

Makhana is harvested during August-September. As the fruit reaches maturity, the pedicel and leaves start rotting, and the follicle bursts out and floats on the water surface. Fruits after maturity fall down and settle at the bottom. The seeds are collected manually in the nets by diving into the pond during August - September. The process of collection is strenuous involving a thorough sweeping of the entire bottom floor of the water area. Sweeping of the floor, making heaps and their retrieval requires several dives inside the water that makes the job really painstaking. One mature plant produces 250-400 g seeds. On an average, 40 kg produce is obtained from one quintal of wet raw seeds. The yield varies from 12 to 20 q/ha.

Processing, Grading and Storage of *Makhana*

After collection, raw *makhana* seeds are threshed manually to break the papery skin covering and are then washed in the pond water, and then cleaned. The cleaned nuts are sun dried to an extent of around 31 per cent moisture content for ease of transportation and temporary storage. Storage of gorgon nuts poses problems to the growers, as it cannot be stored for longer period at ambient conditions. It is necessary to sprinkle water at regular intervals during storage of nuts to keep them fresh. The sun-dried nuts are then categorized into 5 to 7 grades according to their sizes by means of a set of sieves. Grading of gorgon nut facilitates uniform heating of each nut during roasting. The sun-dried nuts are generally heated in earthen pitcher or cast iron pan by placing them over fire and stirring them continuously. After pre-heating of nut, moisture content reduces to approximately 20 per cent. The pre-heated seeds are kept for tempering in basket/pots for 45-72 hours. Tempering of seeds facilitates the loosening of kernels within the hard seed coats.

Roasting and popping are the most painstaking operations of *makhana* processing. *Makhana* seeds are heated in iron pans over the fire at 290°C to 340°C surface temperature with continuous stirring. When crackling sound is heard, 5-7 roasted seeds are scooped quickly by hand and kept on hard surface and sudden impact force is applied on them by means of a wooden hammer. As the hard shell breaks, the kernel pops out in expanded form, which is called *makhana*. Polishing of *makhana* is done immediately after popping since popped *makhana* may absorb moisture and render polishing difficult. It is done by rubbing action of *makhana* pops among themselves in bamboo baskets. Polishing facilitates more whiteness and luster to the *makhana*. After polishing, *makhana* is graded into 2-3 grades namely *Rasgulla, Samundha* and *Thurri*. The graded *makhana* is then packed in gunny bags. A gunny bag with a capacity of 1 quintal of sugar may contain 8-9 kg of good quality *makhana*.

11
Jackfruit

INTRODUCTION

The jackfruit, also known as jack tree, jakfruit, or sometimes simply jack or jak, is a multi-purpose plant species which provide food, timber, fuel, fodder and medicinal and industrial products. The word 'jackfruit' comes from Portuguese word *jaca*, which in turn, has been derived from the Malayalam language term, *chakka*. The primary economic product of jackfruit is the fruit which is used both when mature and immature. When unripe (green), it is remarkably similar in texture to chicken, making jackfruit an excellent vegetarian substitute for meat. In fact, canned jackfruit (in brine) is sometimes referred to as 'vegetable meat'.

COMPOSITION AND USES

Composition

It is a nutritious fruit rich in carbohydrates, proteins, potassium, calcium, iron, and vitamins A, B, and C (Table 11.1). Due to high levels of carbohydrates, jackfruit supplements other staple foods in times of scarcity in some regions. The flesh of the jackfruit is starchy and fibrous, and is a source of dietary fibre. The presence of isoflavones, antioxidants, and phytonutrients in the fruits indicate that jackfruit has cancer-fighting properties. It is also known to help cure ulcers and indigestion.

Uses

In India, jackfruit is used both as vegetable and/or fruit. Unripe fruits are cooked as vegetable especially in North India whereas ripe fruits are considered as a delicacy in some parts of Eastern India, especially, Bihar. Its seeds (nuts) can be roasted like chestnuts, or boiled. The fruit pulp is sweet and tasty and used as dessert or preserved in syrup. The fruits and seeds are also processed in a variety of ways for food and other products. Jackfruit value added products include chips, papads, pickles, icecream, jelly, sweets, beverages like squash, nectar, wine and preserved flakes, *etc*. Additionally, jackfruit leaves, bark, inflorescence, seeds and latex are used in traditional medicines. The wood of tree is also used for various purposes.

Table 11.1: Nutritive Value of Jackfruit per 100 g Edible Portion

Constituent	Pulp		Mature Seed
	Tender	Ripe	
Moisture (per cent)	84.0	77.2	64.5
Carbohydrate (g)	9.4	18.9	25.8
Protein (g)	2.6	1.9	6.6
Fat (g)	0.3	0.1	0.4
Fibre (g)	4.4	1.1	1.3
Total mineral matter (g)	0.9	0.8	1.2
Calcium (mg)	50.1	20.0	21.0
Phosphorus (mg)	97.0	30.0	28.0
Iron (mg)	1.5	500.1	0.8
Potassium (mg)	206.0	350.0	246.0
Vitamin A (IU)	0.0	540.0	17.0
Thiamin (mg)	0.2	30.0	0.2
Riboflavin (mg)	0.1	0.1	0.1
Nicotinic acid (mg)	0.2	0.4	0.3
Vitamin C (mg)	11.0	7.0	11.0
Calorific value	50.0	84.0	139.0

ORIGIN, HISTORY AND DISTRIBUTION

The jackfruit is indigenous to the rain forests of the Western Ghats of India and is cultivated throughout the tropical lowlands in South and South-East Asia, parts of Central and Eastern Africa and Brazil. Major jackfruit producers are Bangladesh, India, Myanmar, Thailand, Vietnam, China, the Philippines, Indonesia, Malaysia and Sri Lanka. It is a popular and relatively cheaper fruit in Southern Asia and other warm countries of both the hemispheres. In Europe, the fruit is sold canned with sugar syrup. Jackfruit is also found across Africa (*e.g.*, in Cameroon, Uganda, Tanzania, Madagascar, and Mauritius), as well as throughout Brazil and in Caribbean nations such as Jamaica. Jackfruit is the national fruit of Bangladesh. Away from the Far East, the jackfruit has never gained the kind of acceptance that is accorded to the breadfruit (except in settlements of people of East Indian origin). This is largely due to the odour of the ripe fruit and traditional preference for breadfruit. In India, jack fruit is grown from hilly states like Assam, Tripura, Meghalya, Himachal Pradesh to Eastern, Western, Northern and Central states like Odisha, Jharkhand, Madhya Pradesh, Chhattisgarh to Southern states like Karnataka, Kerala and Tamil Nadu (Table 11.2).

Table 11.2: Major Jackfruit Growing Belts in India

State	Districts
Assam	Cachar, Goalpara, Darrang, Nagaon, Kamrup, Karbi Anglong, Dima Hasoa
Chhattisgarh	Raipur, Gariaband, Baloda Bazar, Mahasamund, Dhamtari, Durg, Rajnandgaon, Bemetara, Kabirdham (Kawardha), Mungeli, Bilaspur, Janjgir-Champa, Raigarh, Surguja, Korba, Balrampur, Surajpur, Koriya, Jashpur, Bastar (Jagdalpur), Kondagaon, Narayanpur, Dantewada, Sukma, Bijapur, Kanker
Himachal Pradesh	Solan (Nalagarh, Kunnihar), Sirmour (Nahan, Dhaula Kuan), Bilaspur (Ghumarwi)
Jharkhand	Ranchi, Khunti, Chatra, Ghumla, Deoghar, Latehar, Sahibganj, Pakur, Dumka
Karnataka Mandya	Mysore, Bangaluru (Rural), Uttara Kannada, Tumkur, Kolar, Chikkaballapur,
Kerala	Idukki, Kannur, Thiruvananthapuram
Madhya Pradesh	Balaghat, Shahdol, Sidhi and part of Mandla and Dindori
Maharashtra	Thane, Raigad, Ratnagiri, Sindhudurg, Mumbai, Kolhapur, Karad (Sangli), Pune
Meghalaya	Ri-bhoi, Garo Hills, East Khasi Hills
Odisha	Kandhamal, Koraput, Ganjam, Kalahandi
Tamil Nadu	Ariyalur, Cuddalore, Dindigul, Kanyakumari, Namakkal, Perambalur, Pudukottai
Tripura	All Districts

TAXONOMICAL AND BOTANICAL DESCRIPTION

Taxonomy

Jackfruit (*Artocarpus heterophyllus* Lam.) belongs to *Artocarpus* genus of the mulberry family (*Moraceae*), having somatic chromosome number, 2n = 56. Several other important fruit trees such as fig (*Ficus* spp.), mulberry (*Morus* spp.) and osage orange or hedge apple (*Maclura pomifera* Schneid) also belong to Moraceae and yield edible fruits. This family encompasses about 1,000 species in 67 genera, mostly tropical shrubs and trees, but also a few vines and herbs. The genus *Artocarpus* comprises about 50 species, 11 of which are known to produce edible fruits. Even at the species level, a high degree of genetic variability exists. This is true in the case of jackfruit (*A. heterophyllus*), breadfruit (*A. altilis*) and marang (*A. odoratissimus*). At least two species, jackfruit and champedak (*A. integer*) hybridize freely in nature. Furthermore, some species are graft-compatible. The seedless breadfruit, for example, is graft-compatible not only with the seeded strains but also with the kamansi (*A. camansi*) and pedalai (*A. sericicarpus*). The monkey jackfruit (*A. rigidus*) is another promising species because it is very productive, produces perhaps one of the smallest fruits in the genus, has sub-acid taste and possesses good flavour, better than jackfruit, champedak and marang.

Botany

The jackfruit tree is handsome and stately. In the tropics, it grows to an enormous size, like a large eastern oak. All parts contain sticky, white latex. The leaves are oblong, oval, or elliptic in form, 4 to 6 inches in length, leathery, glossy, and deep green in color. Juvenile leaves are lobed. Male and female flowers are borne in separate flower-heads. Male flower-heads are on new wood among the leaves or above the female. They are swollen, oblong, from an inch to four inches long and up to an inch wide at the widest part. They are pale green at first, and then darken. When mature, the head is covered with yellow pollen that falls rapidly after flowering. The female heads appear on short, stout twigs that emerge from the trunk and large branches, or even from the soil-covered base of very old trees. This habit of fruiting is called as cauliflory. They look like the male heads but without pollen, and soon begin to swell. The stalks of both male and female flower-heads are encircled by a small green ring. Botanically, its fruit is a multiple fruit, called as sorosis, which develops from spadix inflorescence bearing sessile flowers. Jackfruit is the largest tree-borne fruit in the world, reaching 80 pounds in weight and up to 36 inches long and 20 inches in diameter. The exterior of the compound fruit is green or yellow when ripe. The interior consists of large edible bulbs of yellow, banana-flavored flesh that encloses a smooth, oval, light-brown seed. The seed is long and white and crisp within. There may be 100 or up to 500 seeds in a single fruit, which are viable for no more than three or four days. When fully ripe, the unopened jackfruit emits a strong disagreeable odour, resembling that of decayed onions, while the pulp of the opened fruit smells of pineapple and banana.

SOIL AND CLIMATIC REQUIREMENTS

Soil

The jackfruit can be grown on a variety of soils as long as they are well-drained, but it performs best on deep, well- drained alluvial, sandy or clay loam soils having pH 6.0 – 7.0. It can also be grown in open textured light or lateritic soils provided such soils have sufficient nutrients.

Climate

Although the jackfruit is essentially a tree of the tropical lowlands, it is adapted to a wider range of conditions. It can tolerate higher altitudes and cold better than the breadfruit. Jackfruit prefers a warm humid tropical climate, mostly lowland coastal areas below 1000 m with more than 1500mm annual rainfall. It can't tolerate cold, drought and flooding situations, but can tolerate moderate winds and salinity.

IMPORTANT CULTIVARS

There are two main varieties. In one, the fruits have small, fibrous, soft, mushy, but very sweet carpels with a texture somewhat akin to raw oysters. The other variety is crisp and almost crunchy though not quite as sweet. This form is the more important commercially and is more palatable to western tastes. So far, there is no well-defined variety in jackfruit and different types are known differently in different localities. Local selections have been named as *Gulabi* (rose scented), *Champa* (flavour like that of *Michelia* sp.), *Hazari* (bearing more number of fruits in a tree).

As a result of local survey, some better types have been collected. Since raw jackfruits have good demand in India as vegetable for culinary purpose, emphasis is also given on fruit characters like thickness of rind and softness of flesh at premature stage of fruit development. Survey conducted in Assam showed that both soft flesh (*Pakikhua*) and firm flesh (*Khoja*) varieties are available while in Western Ghats, varieties such as Muttam Varikka and Rudrakshi were identified. In South India (Kerala, Tamil Nadu and Karnataka), different forms of jackfruit (Varikka, Koozha, Navarikka) are available. A jackfruit variety known as Singapore or Ceylon Jack, introduced from Sri Lanka is a popular variety. Certain other varieties named as Velipala, Hybrid Jack, Panruti Selection, Burliar 1 and Muttam Varikka are also popular in different localities. Variety Konkan Prolific was developed by Regional Fruit Research Station, Vengurla, Maharastra in 2004. It has a yield potential of 420 kg/tree. Fruits are medium size, oblong shape, firm flesh even in monsoon season.

PLANT PROPAGATION

For propagation of jackfruit, the use of seeds is generally preferred because vegetative propagation is quite difficult and cumbersome. However, trees may not exhibit the true characters of the parent plant, take longer time to start flowering, and are generally tall. The seeds should be collected from healthy, mature plants which are prolific producers of fruits with desirable characteristics. Only large seeds are used. Immediately after extraction from the fruit, the seeds are washed in water to remove the slimy coating around the seeds. The horny part of the pericarp is also removed to hasten germination.

As a general rule, the seeds are sown immediately without drying because they are recalcitrant. However, the seeds can be stored in air-tight plastic containers at 20°C to maintain their viability for about 3 months. In sowing, the seeds are laid flat or with their hilum facing downward. Germination starts within 10 days of sowing. About 80-100 per cent of the seeds germinate within 35 to 40 days after sowing.

Jackfruit can be vegetatively propagated using stem cuttings and by air layering or marcottage. However, special techniques are necessary, including the use of rooting hormones at the right concentrations. The Forkert or patch budding as well as cleft grafting and wedge grafting likewise proved successful. In Thailand, suckle grafting is extensively applied in growing jackfruit. It is a form of inarching in which young potted rootstocks are decapitated and inserted in twigs of the mother trees.

PLANTING AND ORCHARD ESTABLISHMENT

Like other fruit crops, proper land preparation is important for growing jackfruit. In sloppy lands or where only a few trees are to be planted, land preparation involves the slashing of the vegetation and round weeding of the immediate peripheries of the hills. If a large number of trees are to be planted, it is best to prepare the land thoroughly by ploughing. Holes are then dug 0.5-1 meter deep and wide. To ensure supply of nutrients, pits should be refilled with top soil mixed with 1/3 proportion of compost. If raw manure or any organic substrate is used, planting should be delayed for at least 15 days to allow decomposition. Field planting can be done by direct seeding or by transplanting using nursery grown potted seedlings. Potted seedlings should be out-planted usually before they are one-year-old or before the roots leak out of the pot because the seedlings are sensitive to root disturbance. Bare root transplanting is inapplicable to jackfruit. Jackfruit can be planted with a spacing of 8-12 meters in square, rectangular or triangular pattern. This is equivalent to a population density of about 70 to 156 plants per hectare (28-63 plants per acre) in the square system and 80 to 180 per hectare (32-72 per acre) in the triangular or hexagonal system. The exact population, however, can only be determined by preparing a planting lay-out plan showing the positions of the hills, plant-to-plant spacing, and the distances of rows to the boundaries. This lay-out plan is similar to a construction blueprint which should be made before actually starting the farm activities in growing jackfruit.

INTERCULTURAL OPERATIONS

Training and Pruning

The height of jackfruit, especially those raised from seed, can be regulated by cutting the main trunk about 2-3 meters from the ground. Early cutback of the main trunk can also be done to induce production of branches, allowing 4 or 5 branches to develop which are evenly distributed when viewed from the top. Properly trained, jackfruit grows with an open center which allows better light penetration. Weak, dead, diseased and overlapping branches should be removed. This is to promote light penetration and air movement, and to prevent build up of insect pests and disease pathogen population. Branches are also removed if they hinder access to the fruits during wrapping and harvesting.

Nutrition Management

Fertilizer application is always a component of growing jackfruit or any crop on a continuing basis. Farm yard manures are applied in increasing doses from 10 to 30 kg per year as the tree matures. To ensure maximum yields of fruiting trees, complete fertilizer is applied at the rate of 1-3 kg per tree per year. Addition of muriate of potash is also generally recommended for fruiting trees. The rate is split into two equal doses, the first application preferably during the onset and the second just before the end of the rainy season. If irrigation is available, fertilizer application can be programmed every 6 months.

Water Management

Regular watering should be done, unless rainfall is sufficient, from planting until the seedlings are fully established. Sufficient water is likewise needed during dry months when the trees are in the flower bloom and fruit development stages.

Weed Management

Ring weeding is practiced to keep the immediate periphery of the tree free of weeds. This operation is a regular necessity in growing jackfruit at least during the first 3 to 4 years after planting. The weeds can be piled around the tree to serve as mulch which will conserve moisture and prevent the germination of weed seeds.

FLOWERING, POLLINATION AND FRUIT SET

Jackfruit has long pre-bearing period and its plants start bearing after 8-9 years of planting. Flowering starts in December and continues up to March. Jackfruit is a monoecious fruit plant and male and flowers appear on the same plant but at different locations. Infloresecence is a spike which is covered by two spathes. The female spikes are borne on footstalks while the male spikes appear both on footstalks as well as on the terminal branchlets. Footstalks bearing female spikes are much more vigorous than male pikes.

The distribution of sex in jackfruit is very interesting. Usually a jackfruit plant bears more number of male spikes (40-90 per cent) than female spikes. The footstalks bearing female spikes appear on trunk and main branches adventitiously in the central region of the plant whereas male spikes appear both in the central and peripheral regions of the plant. Sex of the spike can be easily identified initially as the female spikes are much lengthy than male spikes. Similarly, surface of a young male spike is smooth and that of a female spike is granular.

Jackfruit is anemophilous (wind-pollinated). Pollination is a major problem in jackfruit. Usually the flowers on the inner sides of the tree don't pollinate or remain under-pollinated. As a result, the shape of the fruit is irregular, and due to lack of pollination, there is very high drop of female spikes. After successful pollination and subsequent fertilization, fruit develop in spring and summer and become ready for harvesting in June or early-July.

PLANT PROTECTION

Major Insect-Pests and their Management

It is important that one who engages in growing jackfruit should also be familiar with pests and diseases that affect the crop. The two major pests of jackfruit are the shoot and fruit borer (*Diaphania caesalis*) and the brown bud weevil (*Ochyromera artocarpi*). Caterpillars of the shoot borer tunnel into buds, young shoots and fruit. The grubs of the bud weevil bore into young buds and fruits while the adults feed on the leaves. To prevent damage, the fruits are wrapped with plastic bags when still young. Fallen, overripe and damaged fruits should be collected and buried under the ground.

Major Diseases and their Management

Blossom rot, also called fruit rot and stem rot, is a serious disease of jackfruit, which is caused by the fungus *Rhizopus artocarpi*. It may lead to 15-32 per cent crop losses. The disease affects the inflorescences or the tips of the flowering shoots. The inflorescences turn black, rot and drop. Another disease is the pink disease, which is also called as pink limb blight and is caused by the fungus, *Corticium salmonicolor*. This disease infects many farm crops, including rubber, Citrus, mango, durian, coffee and cacao. Control measures include thorough collection and disposal of the affected parts and fallen fruits.

MATURITY, HARVESTING AND POSTHARVEST MANAGEMENT

Maturity

Young fruits can be harvested for vegetable 2-3 months after fruit set or when the seeds are hardened. For mature fruits, selection is based on the following indices: (1) hollow sound when the fruit is tapped; (2) change in the colour of the skin from pale-green to greenish-yellow or brownish-yellow; (3) emission of a strong aroma; and (4) flattening of the spines with wider spaces. The stalk (peduncle) of the fruit should be cut with a sharp knife and the fruit is gently brought down to the ground.

The tender fruits are generally handled by vegetable dealers and shopkeepers. Hence, care must be taken to avoid damage to the skin which causes browning, resulting in poor external appearance. The cut stalk of the fruit exudes latex, which may permanently stain clothing. When latex exudation stops, the fruits should be wrapped individually in newspapers and packed in a suitable container.

The mature fruits are subjected to mechanical damage, exposure to sunlight and rough handling during transport which reduces the fruit quality. However, induced ripening is not necessary for matured fruits as in case of other climacteric fruits.

Processing

In view of its important properties, ripe jackfruit bulbs (flakes) are consumed worldwide as a dessert fruit or processed in various forms like canned segments (with syrup and honey), jackfruit flavours, drum-dried powder, osmo air-dried segments, enzyme liquefied juice, candy, jam, spread, jelly, ready to serve beverage (RTS), squash, syrup, nectar, slab or bar and chips/papad are also prepared by frying the ripe and semi ripe flakes in margarine. The pulp is also used to flavour ice-cream and beverages, made into jackfruit honey, reduced to concentrate or powder, and used for preparing drinks. Pickles and dehydrated leather are its preserved delicacies. The seeds can be eaten boiled, roasted or dried and salted as table nuts, or they can be ground to make flour and blended with wheat flour for baking. A yellow dye also can also be extracted from the wood particles and used to dye cotton. The fruit has a delicious taste, captivating flavour, attractive colour and excellent quality, which make it suitable for processing and value addition. However, some researchers reported difficulty in the collection of fruits, separation of bulb from the rind, uncertainty and variability in the yield and quality are the major problems involved in the utilization of jackfruit. Hence, there is a need to process the fruit at commercial level.

12

Jamun

INTRODUCTION

Jamun, jambul, jambolan, or jamblang is an evergreen tropical tree in the flowering plant family Myrtaceae, to which several plants of economic importance such as guava, eucalyptus and clove also belong. The name of the fruit is sometimes mistranslated as blackberry, which is a different fruit in an unrelated family. It grows as a tall, evergreen tree under humid tropical conditions.

ORIGIN, HISTORY AND DISTRIBUTION

The *jamun* is indigenous to India. It is also found growing in Thailand, Philippines, Myanmar, which are probably the secondary centre of its origin. *Jamun* is also cultivated in drier parts of Israel, Algeria, Malagassy, West Indies and South Africa. In deciduous forests of subtropical region, it sheds leaves during the spring season. It is believed that the tree was introduced to Florida, USA in 1911 by the USDA, and is also now commonly grown in Suriname and Trinidad and Tobago. In Brazil, where it was introduced from India during Portuguese colonization, it has dispersed spontaneously in the wild in some places, as its fruits are eagerly sought by various native birds. This species is considered an invasive in Hawaii, USA.

COMPOSITION AND USES

Composition

The colour of *jamun* fruit is due to the presence of anthocyanin (under high pH, the pigment gives this characteristic colour). Its fruits are nutritious and are rich in minerals, fibre, carbohydrates and vitamins A and C (Table 12.1).

Table 12.1: Nutritional Value of Ripe *Jamun* Fruit per 100 g

Energy	60 kcal	Protein	0.995 g	Thiamine (vit. B₁)	0.019 mg
Carbohydrates	14 g	Water	84.75 g	Niacin (vit. B₃)	0.245
Vitamin C	11.85 mg	Calcium	11.65 mg	Iron	1.41 mg
Magnesium	35 mg	Phosphorus	15.6 mg	Potassium	55 mg

Uses

Its fruits, seeds and bark have medicinal value. The seeds of jamun are an effective medicine against diabetes and their powder is widely used in India to control diabetes. *Jamun* trees are mainly planted as windbreak and as roadside tree.

TAXONOMICAL AND BOTANICAL DESCRIPTION

Taxonomy

The *jamun* (*Syzygium cuminii*) belongs to genus *Syzygium*, of family Myrtaceae, having somatic chromosome number, 2n = 66. This genus has 8 other species which are found grown as wild in Nilgiris and Western Ghats. These species produce small, acrid fruits with thin pulp and large seeds. These fruits are astringent due to tannins in the pulp. These species have high timber value too. Some of the important species are *S. javanicum* (water apple), *S. jambos* (rose apple) and *S. uniflora* (Suriname cherry).

Botany

The *jamun* is a slow fast-growing tree, reaching full size in about 40 years and can live more than 100 years. It usually forks into multiple trunks at a short distance from the ground. The bark on the lower part of the tree is rough, cracked, flaking and discolored; further up it is smooth and light-gray. Its dense foliage provides shade and is grown just for its ornamental value. The leaves which have an aroma similar to turpentine, are pinkish when young, changing to a leathery, glossy dark green with a yellow midrib as they mature. The fragrant flowers come in clusters, and have a funnel-shaped calyx. The fruit, in clusters of just a few or 10 to 40, is round or oblong, often curved and usually turns from green to light-magenta, then dark-purple or nearly black as it ripens. A white-fruited form has been reported in Indonesia. Its skin is thin, smooth, glossy, and adherent. The pulp is purple or white, very juicy, and normally encloses a single, oblong, green or brown seed, though some fruits have 2 to 5 seeds tightly compressed within a leathery coat, and some are seedless. The fruit is usually astringent, sometimes unpalatably so, and the flavour varies from acid to fairly sweet.

SOIL AND CLIMATIC REQUIREMENTS

Soil

The *jamun* trees can be grown on a wide range of soils-calcareous, saline sodic soils and marshy areas. Deep loam and well-drained soils are, however, the most ideal. It does not like very heavy and light sandy soils.

Climate

Since *jamun* is a hardy fruit, it can be grown under adverse soil and climate conditions. It thrives well under both tropical and subtropical climate. It requires dry weather at the time of flowering and fruit setting. Early rains are beneficial for better growth, development and ripening of fruit. Young plants are susceptible to frost. Usually, dry atmosphere favours good fruiting in *jamun* but fruit ripening is hastened in hot humid conditions.

IMPORTANT VARIETIES

There is no improved variety for commercial cultivation. Seedling trees of *jamun* are found all over India. Some seedling selections have been made on the basis of fruit shape (oblong, obovate), fruit size (big, small), fruit colour (purple, deep purple, red, bluish, black) and pulp (sweet, slightly acrid). Some such selections grown in North India are given below:

- ☆ **Ra-Jamun:** It produces big sized fruit with average length of 2.5 - 3.5 cm and of diameter 1.2 - 2.0 cm. Fruits are oblong in shape, deep purple or bluish black in colour at fully ripe stage. The pulp colour of ripe fruit is purple pink and the fruit is juicy and sweet. The stone is small in size. It ripens in the month of June-July. The variety is very common among the people.

- ☆ **Badama:** Fruits are large and very juicy.

- ☆ **Kaatha:** Fruits are small and acidic.

- ☆ **Jathi:** Fruits ripen in May-June.

- ☆ **Ashada:** Fruits ripen in June-July.

- ☆ **Bhado:** Fruits ripen in August.

A type having large- sized fruits is known as *Paras* in Gujarat. Another type found in Varanasi has no seed. Among other named cultivars, 'Early Wild', 'Late Wild', 'Pharenda'; and, secondarily, 'Dabka' are important. A selection with desirable traits has been located by the Acharya Narendra Dev University of Agriculture and Technology at Faizabad. It has been named by them as Narendra *Jamun* 6. Similarly, Rajendra Jamun 1 from Bihar Agricultural College, Bhagalpur, Bihar; CISH J-42 (Seedless type) and CISH J-37 from ICAR-Central Institute for Subtropical Horticulture (CISH), Lucknow, U.P. have been developed.

Some other improved varieties developed by the State Agricultural Universities (SAUs) and ICAR institution are given below:

- ☆ **Goma Priyanka:** It was released in 2010 from CHES (ICAR-CIAH), Godhara. Fruits are good in test having 16.86 TSS°Brix. Fruit are rich in Vitamin C (45.44 mg/100g). Yield potential is 30 kg/plant.

- ☆ **Konkan Bahadoli:** It was developed in 2004 at Regional Fruit Research Station, Vengurla, Maharashtra. It has a yield potential of 125-150 kg/tree. Big size fruits and longer keeping quality, semi spreading, dome shape, dark green foliage, regular bearer.

PLANT PROPAGATION

Jamun is propagated both by seeds and vegetative techniques, the most common being by seeds as due to poly-embryony, it could provide true to type plants in *jamun* and could be a commercial practice to raise hardy plants, especially, for road side and windbreak plantations. Freshly harvested seeds should be sown during June-July in North India and during August in South India as *jamun* seeds lose viability after prolonged storage. The seeds germinate within 8-10 days. The seeds should be sown at a distance of 15 cm in rows which are 25 - 30 cm apart. The seedlings can also be raised in polythene bags of 22.5 - 30 cm size. Plants grown from seed become transplantable during next spring season. But it is advisable to keep them in nursery upto next rainy season which is the best time of its plantation.

Vegetative propagation methods like cutting, air layering, inarching, kernel grafting and budding have been tried with varied success. Semi-hardwood cuttings, treated with growth-promoting hormones have given 20 per cent success and have grown well. Budding onto seedlings of the same species has also been successful. Veneer-grafting of scions from the spring flush has yielded 31 per cent survivors. The modified Forkert method of budding may be more feasible. Approach-grafting and inarching are also in practice. Air-layers treated with 500 ppm indole butyric acid have rooted well in the spring (60 per cent of them) but have died in containers in the summer. *S. densifolium* can be used as rootstock because it is resistant to the attack of termite.

PLANTING AND ORCHARD ESTABLISHMENT

Jamun can be transplanted during spring (February-March) or during monsoon (August-September). However, the later season of planting is considered better because the plants easily get established during the rainy season. Pits of 1m × 1m × 1m size are dug 10m apart for seedling trees and 8m apart for budded plants in a properly cleaned field. Pit digging should be completed before the onset of the monsoon or spring season. They should be filled with a mixture of top soil and well-rotten farmyard manure or compost in a 3:1 ratio. In poor soils, *jamun* can be planted closer (6 × 6 m). Planting should be done in the already prepared pits of 1 x 1 x 1 m filled with a mixture of soil and 20 kg FYM. Planting can also be done during the spring but the plants would need special care during summer months.

INTERCULTURAL OPERATIONS

Training and Pruning

Young plants need training for the development of framework. *Jamun* plants should be trained according to the modified leader system. Regular pruning is not required in *jamun* plants. However, in later years, the dry twigs and crossed branches are removed. The main stem or trunk should be kept clean up to a height of 60–90cm from the ground level by removing the basal branches and sprouts. The tree bole should be kept straight upto 1 m height without side branches. Sprouts emerging from the rootstock portion should be periodically removed. *Jamun* develops narrow crotch angle. Since the branches are brittle, these often break due to fruit load. Therefore, wider crotches should be developed at least in the main branches. *Jamun* plants do not require any pruning except removing diseased, dry and crisscross twigs in later years.

Nutrition Management

Jamun trees are seldom manured. However, annual dose of FYM @ 20 kg/ tree during pre-bearing stage and @ 40 kg/tree at bearing stage should be applied. Application of 1000 g N, 500 g P_2O_5 and 500 g K_2O to a 10-15 year old trees is considered useful under irrigated conditions during February-March.

Water Management

Jamun is a deep rooted tree and can draw moisture from lower strata of the soil. Therefore, once established, the trees can withstand much adverse weather conditions even under rainfed arid conditions. Young plants require 6–8 irrigations for better growth. In bearing trees, irrigation should be given from September to October for better fruit bud formation and from May to June for better development of fruits. Normally 5–6 irrigations are required. Under arid environment, young trees of *jamun* should be protected from frost during winters by light irrigations. Mulching with weeds and crop residues helps in conserving the moisture and temperature status of the soil.

Intercropping

To supplement the income from pre-bearing period of *jamun*, intercropping should be practiced judiciously. Intercropping *karonda* or *phalsa* or seasonal vegetable crops in initial years between the rows or interspaces can be done. Sprouts arising from base of its plants should be removed timely and the plantation should be kept weed-free. Under arid and semi-arid environment, intercrops improve the water holding capacity of soils. The tree is suitable for silvi-horti cropping system.

FLOWERING, POLLINATION AND FRUIT SET

The main fruiting season (tree blooms principally in March and April) extends through late May, June and July. The trees are in full bloom in the second week of April. The inflorescence in *jamun* is generally borne in the axils of leaves on branchlets. The flowers are hermaphrodite, light yellow in colour. It is a cross-pollinated tree and pollination is done by honeybees, houseflies and wind. Flower buds take 15-20 days to open. The maximum anthesis and dehiscence were recorded between 10 A.M. and 12 Noon. The pollen fertility was higher in the beginning of the season. The maximum receptivity of stigma was observed one day after anthesis. Stigma remains receptive for only 24 hours and about 60 per cent of the flowers drop. Flower drop can be reduced by two sprays with 60 ppm GA_3 first at full bloom and the second after 15 days. Since, *jamun* is a cross- pollinated crop, hence keeping of honeybees near the plantation is beneficial for maximum fruit set and productivity. Fruit development pattern is single sigmoid. Three distinct phases are observed in fruit growth, first between 15-52 days after fruit set when the growth rate remains slow, second about 6 days after the completion of first phase when development is very rapid and the third within 3 days when growth rate is slow with little increase in weight. The whole process of fruit growth and development is completed within 62 days after fruit set in North India. Thus the fruits mature in 3-4 months after full bloom. Spraying 25 ppm 2,4-D at fruit set induces about 83 per cent seedlessness in fruits.

In *jamun*, the flower and fruit drop start just after opening of flowers and continue up to maturity. Only 12 - 15 per cent flowers reach maturity. The flower and fruit drop occurs at 3 stages. Of which, the first drop takes place during bloom or shortly there after, which accounts for about 52 per cent of the flowers drop off after 4 weeks from flowering. The second drop starts about 35 - 40 days of full bloom and apparently there is no difference between the developing and aborting fruits. The third drop takes place after 42 - 50 days of full bloom and continues till 15th July. The extent of flower and fruit drop in *jamun* may be reduced by two sprays of 60 ppm GA_3, one at full bloom and the other 15 days after initial setting of fruit.

PLANT PROTECTION

There is no major pests and disease. Leaf eating caterpillar (*Carla subtilis*) eats the tender leaves. It can be controlled by two sprays 0.2 per cent carbaryl or 0.04 per cent monocrotophos at 15-20 days interval. White fly (*Dialeurodes eugenia*) is another pest giving the fruit a wormy appearance. It can be controlled by proper cultural operations maintaining field sanitation. Buds, rodents and bats also causes damage to *jamun* fruit.

Anthracnose is a common disease which affects the leaves and young fruits and cause fruit rot. Spraying of mancozeb (0.2 per cent) or Bordeaux mixture (1 per cent) checks its incidence.

MATURITY, HARVESTING AND POSTHARVEST MANAGEMENT

Seedling trees start bearing at the age of 9-10 years, whereas budded ones take 5-6years. It takes about 3-5 months to ripen after full bloom. Fruits change their colour from green to deep red or bluish black. *Jamun* fruit is botanically a berry and is non-climacteric fruit hence it does not ripen after harvesting. Fully ripe fruits are harvested daily by hand picking or by shaking the branches and collecting the fruits on a polythene sheet. It has delicate skin; therefore, the ripe fruits are picked singly by hand and in all cases care should be taken to avoid all possible damage to fruits. A full grown tree produces 70-80 kg fruits per year. Harvesting season varies according to climatic conditions. In North India, it is harvested in June-July, while in Karnataka and Andhra Pradesh in May and July-August in Tamil Nadu.

Jamun is a highly perishable fruit and cannot be stored for more than 24 hours at room temperature. However, pre-cooled fruits are packed in polythene bags for storage for 3 weeks at 8-10 °C and 85-90 per cent relative humidity.

There is no standard practice for grading of *jamun* fruits. Blemished or bruised fruits must be sorted out before packing. Fruits are normally packed in bamboo baskets and transported to local markers. The fruits prepacked in leaf cup covered with perforated polythene bags have little or no damage, during handling. In addition, handling of fruits during transit from market to home is also easier in this container.

Fresh salted fruits are eaten. *Jamun* fruits can also be processed into excellent quality fermented beverages such as vinegar and cider, and non-fermented ready-to-serve beverages and squashes. A good quality jelly can also be prepared from its fruits. The seeds can be processed into powder which is very useful to cure diabetes. These have a ready market. A natural dye is also prepared from the fruit.

13

Kiwifruit

INTRODUCTION

The kiwifruit, also known as Chinese Gooseberry (*Actinidia deliciosa* Var. *Delicious*) is a deciduous fruit of oblong shape, having rusty brown hairs and look like a sapota fruit. It is also known as 'China's miracle fruit' or the 'Horticultural wonder of New Zealand'. It has gained enormous popularity throughout the world during the last 3-4 decades. Uptil 1960, the fruit was known as Chinese gooseberry even in New Zealand, but to promote sale, it was named as kiwifruit because of its brownish colour and hairy appearance like flightless bird 'kiwi', which is the national bird of New Zealand. It has been assessed as one of the important future fruit of our country because of its wider adaptability, high nutritive and medicinal value, precocity in bearing and higher yields and has no serious pests and diseases.

COMPOSITION AND USES

Composition

Its fruits are highly acclaimed for its nutritive value as it is a rich source of vitamin C and vitamin K and a good source of dietary fibre and vitamin E. Fruits also contain appreciable amount of minerals like phosphorus, potassium and calcium (Table 13.1).

Uses

It is sweet sour in taste and has delicate distinct melon like flavor similar to strawberry, rhubarb and gooseberry. The fruits are eaten fresh after removing the hair present on fruit peel. The greenish pulp of the fruit is generally used fresh as dessert. It can also be consumed whole or as sliced segments. Various products like wine; liquor, highly aromatic jam or unattractive marmalade is also made from this fruit. Such products retain higher amount of vitamins even after processing in comparison to other fruits. It also contains various valuable enzymes, one of which Actinidin (proteolytic) prevents jelling and when separated in a powder form, it is useful commercial meat tenderizer. Hard kiwi-fruits are very acid in taste and not palatable.

Table 13.1: Chemical Composition of Ripe Kiwifruit per 100 g Fruit

Content	Composition	Content	Composition	Content	composition
Energy	217 kJ (51 kcal)	**Minerals**		**Organic Acids**	
Water	83.8 g	Sodium	4 mg	Malic acid	500 mg
Protein	1.0 g	Potassium	295 mg	Citric acid	990 mg
Carbohydrate	9.3 g	Magnesium	25 mg	Salicylic acid	320
Vitamins		Calcium	40 mg	**Carbohydrates**	
Vitamin B2	50 µg	Iron	800 µg	Sucrose	1250 mg
Nicotinamide	410 µg	Phosphorus	30 mg	Glucose	4490 mg
Vitamin C	20-300 µg	Chloride	65 mg	Fructose	3540 mg

ORIGIN, HISTORY AND DISTRIBUTION

The kiwifruit is of Chinese origin where the woody vine is known as Yangtao but its full economic potential was exploited by New Zealand. It has made a lead role to change the National horticultural scenario of New Zealand and they named it Kiwifruit. Now New Zealanders have spread the commercial fruit varieties all over the world. It has been introduced to Japan, Russia, Europe, USA and New Zealand at the turn of this century. However, its cultivation gained momentum after 1960. Now, it is being cultivated commercially in New Zealand, Chile, Greece, France, Italy, Turkey, Japan, China, USA, France, Belgium, Germany, Australia, and Spain. At present, total world production of kiwifruit is 1,412,35 tones and over 70 per cent of kiwi production is in Italy, New Zealand, and Chile (Table 13.2). Italy produces roughly 10 per cent more kiwifruit than New Zealand, and Chile produces 40 per cent less. Outside of Australasia, all New Zealand kiwifruits are now marketed under the brand-name label *Zespri*. Kiwifruit has now been declared a national fruit of China. Until recently, China was not a major producing country of kiwifruit, as kiwifruit was traditionally collected from the wild.

In India, it was planted in the Lalbagh gardens at Bangalore in 1960 where it did not fruit and remained as an ornamental tree. Later, plants were introduced at ICAR-NBPGR, Shimla where plants fruited due to fulfilment of chilling requirements. With the extensive research and development, it is now commercially cultivated in mid-hills of H.P., J&K, Uttarakhand, Sikkim, Meghalaya, Arunachal Pradesh and Nilgiri hills in South India.

Table 13.2: Kiwifruit Producing Countries of the World

Country	Production (Tones)	Country	Production (Tones)
Italy	384,844	Turkey	36,781
New Zealand	376,400	Iran	32,000
Chile	240,000	Japan	28,000
Greece	161,400	USA	26,853
France	65,253	Portugal	25,000

TAXONOMY AND BOTANICAL DESCRIPTIONS

Taxonomy

Kiwifruit belongs to genus *Actinidia* of family Actinidiaceae. The genus *Actinidia* contains around 60 species. Though most kiwifruit are easily recognized as kiwifruit (due to basic shape) their fruit is quite variable. The skin of the fruit can vary in size, shape, hairiness, and color. The flesh can vary in color, juiciness, texture, and taste. Some fruits are unpalatable while others taste considerably better than the majority of the commercial varieties. The most common kiwifruit is the fuzzy kiwifruit and comes from the species *A. deliciosa*. Other species have fruits that are commonly eaten; some examples are golden kiwifruit (*A. chinensis*), Chinese egg gooseberry (*A. coriacea*), baby kiwifruit (*A. arguta*), Arctic kiwifruit (*A. kolomikta*), red kiwifruit (*A. melanandra*), silver vine (*A. polygama*), purple kiwifruit (*A. purpurea*). The exact scientific name of fuzzy kiwifruit is: *Actinidia deliciosa* (A. Chev.) C.F. Liang et A.R. Ferguson. The cultivated kiwifruit, *Actinidia deliciosa*, is a hexaploid species with 6x=174 chromosomes, with basic chromosome number, x=29. However, the majority of *Actinidia* species are diploid, with 2x=58 chromosomes.

Botany

The kiwifruit is borne on a vigorous, woody, twining vine or climbing shrub reaching 30 ft (9 m). Two types of shoots are produced – terminating and non-terminating. The former are short, 3-6 leaved shoots, which frequently form flower buds the following year. Non-terminating shoots may grow 10-15 ft in a season, producing smaller leaves along long internodes, with the distal portion of the shoot often coiling on contact with other shoots or solid objects. Non-terminating shoots perform the function of tendrils, which are absent in kiwi. Its alternate, long-petioled, deciduous leaves are oval to nearly circular, cordate at the base, 3 to 5 inch long. Young leaves and shoots are coated with red hairs; mature leaves are dark-green and hairless on the upper surface, downy-white with prominent, light-colored veins beneath. Flowers are relatively large, white changing to golden-yellow, flowers are borne on long peduncles in axils of leaves on current season's growth. Although flowers are produced in few-flowered cymes, most lateral flowers do not develop, giving 1 flower per axil. A whorl of many stamens surrounds the ovary, although they are smaller and produce non-viable pollen in pistillate flowers. Female flowers have 30 styles fused at the base into a superior ovary with many carpels, each containing 10-20 ovules. These flowers are usually larger than males. Male flowers have rudimentary ovaries and reduced styles, but a pronounced whorl of many stamens which are longer than in flowers. Flower buds appear shortly after budbreak, but anthesis is delayed about 2 months since flower development takes place entirely in spring. The fruit is a many-seeded berry with a brown, hispid exocarp (peel). The flesh is green due to chlorophyll, which usually does not degrade during ripening. The core or central axis is white, edible, having 1-3 rows of small black seeds radiating around it. There is no fruit drop in kiwifruits, all fruit that set will mature.

IMPORTANT VARIETIES

Kiwifruit is a dioecious plant, which bears staminate and pistillate flowers on separate plants. The commercially grown pistillate cultivars are Hayward, Abbott, Allison, Bruno and Monty. Similarly important staminate cultivars are Tomuri, Matua and Allison.

Pistillate Varieties

☆ **Abbott:** This is an early maturing cultivar. The fruits are oblong in shape, medium in size, and covered with dense hairs. Fruits are sweet in taste with low ascorbic acid content. Fruits are slightly tapering at distal end.

☆ **Allison:** Fruits are slightly longer than the Abbott and slightly tapering at both ends. It has more production potential. Ascorbic acid content is low. It bears profusely and is suitable for mid and high hill regions. Fruits are larger in size than others cultivars available in India. It more or less dumbbell elongated shaped.

☆ **Bruno:** It is a good fruit bearing cultivar. Its fruits are medium in size and easily distinguished from other cultivars by its slightly tapering shape towards stem end broader at distal end. Fruit contains more TSS than other cultivars. Its chilling requirement is also low.

☆ **Hayward:** It is much preferable cultivar in the world because of its large size, attractive oval shaped fruits with very good keeping quality. It is also considered superior in flavour and sugar content and has fairly high ascorbic acid contents. It is a late maturing cultivar. It requires more chilling and therefore highly suitable for planting in high hill conditions. The cultivar was selected as chance seedling from the Kiwi nursery by Hayward Wright in New Zealand.

☆ **Monty:** This is a late flowering cultivar but the maturity period is short. It resembles with the Allison and Abbott cultivars. The fruits are oblong, medium in size and slightly tapering and flat at both the ends. It has profuse bearing. The fruits have higher acidity and medium sugar content.

☆ **Red Kiwi:** Fruits are very small with smooth greenish skin. Start flowering in 2nd week of May and fruits ripen in mid August.

Staminate Varieties

☆ **Tomuri:** It is a good pollinizer cultivar. The flowers usually appear in groups of five, bold and healthy in comparison to pistillate flowers.

☆ **Red Kiwi:** White small flowers, which usually appear in group, light purplish at perianth.

☆ **Matua:** Very good pollinizer for pistillate cultivars.

SOIL AND CLIMATIC REQUIREMENTS

Soil

Kiwifruit can be grown in almost all types of soils provided adequate soil moisture is available. Well drained sandy-loam soil, which is fairly fertile with good amount of organic matter and with a minimum depth of 1-3 m is considered as most ideal. Ideal soil pH for kiwifruit is 5.0 to 6.0. Kiwifruit has higher water requirement but susceptible to water logged conditions. There, provision of adequate drainage is essential in kiwifruit orchards.

Climate

The areas, which are warmer for apple and colder for subtropical fruits are suitable for kiwifruit. It can be grown in sub-tropical regions lying between 900-1,800 m elevations, which provide 600-800 chilling hours below 7°C required for breaking dormancy in the winter. Spring frost is harmful as it destroys flower buds and actively growing young shoots. In summer, hot weather is beneficial but a temperature higher than 35°C and low humidity is harmful, which cause scorching of leaves and sun burn of fruits. Well distributed annual rainfall of 150 cm is considered good for growth and development.

PLANT PROPAGATION

Kiwifruit is propagated through several ways, but propagation through cuttings is rapid, easiest and suitable method of propagation. Although, different types of stem cuttings (softwood, semi-hard or hardwood) are used but best results are obtained from hardwood cuttings, which are prepared during dormant season from one-year-old wood. Such cuttings should be 15-20 cm long and of pencil thickness (0.5 to 1.0 cm diameter) and should have at least 4 healthy buds. Soft wood cuttings root easier and quicker than hardwood cuttings but they require mist chamber for creating humid conditions. Such cuttings should be 10-15 cm long and 0.5 to 1.0 cm thick, with at least 4 nodes/buds and should be prepared from middle portion of the shoot. After preparation, these cuttings should be kept turgid and are dipped in 4,000-5,000 ppm IBA solution for 10-15 seconds and planted in mist propagation chamber. Kiwifruit can also be propagated by budding and grafting, but success is limited.

A prototype of bottom heating propagation tray has been designed at ICAR-IARI, Regional Station, Shimla to induce rooting in hardwood cuttings to save the time and space. The procedure consists of callusing and root initiation of cutting by dipping them in rooting hormone (IBA @3000 ppm) for 30 seconds and maintaining 20°C temperature at the root zone of cuttings. Such cutting root better and profusely and give more than 80 per cent success in the field.

PLANTING AND ORCHARD ESTABLISHMENT

Flat land with gentle slop is ideal for planting of kiwifruit orchard. The land should be thoroughly ploughed and then pits of 1x1x1 m size are dug out in December and filled with mixture consisting of 30-40 kg FYM, 1 kg super phosphate and top soil. Planting is done in January-February in warmer area, while in colder areas, planting should be done during early spring to reduce the risk of frost damage. Planting distance depends on several factors like cultivar, soil type, and training system. However, in all, planting distance of 5x3 m for Hayward and 6x4 m for vigorous cultivars is suitable. In 'T' bar system of training, 4x6 m and in pergola system of training, a spacing of 6x6 m is recommended for the production of quality fruits.

Kiwifruit is a dioecious fruit palnt and hence, male plants should also be planted in appropriate proportion. This can be achieved by 1:6, 1:8 or 1:9 male to female plant ratio (Figure 13.1).

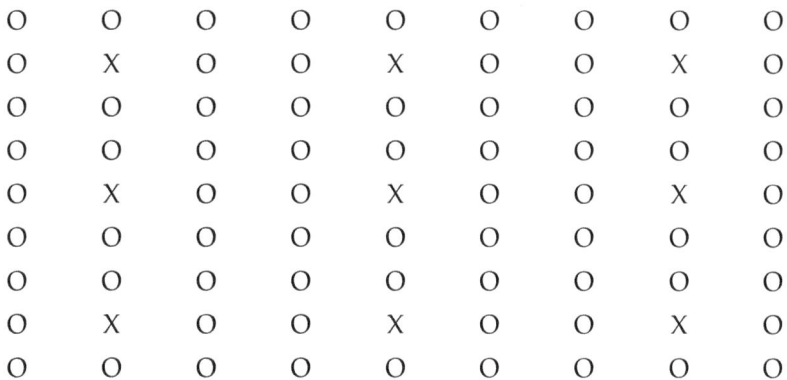

O	O	O	O	O	O	O	O	O
O	X	O	O	X	O	O	X	O
O	O	O	O	O	O	O	O	O
O	O	O	O	O	O	O	O	O
O	X	O	O	X	O	O	X	O
O	O	O	O	O	O	O	O	O
O	O	O	O	O	O	O	O	O
O	X	O	O	X	O	O	X	O
O	O	O	O	O	O	O	O	O

Figure 13.1: A Layout Planting Plan of 1:8 Male : Female Plants in Kiwifruit Orchard.

INTERCULTURAL OPERATIONS

Training and Pruning

Training is very important aspect in kiwifruit and it requires constant attention. In fact, training structures should be erected well before planting. Kiwifruit is usually trained to 'T' bar or pergola systems. In 'T' bar, the pillars of iron or concrete about 1.6 m in height above the ground level are erected at a distance of 6 m in a row. A cross arm (1.5 m) is fixed on each pole, which carries 5 outriggers wires at a distance of 45 cm each (Figure 13.2). Then laterals arising from the main branch are trained on these five wires. Initially, a strong growing shoot is selected as main trunk to carry vine upto the wire. The vine is staked to provide support and is tied with tread at

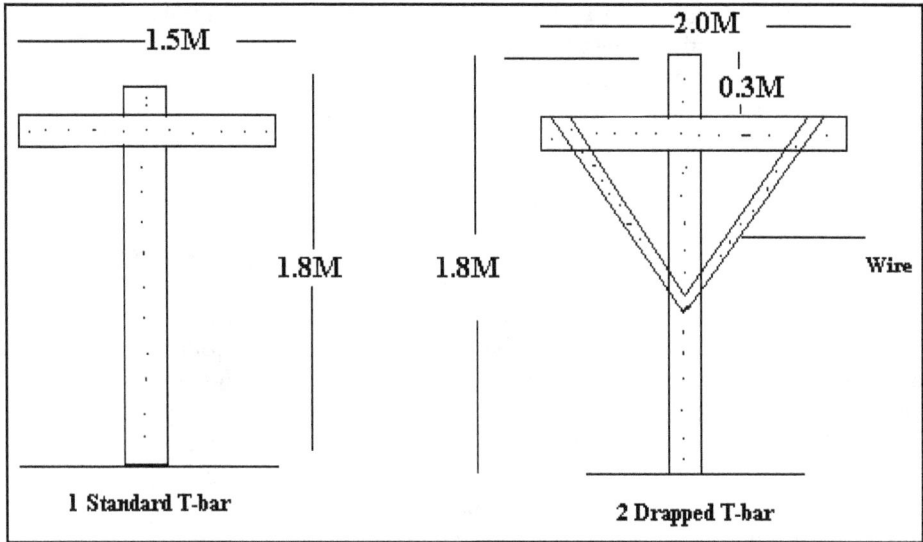

Figure 13.2: A View of T-bar Training System in Kiwifruit.

frequent intervals to prevent wind damage and to avoid the twisting of vine. As the vine attains the height of 2 m or reaches the wire, one permanent leader/arm is allowed to grow out in each direction along the central wire. The main leader can be cut just below the wire to force the production of two leader growth, which can be trained as leaders in two opposite directions. From the permanent leaders, temporary fruiting arms 20-30 cm apart are selected, which bear the crop.

Pruning

In kiwifruit, vine management's is the most important aspect for obtaining higher yields of quality fruits. Pruning prevents dense and tangled vine growth and allows access to bees during flowering periods, penetration of light, insecticide and fungicide sprays, air movements to minimise the fungal diseases. Adequate light penetration through open vines provides to ripen the fruits and increase in its quality. The pruning operation in Kiwi fruit is carried out in two seasons *i.e.* during summer and winter.

(a) **Summer pruning:** It is the more essential to maintain the order; spacing but light penetration is achieved in the winter pruning. The fruiting shoots are headed back beyond 6-8 buds from the last fruits. If there is an excessive growth on this shoots, it should be removed during summer pruning. The summer pruning must be completed each year by the middle of June to July under hill conditions, because of the fast growth during these months.

(b) **Winter pruning:** The main purpose of dormant pruning is to cut the fruiting laterals of two vegetative buds beyond the last fruits. In the second year, these vegetative buds produce the vegetative shoots, which are pruned again. The arms on the lateral shoots are allowed to fruits during third or fourth years. After fifth year onwards, the laterals are removed leaving sixth to seventh fruiting spur buds from the main branch and other laterals are selected and pruned accordingly so that the balance between vegetative and reproductive growth is maintained for the continuity in fruit production. Dormant pruning must be completed by mid - February each year.

Nutrition Management

For proper growth and yield of quality fruits, kiwifruit also requires plentiful supply of manures and fertilizers. Although, manurial and fertilizer requirements are affected by several factors, but a basal dose of 20 kg FYM coupled with 0.5 kg of NPK fertilizer mixture containing 15 per cent N be applied in 2 dressings each year/vine. After five years of age, 40 kg FYM, 850-950 g N, 500-600 g P_2O_5 and 800-900 g K_2O/vine is applied every year. N should be applied in two split doses first during January-February and remaining half in last week of April or upto mid-May. Full dose of P and K should be applied along with FYM in December-January. The fertilizers should be applied 30 cm away from tree trunk and mixed thoroughly.

Water Management

Water requirement of kiwifruit is high because of vigorous vegetative growth and larger leaf surface. Kiwifruit is also sensitive to water stress, especially, during April to mid-May when there is high demand for water due to rapid vegetative and fruit growth. The frequency of irrigation depends on several factors like water holding capacity of soil, rainfall distribution, age of plant and mulch material used. However, in general, kiwifruit should be irrigated at 8-10 days interval.

FLOWERING, POLLINATION AND FRUIT SET

Kiwifruit starts bearing sizeable crop after 4-5 years of planting. Flowering starts in April and fruits are ready for harvesting during last week of October or first week of November, depending on climatic conditions and cultivar. Being dioecious, pollination is a problem, if male plants are not in suitable proportion. Without good pollination, fruits will be small and not upto marketable size. Kiwifruit is pollinated by insects, mainly by honeybees, however wind also plays vital role. Artificial pollination by rubbing the freshly opened male flowers on the stigma of female flowers is always beneficial.

Kiwifruit bears profusely; as a result, there is severe completion for nutrients and photosynthates among the fruits. The fruits remain small in size and thus fetch low price in the market. Therefore, it is necessary to thin out the crop. Only 4-5 fruits should be retained/cane. Fruit thinning can be done by hands. However, dipping fruits in CPPU (5-10 ppm) gives best results. Thinning should be carried out after about a month after fruit set.

PLANT PROTECTION

No serious inset-pest and disease have so far been reported in kiwifruit. However, with the establishment and spread of kiwifruit orchards, damage may occur due to pests like leaf roller, greesy scale, two spotted mite and thrips and diseases like root rot, crown rot, field rot and storage rots. These pests and diseases can be effectively controlled by following control measures recommended for such pests or diseases for other fruits.

Physiological Disorders and their Management

In addition to fruit rejections due to poor size and pest or disease infestation, there are a number of rejection categories resulting from non-pathogenic causes which are important in the fruit grading operation. These categories include:

☆ **Flats**: A flat is a fruit whose width is greater than its length. Flats are difficult to pack in the standard single-layer tray. Flat fruits are more likely to be produced on the two proximal flowering buds of a shoot. Various influences that affect vegetative and fruit growth include tipping during normal summer pruning and the application of growth retardants.

☆ **Fans**: A fan is a fruit formed from a fasciated or fused flower. Although, many people enjoy eating such fruits because of their larger size. Such fruits undoubtedly has less commercial value than a symmetrical fruit of a cultivar.

☆ **Dropped shoulder**: The fruit appears lop-sided with the shoulder at one side sloping away abnormally. The cause of this condition, which can occur on both large and small fruit, is not completely known. It may be the result of inadequate pollination of the top most locules on one side of the fruit because of a deformed flower or may be due to bud rot infection.

MATURITY, HARVESTING AND POSTHARVEST MANAGEMENT

Maturity

Kiwifruit comes into commercial bearing after 7-8 years of planting. Judgment of optimum harvest maturity is difficult, because no perceptible changes occur in skin or flesh of fruit at the time of maturity. Hence, the most important maturity index is total soluble solids. Internationally, a maturity index of 6.2° B TSS is considered satisfactory for fruit harvest but it is usually harvested between 6.5-8.0°Brix in different kiwifruit producing countries of the world including India. In addition, days after full bloom (DAFB) is another useful maturity index for harvesting kiwifruit at a right stage of maturity (Table 13.3).

Table 13.3: Optimum Harvest Time for Mid Hills of Himachal Pradesh

Cultivar	DAFB	Harvest time
Allison	188-195	4-6 November
Abbott	192-200	7-12 November
Bruno	190-199	1-3 November
Monty	189-197	6-10 November
Hayward	194-202	15-20 November

Harvesting

The fruit is harvested hard and under ambient conditions, it ripens after 8-10 days of harvesting. The harvesting period varies from area-to-area. The fruits mature earlier at lower altitude and later at higher altitudes, because of variation in temperature during fruit development. Under mid-hill conditions (1200- 1500m) conditions of India, the fruits are ready for harvest from October-end to third week of November depending upon cultivars, whereas, under high-hill conditions (2000 m), the fruits are harvested up to last week of November. Unfortunately, in India the kiwifruit is harvested at immature stage which does not give good taste and aroma typical of the fruits. This is mainly due to lack of awareness about its ripening behaviour among the farmers and consumers.

Method of Harvesting

The fruits are easily harvested by snapping off the fruit at the base of the stalk. After harvesting, the fruits are rubbed with a coarse cloth to remove stiff hairs found on their surface. Hard fruits are transported to the market. Subsequently, they lose their firmness in 2 weeks at room temperature and become edible. On an average, kiwifruit yield varies from 60 to 70 kg/vine.

POSTHARVEST MANAGEMENT

Pre-cooling

Rapid removal of field heat from freshly harvested commodities before shipment, storage or processing. Like other fruits, it is essentially required for kiwifruits as well, although the temperature at the time of its harvesting is quite low. Usually, it is done in conventional cool stores for 7-10 days at 2°C. In forced air pre-cooling, 8 hours pre-cooling is recommended.

Sorting and Grading

After harvesting, fruits should be sorted out and the diseased or culled fruit should be removed. After sorting, fruits should be graded in to different grades. There are 8 designated fruit sizes. In International market, fruit weighing 100 g is a preferred. In India, fruits weighing 80 g and above are graded as 'A' grade; between 50 and 80 g are graded as 'B' grade and less 50 g are regarded as C grade which are mainly used for processing purpose.

Packaging

Packaging plays a very important role in the postharvest life of a fruit. Like other fruits, kiwifruits should be packed in a proper box to make it attractive and gain maximum benefits. Kiwifruits can be packed manually or by semi-automatic systems in plastic pocket trays. The average weight: 3.6 kg, which contains 25-46 fruits/tray. Since, in India there is no standard package for kiwifruit, cardboard boxes of 3 kg capacity are generally used for packaging. Recently, few progressive farmers in HP have designed small beautiful packing boxes of 1 kg capacity which can be seen in the market during festive season of Diwali. Kiwifruits can be packed in trays using heat shrinkable films. Using cryovac heat shrinkable films, kiwifruits can be stored for about 20 days at room temperature.

Storage

The kiwifruit have an excellent keeping quality. The fruits can be kept in good condition in a cool place without refrigeration up to 8 weeks. It can be stored for 4 to 6 months in a cold store at $O^\circ C$ with 80-90 per cent relative humidity. If stored at $2^\circ C$, its storage life reduced by 1-2 months and at $5^\circ C$, respiration rate doubles, which further decreases its life significantly. In CA relative storage, life of kiwifruit is increased by 100 per cent than at room temperature and about 33 per cent than at optimal refrigeration temperature. The most commonly used mixture of gases is: 5 per cent CO_2 and 3 per cent O_2. In low O_2 storages, 2 per cent O_2 and < 1 per cent CO_2 levels are recommended. In ultra low (UL) O_2 storage, 1-1.5 per cent O_2 and <1 per cent CO_2 level have been recommended.

Processing and Value Addition

C grade fruits or cull fruits should be used for processing. Fruits can be processed into jam, juice and squash.

14

Loquat

INTRODUCTION

Loquat (*Eriobotrya japonica* Lindl. syn. *Mespilus japonicus* Thunb.) is an evergreen, subtropical fruit, of the rose family, Rosaceae. It has been called by various names like Japan plum, Maltese plum and Japanese Medlar. The name loquat derives from *lou gwat*, the Cantonese pronunciation of its old classical Chinese name (literally "reed orange"). This fruit species is unique in sense that it has a reversed annual cycle; it rests during summer and matures in early spring, being the first fruit crop reaching the market. It is available in the market in North India during mid-March –May when very few fruits are available in the market. The early arrival in the market renders growers to fetch high prices for their produce. Though loquats are not popular in many countries, they are one of the most dearly loved fruits in countries like China and Japan. In these cultures, legends and mythology surround its origins. For instances, Chinese folklore holds that loquats stemmed from the base of a waterfall and when a carp consumed the fruit, it transformed into the mythical dragon. Considering it as a miraculous fruit, the royal family prohibited anyone except their family from eating loquats. Loquat is grown commercially throughout the subtropical and Mediterranean areas of the world for its fresh fruits; however, at some places it is grown as an ornamental plant. Being a plant from the family Rosaceae, loquats have similar taste and flavor as that of apples; tart, and sweet with pleasant aroma. But they are soft and juicy in texture instead of crispy in apples.

COMPOSITION AND USES

Composition

Presently, most production is for the fresh market. Loquat fruits are consumed largely as fresh fruit, although small amounts are used in jams, jellies, syrups, and pies. Loquat may be eaten fresh without the peel, combined with other fruits in fruit salads, used as a pie filling, and made into sauces and gelatin desserts, jams, and jellies. Fruit may also be canned, dried, frozen, and made into syrup. Loquat fruits are a good source of vitamins and minerals as depicted in Table 14.1.

Table 14.1: Nutritive Value per 100 g of Edible Portion of Loquat Fruits

Constituent	Content	Constituent	Content
Calories	47 Kcal	Magnesium	13 mg
Carbohydrates	12.14 g	Sodium	1.0 mg
Protein	0.43 g	Vitamin A	1528 IU
Fat	0.20 g	Thiamin	0.019 mg
Dietary Fiber	1.7 g	Riboflavin	0.024 mg
Calcium	16 mg	Niacin	0.180 mg
Phosphorus	27 mg	Folates	14 µg
Iron	0.28 mg	Ascorbic Acid	1mg
Selenium	0.6 µg	Pyridoxine	0.1mg

Uses

The carotenoids of loquat fruit are mainly responsible for pulp and skin colour, which varied from yellowish white, yellow to deep orange. Total carotenoid values, especially carotene, varies widely in fresh fruit peel and pulp. Total carotenoid values in the peel are several times higher than the pulp. The content of carotene in yellow-orange fruit is 5-10 times higher than in the yellow-white fruits, while the contents of zeaxanthin, lutein, and violaxanthin in yellow-orange fruit are much lower. Since fruits are rich in insoluble dietary fiber, pectin that helps retaining moisture in the colon, it functions as bulk laxative. Further, loquats are considered a functional food for the prevention of diabetes and hyperlipidemia. Loquat preserve can be used to top pancakes, cakes, ice cream and pastries. Baked loquats with nuts, sugar, garlic and onions can be used to serve atop rice and grain dishes such as tagine, biryani, and couscous. The seeds, which have an almond like taste, are used to flavor drinks and cakes.

Leaves are used for the treatment of diabetes mellitus, skin diseases and are used as a folk medicine for the treatment of chronic bronchitis, coughs, phlegm, high fever and ulcers. Loquat leaf is used for the reduction of skin inflammation, edema and histamine-induced skin contraction. Leaf poultices are applied on swellings. Further, flowers are used as expectorant. Due to their effectiveness as a diuretic by increasing the production of urine and promoting the elimination of excess uric particles, loquats are highly recommended in cases of excess uric acid, kidney stones, kidney failure, and gout. In addition, loquat helps decongest the volume of an enlarged liver (hepatomegaly). Researchers have suggested that leaf extracts of loquat may protect the body against cellular aging. Loquat extracts offer considerable protection against memory impairment and neurological oxidative stress. In apiculture too loquat finds its uses. Bees are easily attracted to the fragrant, white flowers. The honey; thereby produced, is amber colored with an agreeable flavor. Since the wood has very little tendency to split, much harder, and takes a good polish, it is suitable for poles and posts, carving and drawing materials such as rulers, and is in demand for making stringed musical instruments.

ORIGIN, HISTORY AND DISTRIBUTION

The loquat is indigenous to Dadu River Valley in South-Eastern China and possibly Southern Japan, though it may have been introduced into Japan in very early times. The fruit was brought from China to Japan, although others contend that it's the other way around. The fruit was introduced from China to Japan as early as 700 AD. In 914, the first Chinese medical textbook was translated to Japanese and mentioned how to use loquat to obtain clear lungs. Japanese law books in the early 900s stated the proper way to present loquat as an offering to the Shinto gods. Records show that the fruits were present on Japan's soils as early as 1100AD. The Japanese used to call loquats with a name resembling its Chinese term, 'biwa' indeed, during the Tang Dynasty from 618 to 907, loquats were called 'Tang biwa'. It was introduced to Europe from Japan in the 18th century as an ornamental tree. The fruit spread to other countries by way of German physician and botanist Engelbert Kaempfer, who recorded his loquat sightings in Japan in his book, Amoenitatum Exoticarum in 1712. Thunberg, who saw it in Japan in 1712, provided a more elaborate description under the name *Mespilus japonica*. Since some primitive types of *E. japonica* occur in several prefectures in Japan, some Japanese authors consider the origin to be both China and Japan. Most authors around the world now believe loquat originated in China.

It was planted in the National Gardens, Paris, in 1784 and plants were taken from Canton, China, to the Royal Botanical Gardens at Kew, England, in 1787. Soon, the tree was grown on the Riviera and in Malta and French North Africa (Algeria) and the Near East and fruits were appearing on local markets. In 1818, excellent fruits were being produced in hothouses in England. Chinese immigrants are presumed to have carried the loquat to Hawaii. It was common as a small-fruited ornamental in California in the 1870's, and the improved variety, Giant, was being sold there by 1887. This species was introduced to Anatolia (Asia Minor) 150 to 200 years ago, possibly from Algeria or Lebanon. Loquat has also become naturalized in India and many other areas. Though it is naturalized in India, how the fruit ultimately came to its soils remains unclear. Most sources claim that the fruit didn't arrive until long after the US had received it around 1870; however, the 1857 publication, "India under the British Empire," references loquat as already being present. Given Afghanistan's long-standing love affair with the fruit, perhaps loquats arrived by way of the Muslim rulers in the North. Later, in the 19th century, selections of cultivars with large fruits were used for fruit production. The exact date of its introduction in India is not known. However, Government Botanical Gardens (now Horticultural Experiment and Training Centre), Saharanpur, Uttar Pradesh is said to be the first garden to grow it, from where it has been distributed all over the country. Some species of loquat are found growing naturally on the hills of India. *Eriobotrya angustissima*, an evergreen shrub is found at an elevation of 1,300-1,700 m in Assam, while another species *E. dubia*, a small tree that bears edible fruits, has been found growing at slightly higher elevation in the Eastern Himalayas.

Loquat is cultivated mainly in China, Japan, India, Pakistan, Madagascar, Reunion Island, Mauritius Island, the Mediterranean countries (Spain, Turkey, Italy, Greece, Israel), United States (mainly California and Florida), Brazil, Venezuela, and Australia. This species has adapted well to the Mediterranean climate and grows in the same areas where citrus species are cultivated. Generally, loquats are found between latitudes 20° and 35° North or South, but can be cultivated up to latitude 45° under maritime climates. In India, its commercial cultivation is mostly confined to Uttar Pradesh, Delhi, Punjab, Himachal Pradesh (Table 14.2) and to a small extent in Assam, Maharashtra and hills of South India (Nandi hills).

Table 14.2: Major Loquat Growing Belts in India

State	Districts
Himachal Pradesh	Kangra (Palampur, Baijnath), Mandi (Sunder Nagar), Solan (Nalagarh)
Punjab	Amritsar, Hoshiarpur, Gurdaspur, Patiala
Jammu and Kashmir	Kathua
Uttar Pradesh	Saharanpur, Muzaffarnagar, Meerut, Farrukhabad, Kanpur, Bareilly
Uttarakhand	Dehradun

TAXONOMY AND BOTANICAL DESCRIPTIONS

Taxonomy

The generic name is derived from the Greek words *'erion'* (wool) and *'botrys'* (cluster), from the woolly appearance of the spiked inflorescence. The specific epithet *'japonica'* was based on Thunberg's belief that the origin of loquat was Japan. Loquat belongs to Class Magnoliopsida, Order Rosales and family Rosaceae. The number of chromosomes in loquat cultivars cultivated China are all $2n = 2x = 34$. Tetraploidy was also attained from colchicine treatment, and triploidy was derived from hybridization (diploid x tetraploid) and by endosperm culture. Hybrids among cultivars are cross compatible. Most of the cultivars are self-fertile.

Botany

It is a large evergreen shrub or small tree 6-8 m high; with a ovoid or globular crown, short trunk, bark being grey and shallowly fissured, on young branches it is pale brown and hairy. Leaves are whorled towards the end of the stout, woolly branchlets, large, alternate, subsessile, stiff, coriaceous, elliptic, lanceolate to obovate, lanceolate in outline, 21-32 cm in length, with remotely toothed to sharply dentate margins; dark, glossy, green above and rusty-tomentose below: base green, obtuse or narrowed into a very short, stout, woolly, stipulate petiole. A hermaphroditic species, the self-incompatibility of *E. japonica* is gametophytic. Loquat is pollinated by insects.

Sweetly fragrant flowers with inferior ovaries, borne in rusty-hairy, terminal panicles of 30 to 100 blooms, are white, 5-petalled, 1.25-2 cm wide. Fruit are pomes, held in clusters of 4 to 30, oval to rounded to pear-shaped, 2 to 5 cm long and weigh an average of 30 to 40 g; some cultivars up to 70 g. The peel is smooth to slightly downy, light yellow to orange. The pulp is white to light yellow to orange, 6.7 to 17°Brix, sweet to sub acid, and juicy. There may be 1 to 10 dark brown seeds. Birds and bats disperse the fruits. Loquat like other rosaceous fruits utilizes sorbitol as the main metabolite of photosynthesis. The edible portion in loquat is torus (fruit cortex or fused hypanthium), consisting of pith and cortical areas. The development of edible portion consists of rather uniform growth of receptacle tissue throughout the fruit. The toral cells of mature fruit are large, thin walled and very juicy. The growth pattern of loquat fruit is neither sigmoidal, as in most small-seeded pome fruits, or double sigmoidal, as in stone fruits that have a large seed, but is exponential with a rapid growth toward the end of fruit development, in spring, until ripening. Malic acid content in the pulp decreases and the color increases as the fruit reaches towards maturity. The loquat is a non-climacteric fruit and shows no respiration climacteric rise and no peak of ethylene production either on the tree or after harvest. The fruit does not abscise after ripening but shrinks on the tree.

SOIL AND CLIMATIC REQUIREMENTS

Soil

Loquat trees grow well in a range of well-drained soils, from fertile loamy soils to clay to gravelly limestone-based soils. Loquat trees are not tolerant of flooded soil conditions. Similarly, subsoil should also be free from hard pan. Stagnation of water in the land, even for a short time may damage or kill even grown up trees and such soils should be discarded. On less fertile soils loquats respond well to NPK fertiliser.

Climate

Loquat is specific in its climatic requirements; however, it adapted well to the subtropical climate of North India. It is grown nearly throughout the country up to an elevation of 1525 m above sea level. It can be grown throughout the tropics where there are elevations of a few thousand feet. Further, it is performing well in tropical climate and lower hills, where temperature rarely falls below 0°C. It needs about 90 cm well distributed rainfall throughout the year. It cannot tolerate severe frost in winter. Frost is a limiting factor for its successful cultivation. At certain places, the crop may be destroyed by moderate winter frost, since flowering takes place from October to late January. At -3°C smaller fruits (diameter <9.5mm) are more susceptible to cold injury than larger ones. The fruit of loquat is most susceptible to frost injury when it just starts colouration. In addition, heavy damage is noticed in early ripening varieties. Loquat is wind tolerant; however, its cultivation is problematic in areas where summer sets in early along with hot scorching winds. Hot scorching winds results in to development of small sized, unevenly ripened fruits with poor juice contents. There may be a heavy loss of crop because of sun burning as sun burnt fruits fetch low price in the market. Loquat grows best in full sun, but also does well in partial shade. Likewise, it likes warm and dry climate at fruit ripening time. It is tolerant to drought and heat as the thick, leathery leaves are well adapted to withstand seasons of neglect without serious injury. The environmental factor or factors responsible for flower induction are not known, although a cessation of growth prior to the fall/winter flowering is essential. The optimum climate appears to be where trees stop vegetative growth during the early fall, perhaps due to cool temperatures and/or dry soil conditions. This is followed by continued cool but non-freezing temperatures during the winter, and warming temperatures during the spring. Therefore, loquat should preferably be cultivated in the submontaneous regions or other areas with mild climate, free from severe hot or cold weather conditions.

IMPORTANT VARIETIES

Based on origin, two groups are distinguished: the Chinese groups with large, pyriform, deep orange fruit, which can be kept for 1-2 weeks, and the Japanese group with small, slender, light-coloured fruit, maturing early and having a shorter shelf life. The important loquat varieties are:

- ☆ **Pale Yellow:** It has large fruits, which are slightly conical to roundish in shape and pale yellow in colour. Pulp is white and tastes sour-sweet. Each fruit contains 2 or 3 medium-sized seeds.

- ☆ **Improved Pale Yellow:** Fruit medium, oblong pyriform, pulp medium thick, cream colour, smooth and soft, pleasant taste, sub acid and moderately seeded.

- ☆ **Golden Yellow:** It has medium sized, egg-shaped fruits with attractive golden yellow colour. Pulp is yellowish which tastes sour-sweet. Each fruit contain 4 - 5 dark brown, medium sized seeds.

- ☆ **Fire Ball:** Fruit small, oblong to ovate, saffron-yellow, pulp thick, corn husk colour, smooth and crisp, taste mild, sub acid and moderately seeded.

- ☆ **Large Agra:** Fruit medium, oblong to ovate, pulp medium thick, pale orange, smooth and firm, pleasant taste, sub acid and moderately seeded.

- ☆ **Thames Pride:** Fruit medium, pyriform, marble color. Pulp medium, pale orange, coarse and lightly granular, mild taste, sub acid and moderately seeded.

- ☆ **Mammoth:** Fruit small, oblong pyriform, color snow white. Pulp medium, orange, coarse and granular, pleasant taste, sub acid and few seeded.

- ☆ **Matchless:** Pulp medium, orange, coarse and granular, pleasant taste, sub acid and few seeded.

- ☆ **Safeda:** Fruit large, oblong pyriform. Pulp thick, creamy white, smooth and melting, excellent taste, sub acid and moderately seeded.

- ☆ **California Advance:** Fruit medium-sized, conical to round in shape, external colour yellow. Pulp creamy white, sour-sweet in taste. Fruit contains 2 or 3 medium-sized seeds.

☆ **Tanaka:** Fruit medium, 4 - 5 cm long, 3 - 7 cm broad ovate in shape, smooth and moderately pubescent. Rind is medium and firm. Pulp medium, completely filled, sayal brown, coarse firm, juiceness abundant, taste pleasant subacid. Seeds 2 to 4 per fruit and are medium in size.

On the basis of maturity, loquat varieties can be divided into three groups *viz.,* early (which ripen from mid-March), mid (ripens from last week of March) and late season (ripens from mid April).

☆ **Early Varieties:** Golden yellow, Improved Golden Yellow, Pale Yellow, Large Round and Thames pride.

☆ **Mid-Season Varieties:** Large Agra, Mammoth, Improved Pale Yellow, Matchless, Safeda and Fire ball.

☆ **Late-Season Varieties:** Tanaka and California Advance.

Self unfruitfulness has been noted in loquat varieties. Therefore, a pollenizer variety should be planted along with the main variety. On the basis of self fruitfulness, the varieties can be grouped as follows:

i. **Self Incompatible:** Golden Yellow, Improved Golden Yellow, Pale Yellow and Large Agra.

ii. **Partially Self Incompatible:** Large Round, Fire Ball, Thames Pride, California Advance and Tanaka.

The variety California Advance is the best pollenizer for Improved Golden Yellow.

FLOWERING AND FRUIT SET

Vegetative growth is in the form of a series of flushes that occur once each season. Summer shoots are the most abundant; spring shoots, summer shoots, and sometimes autumn shoots will be flower branches; winter flushes depend on tree age and nutrition. Loquat bears flowers at the terminal end of current year's growth. In India, the flowering period of loquat is very long, lasting from mid July to January or sometimes even upto May, depending upon the location. Three reproductive flushes under tropical conditions are noted, out of which the intermediate ones give the higher yield of better-sized fruits. The first flush of flowers can be noticed during July-August, however, due to non-viable pollen grains there is no fruit set. The second flush of flowering takes place during October- November and sets maximum fruits. Another flush of flowers can also be noticed during January-February but hot and desiccating winds in early summer have adverse influence and thereby reduce the fruit size. It was observed that flower bud differentiation occurs early in warmer climate. The time span from flower bud differentiation to anthesis is three months. The number of inflorescence has been found to be higher at the beginning of floral flush emergence, but only a few of them set fruits. The number of inflorescence continues to decrease with the advancement of season, but the percentage of fruit-bearing panicles gradually increases. The fruit size is the inversely proportional to the number of fruits per panicle. Earlier panicles gave best-sized fruit and size reduced afterward. In Northern India, only one continuous reproductive flush appears. No peak hour of anthesis has been observed and the opening of flowers continues throughout the day. The dehiscence of anthers takes place in longitudinal fashion. It takes more than a day to complete dehiscence in all the varieties except improved 'Golden Yellow' and 'Pale Yellow' which requires only a day to complete. The time of dehiscence was found to have hastened with the increase in temperature. In Punjab, the flowering period in loquat is very lengthy. The flowering starts in the first week of October and continues upto third week of December. The number of flowers per cluster may vary from 50 to 100 but in general, not more than 15-20 fruits per cluster are set. Flowering in loquat may extend over 1.5 to 2.5 months and fruits normally ripen about 150 to 200 days from flowering.

Loquat trees grow singly or in small groves, though produce perfect flowers, yield negligible or no crop. This is reported to be due to self-incompatibility in commercial loquat varieties. In India, it has been observed that cross-pollination generally results in 10-17 per cent increased production over self-pollination. 'Tanaka' pollinated by 'Pale Yellow' has a lower yield than when self-pollinated, indicating a degree of cross-incompatibility. However, when pollinated by 'Advance', the normal yield of 'Tanaka' is nearly doubled. When cross-pollinating for the purpose of hybridizing, only flowers of the second flush should be used, as early and late flushes have abnormal stamens, very little viable pollen, and result in poor setting and undersized fruits. Honeybees (*Apis dorsata*) are its pollen vectors. Other species of insects found occasionally included syrphids, houseflies, Myrmeleontidae, Bombinae and *Pieris rapae* (L.). Higher fruit set has been recorded in unbagged than in bagged flowers.

Fruit set was in loquat can be improved considerably with the application of plant growth regulators. Best results with regard to fruit set and fruit quality were obtained with GA at 60 ppm. Mature loquat trees sprayed when fruits were at the pea stage and again one week later with NAA, 2,4,5-T, or GA_3, each at 10, 20, or 40 ppm, ripened about 10 days earlier than with GA at 10 ppm, gave the best fruit retention and greatest fruit volume, weight and pulp content.

FRUIT THINNING

In loquat, each flower panicle can set a large number of fruits, unless these are thinned, the fruit will be small and often scarcely worth picking. Therefore, thinning of flowers and/or fruits is an essential cultural operation in loquat to obtain marketable fruit size. Thinning will increase fruit size from 25 per cent - 100 per cent. Thinning of flowers is necessary not only to promote the growth of fruits but also to reduce the size of the inflorescence for bagging. After flower thinning, one inflorescence bears a maximum often fruits and later fruits are thinned to reduce the number between 4 and 6. Flower thinning is done by removing as many as 30 per cent of the whole inflorescence and further thinning of flowers from the remaining inflorescence. Another procedure is to remove some lower and upper peduncles and further thinning of flower buds to provide sufficient space for the fruits to develop in good shape and size. Chemical thinning can also be done by use of growth hormones. NAA and NAAm (naphthaleneacetamide) applications

(25, 50, or 100 ppm) effectively thinned loquat fruits. Optimum level of thinning was obtained with 25 ppm. Fruits on thinned branches develope more rapidly than non-thinned branches. Fruit thinning is done as soon as the fruits develop and fruits of uniform shape and size are retained. The fruits are bagged simultaneously with fruit thinning to maintain their attractive appearance. Paper bags are usually used for this purpose. A loquat fruit may contain one or more large seeds, which may even reach up to 10 seeds. However, varieties which naturally have relatively large fruit with few and small seeds are preferred. And for achieving it, good management practices are required to be followed to further improve the pulp to seed ratio. Besides, seedlessness can be induced in loquat by application of GA @ 250 ppm after the emergence of floral buds or NAA @ 20 ppm during full bloom. GA_3 applied at 250-500 ppm after emergence of the floral buds in mid-October resulted in production of seedless fruits with high pulp content.

PLANT PROPAGATION

In general, seeds propagation is followed either with the aim of raising tree for ornamental purposes or for raising rootstock. The loquat seeds germinate promptly when sown immediately after extraction from the fruit. It is advisable not to let seed get dried after extraction as exposure to heat and light may lead to poor germination and subsequently stunted seedling growth. Loquat seeds remain viable for 6 months if stored in partly sealed containers under high humidity at room temperature; the best temperature for storage is 5 °C. For rootstock, the seed are washed and planted in flats, soon after extraction from fruit, during April-May in moist sand for germination. When the seedlings are 4 - 5 cm tall, they are transplanted in the nursery under the mother trees for inarching. If the mother plants are high headed, the seedlings are transplanted in the pots and brought in contact with the mother plants by raising platforms when they attain inarchable size (*i.e.* 1/2 inch thick at the base). The seedlings grow rapidly and are fit for grafting in the following rainy season. The rootstock generally is *E. japonica* itself, although *E. deflexa* and other species, even *Photinia serrulata*, have been used for rootstocks. The use of loquat seedling rootstock usually results in a comparatively large tree with a high canopy. Cultivars grown on quince (A, C and BA 29) and pyracantha rootstock produce a dwarfed tree of early bearing character. The smaller tree has no effect on fruit size and gives adequate fruit production with the advantage of easier picking. Further, Quince rootstock tolerates heavier and wetter soils.

Loquats can be propagated by various grafting methods, including inarching, shield-budding or side-veneer grafting and cleft-grafting. Inarching is common of grafting in loquat. The seedlings are inarched during July-August. An improved method, called a young stock cleft graft, was developed in China. Using 1-cm-diameter stock, cleft grafts are made and are held tightly in place with parafilm strips. Because there are several leaves under the graft position, the scion grows quickly and survival rates are higher. Grafted trees will begin to bear fruit in 2 to 3 years, compared to 8 to 10 years in seedling trees. Budding can also be done in January-February. Shield budding, using buds from three month-old branches in January and February gave encouraging success, while poor success was observed during September- October. Loquat cuttings are not easy to root; however, cuttings dipped in IBA solution roots under mist but with poor survival. Likewise, air layering can also be done but with limited success; though, 80 to 100 per cent of the layers root in 6 weeks if treated with 3 per cent NAA (2-naphthoxyacetic acid). Old seedling trees, which are unprofitable, can be top-worked with some improved commercial varieties. The vegetative method of propagation such as inarching, T-budding and bark grafting could be employed. Top-working of inferior loquat tree should be done during May. Ample number of healthy shoots emerges from the headed back plant during the rainy season. Only one or two healthy branches should be retained for grafting purpose. Grafted plants start bearing fruits after third years.

PLANTING AND ORCHARD ESTABLISHMENT

It can be planted in February-March and July- September. However, Monsoon is the best time for planting. Spring planting may be done where adequate irrigation facilities are available. Planting distance may vary with variety, environmental conditions and soil fertility level. The planting distance could be 6 meters to 8 meters. Thus one hectare shall accommodate 156 to 256 plants.

Pits should be prepared at least one month prior to actual plantation by digging out one metre deep and one metre diameter holes. In areas where termites are menace, Chloropyriphos solution @ 10 ml/10 L may be applied to each pit to check them. A dose of 40-50 kg well rotten farmyard manure and 200g single super phosphate should be given. The saplings are planted in mid-August or mid-February. In general, square system of planting is followed. Thorough watering should be done after planting and again every three or four days for the first week. Thereafter, the interval between waterings can be increased over the next several months until the tree is well established. A water ring several inches high and thick, and a couple of feet across, atop the soil around the newly planted tree may be constructed for the ease of watering. Ring can be filled with water as per need. By the time the ring melts into the surrounding soil, the plant establishes itself.

INTERCULTURAL OPERATIONS

Training and Pruning

If left unpruned, loquat trees grow upright and attain unmanageable heights, which make intercultural operations difficult to be followed. Further, such tall trees are often prone to damage by strong winds. Pruning is essential to reduce the number of bearing shoots and to secure sufficient flower buds for a satisfactory crop load. Central leader or open system is usually followed to train the loquat. It may also be trained into a vase-shape. During initial stages, stem up to height of 50 cm should be kept clean. Overgrown branches of the tree crown are removed with shears or handsaws, and sprouts are removed or cut back. Thin dense growth to make centre of canopy more open to light. Later on, pruning is confined to removal of dead and diseased branches in the full grown tree. The bearing tree benefits from annual pruning to regulate the crop. Since, flowers are born on current year's growth in loquat; therefore, a timely and judicious pruning is necessary. The best time for pruning is during summer after the crop has been harvested, otherwise terminal shoots become too numerous and cause a decline in vigor. Pruning should be made by snipping off 5 cm below the tips towards the May end. Harvesting of mature bunch is also a kind of pruning as it encourages new growth. Heavy pruning should be avoided, as it seriously hampers the yield.

Nutrient Management

Regular and light applications of nitrogenous fertilizers benefit loquats; however, too much nitrogen will promote excessive vegetative growth, which in turn will reduce flowering. It is generally accepted that loquat is a voracious feeder and tends to exhaust the soil. Therefore for realization of satisfactory regular cropping, it needs to be adequately fertilized. Given below is a broad nutritional schedule being recommended by the Panjab Agricultural University, Ludhiana.

Table 14.3

Age of Tree (years)	Farmyard Manure (kg/tree)	CAN (kg per tree)	Superphosphate (kg per tree)	Muriate of Potash (kg per tree)
1-3	10-20	0.3-1.0	0.2-0.5	0.15-0.4
4-6	25-40	1.1-1.5	0.6-1.5	0.6-1.0
7-10	40-50	1.6-2.0	1.5-2.0	1.1-1.5
10 and above	60	2.0	2.0	1.5

Full doses of phosphorus and potash should be applied in September along with farmyard manure. However, one-half dose of nitrogenous fertilizer should be applied in October before flowering and remaining half in February-March *i.e.* after the fruit-set.

Water Management

Loquat trees are drought tolerant; however, for a good harvest regular irrigations are required. The best results are augmented when the orchard is irrigated judiciously. Moisture condition of soil should be examined frequently and irrigation decided accordingly. Care must be taken to maintain sufficient moisture in the soil in order to enable the shoots to develop and the mature terminal buds to fill out properly as the trees blossom buds. During fruit growth to maturity 3-4 irrigations are generally advisable. Loquat does not tolerate standing water.

Weed Management

Thorough cultivation should be given to check the weed growth in loquat orchard. Because of the shallow root system of the loquat, care should be taken not to damage the roots, while mechanical cultivation. Mulching with brown, black, or transparent polythene film from November to June in loquat orchard has been found effective to conserve moisture and to check weed growth.

Intercropping

Loquat can be grown under clean cultivation. The cultivation of leguminous cover crops is, however, considered beneficial. In *kharif*, pulses like *moong*, mash, peas and oil seed crop like *toria* can be grown. Growing *rabi* crop is not advisable in loquat orchards since fruit is harvested in March-April. However, winter cover crops may be raised if planted before September with the aim to turn their sufficient vegetative growth under tree canopy before the harvest begins in loquat. The leguminous crops like gram, peas, mash, *etc.* should be preferred as intercrops.

HARVESTING AND YIELD

Loquat trees may start bearing 3 years after planting and reaches the maximum yield at the age of 15 years. Loquat fruits should be allowed to ripen fully before harvesting. They reach maturity in about 90 days from full flower opening. Fully matured and distinctive color fruits, depending on the cultivar, with TSS above 9° Brix should be harvested. Varieties like California Advance, Golden Yellow and Thames Pride should be harvested at 11° Brix TSS. Usually, all the fruits in a cluster almost ripen uniformly; therefore, the entire cluster may be cut at once. However, sometimes when the fruit at the base ripens before than those at the tip of the cluster, the ripe fruits should be picked individually. The fruits are difficult to harvest because of the thick, tough stalk on each fruit which does not separate readily from the cluster; hence, fruits must be picked with stalk attached to avoid tearing the skin. The best method is to harvest bunches is with the help of a sharp instruments like secateurs or knives. Whole clusters are not particularly attractive for the market; therefore, the individual fruits are clipped from the cluster, the stalk is detached from each fruit and the fruits are graded for size and color to provide uniform packs. Separate the clusters as A (large sized unblemished fruits) and B grade (small sized good fruits); however, the poorest fruits (undersized or misshapen) can be used for preparation of jams, jellies and other by-products. Fruit are easily bruised and scratched, and the damaged areas usually turn brown or black; thus, great care should be taken for handling and packaging during and after harvest. On an average, a loquat tree yields 16-20kg/tree; however, a well managed tree may yield 30-40 kg fruits/tree. To save the fruits from sun burn, sprays of 2,4,5-T (20-40 ppm) is advised as it hastens maturity.

Loquat fruits are very delicate; therefore, a careful packing by providing sufficient cushion is required to avoid injury. Paper shreds are placed at the bottom of each box for providing cushion. The wood boxes of 14 kg size can be used for sending fruits to nearby market. However, for distant markets, smaller packages are used to give considerable protection to the fruit. Loquats can be stored for 4-6 days at ambient temperature and for 15 days in cool storage (11°C temperature and 85-90 per cent humidity). After removal from cool storage, the shelf-life may be only 3 days. Loquat fruit responds well to modified atmospheric (MA) conditions. However, physiological disorders such as internal browning and brown surface spotting may occur during long-term or high-CO_2 storage. The combination of MA and low-temperature storage could be an ideal approach to improve shelf life and reduce fruit disorders. Modified atmosphere packaging (MAP) using films with low permeability to CO_2 under ambient conditions may lead to high-CO_2 atmospheres surrounding the fruit, causing internal browning and increased incidence of decay. Treatment with the fungicide, benomyl, may extend the shelf life by minimizing the decay during storage.

VALUE ADDITION

Loquat is used for canning and preparation of value added products such as jam, juice, syrup, candied fruits, and jellies. Dried fruit has good flavor. Loquats can also be used to make light wine. Loquat wine is very popular in Asia and a liqueur (Bermuda Gold) is made from loquats in the West Indies.

PLANT PROTECTION

Major Insect-Pests and their Management

Bark Eating Caterpillar (*Inderbela quadrinota*)

Bark eating caterpillars cause huge damage to loquat trees. Caterpillars feed on the bark and make tunnels in the trunk. Tunnelling leads to girdling of stem, which may kill the plant. To check this pest, clearing of the holes/tunnels with wire, followed by injecting kerosene oil or chloropyriphos solution @ 50: 50 water in tunnels.

Fruit Fly (*Dacus dorsalis*)

Fruit fly attacks fruits during their maturity. Upon hatching maggots bore into the fruits and feed on the pulp. And as a result, the infested fruits become unfit for human consumption.

Though it is difficult to control the fruit flies, once they enter the fruits; however, their incidence can be reduced by regular spray of fungicide Imidaclorpid @ 0.5 ml per litre at 15 days interval. Similarly, cultural practices like hoeing of the tree basins should be done to expose the pupae to their natural enemies, collection of the infested fruits and destroying them, sanitation in orchard *etc.* may be followed to check their population. To repel the fruit fly, spray Sevin 50 WP (carbaryl) @ 2 g/L twice at 15 days interval during February.

Major Diseases and their Management

Shoot/Fruit Blight and Bark Canker (*Phoma glumerata*)

The cankers are noticed on the bud scars, wounds, twig stubs or in crotches. Small circular brown spots appear around a leaf scar. As the canker enlarges the centers become sunken with the edges raised above the surrounding healthy bark. The fungus perpetuates itself on the trees in bark cankers.

For its management, the affected cankered portions should be removed and the dead bark decorticated along with 2 cm of healthy bark. Pruned dead-wood should be destroyed. The wounds should be covered with Bordeaux mixture immediately after removal of affected portions followed by another application in March and June.

Crown Rot (*Phytophthora* sp.)

This fungus proliferates where moisture stagnate for long time and perpetuates itself in the dead cankers; though, its soil borne. It attacks the bark and manifest into cankers, which extend from the ground level up to the points, where the main stem bifurcates. The rot progresses with the passage of time and girdles the trunk during the next 2 - 3 seasons. Infected plants, flower profusely; however, ultimate fruit retention is abysmally low. Appearance of yellowish green foliage is the typical feature of the crown rot disease. Often the half side of the tree affected by crown-rot may show symptoms. Some branches may be killed every year and ultimately the whole tree may succumb to the disease and dry up completely.

The spread of disease can be minimized by uprooting the severely affected trees and destroying them. Removal of the diseased bark during the dry season by extending the cut an inch beyond the diseased zone on all sides is advisable followed by application of Bordeaux paint after a week. Spray the tree with 2: 2: 250 with Bordeaux mixture immediately after paste application. The spray may be repeated before the monsoon, during the monsoon and post-monsoon, till October.

Root Rot/White Rot (*Polyporus palustris*)

The affected loquat tree begins to show symptoms of wilt, early leaf fall and increase in the fruit-set. The bark and the wood of the root including the root collar is affected in the loquat trees infected by this pathogen. The decayed wood is pinkish to dull violet in colour, whereas in the advanced stages, small, white, elongated pockets appear and they form a mass of spongy white fibers. At advance stages of disease development, the fruiting bodies conks, which may grow up to 30 cm or more in diameter usually appear. They are either hidden by the litter or lie exposed on the surface of the soil.

For disease management, find out the affected trees in early stages by examining the roots and the root-collar region of the tree showing weakening signs. Decayed roots should be dug out and cut completely right from the collar region. Bordeaux paste should be applied on these cut ends. Further, soil drenching, with 2: 2: 250 Bordeaux mixture, should be done in areas from where the dead roots have been dug out. To further, check the spread of this disease, avoid irrigation water to come into contact with the stem. Deep ploughing and interculture are to be avoided to obviate injuries to the roots, through which fungus attacks.

Physiological Disorders and their Management

Physiological disorders such as purple spot and russeting decrease market value of loquat fruit as they severely affect visual quality. In China, skin creasing has also been reported. Bagging of fruits, usually, reduces the incidence of these disorders.

Purple Spot

The origin of purple spot of loquat fruit is related to an alteration of water relationships between the pulp and the rind caused by the simultaneous occurrence at fruit color break of a period of high sugar accumulation in the pulp in addition to a high fruit growth rate. The dehydration process is enhanced by cultivation practices (thinning intensity), and environmental factors (low temperature and sunlight exposure) which affect sugar and mineral assimilation and partitioning in favor of the pulp, increasing the gradient of solute concentration between both tissues. Hand thinning resulted in larger fruit and as well as higher incidences of the purple spot. More fruits get affected with the disorder, when panicles are thinned to a single fruit. Thinning reduces competition among developing fruit, thus increases pulp sugar concentration, which correlates with the incidence of spotting and modifies pulp-rind sugar partitioning.

Application of calcium nitrate, calcium chloride, Ca-EDTA, ammonium nitrate and potassium nitrate at a concentration of 150 mM applied 2 weeks before fruit colour break reduces significantly the proportion of purple-spotted fruit, giving rise to a reduction of water potential of the epidermal tissue that allows it to retain water. The most favourable date of treatment is during the 2 weeks prior to fruit colour break. Further, bagging of fruit is a desirable practice as no fruit are affected when they are covered.

Russeting

Russeting could be detected as skin blemish a month after full bloom, while the fruit is still green. The blemish turns into brown strips toward the phase of colour change. With the fruit maturation, the percentage of russeting increases. Varieties differ in their susceptibility for russeting. The "late" fruit are more susceptible than the "early" fruit. Conditions which delay fruit development during cell division stage tend to reduce the incidence of russeting during the cell enlargement stage.

Skin Creasing

Creasing is more severe in the "early" than in "late" fruit. Creasing incidence in the "early" fruit could be double than in "late" fruit. Further, skin creasing may also result in heavy fruit drop in loquat.

Pulp Browning

Pulp browning is an important postharvest storage and processing disorder of loquat fruit. The fruit turns brown rapidly when peeled or crushed owing to enzymatic oxidation of endogenous polyphenols into quinones, which are then polymerized with other quinones and amines to form brown pigments. During storage, fruit browning occurs from the core area, accompanied by lignification of the pulp tissue. Sulfites, ascorbic acid and its derivatives, and cysteine have been reported to reduce pulp browning in loquat fruit.

Chilling Injury

Loquat fruit are not sensitive to chilling; however, red-pulped cv. 'Luoyangqing' fruit has been found to prone to chilling injury, including stuck peel, firm and juiceless texture (pulp leatheriness) and internal browning, while white-pulped cv. 'Baisha' is reported to be resistant. The symptoms of chilling injury in fruit are tissue browning and lignification, a decrease in juice content and increases in oxidative stress related compounds. The symptoms become more severe once the fruits are moved to 20 °C. Treatments like low-temperature conditioning and application of polyamines, salicylic acid or 1-MCP has been reported to reduce chilling-related disorders in loquat.

15

Passion Fruit

INTRODUCTION

Passion fruit, a perennial woody climber, is a high value and export oriented crop. It is widely cultivated in frost-free areas in over 20 countries around the world for its edible fruits and ornamental flowers. Passion fruit stands out not only for its exotic and unique flavor and aroma but also for its remarkable nutritional and medicinal properties. Unlike popular belief, the name 'passion fruit' is not derived for any aphrodisiac quality of the fruit but was named reportedly by Spanish Catholic missionaries, who landed in South America, saw in the flower, the symbolism of the Passion of Christ where '*Passus*' refers 'suffering' and '*Flos*' refers 'flower'. The English prefix 'passion' derives from the passion of Christ suggested by the prominent four-branched style that appears in the flowers. They believed that the flower symbolized the death of Christ; the five petals represented the disciples (minus Peter and Judas), the corona symbolizes the crown of horns around Christ's head, and other features were a symbol of the wounds, nails, and whips used on Christ. Passion fruit should more correctly be referred to as the 'passion flower fruit'; however, in the trade 'passion fruit' is in vogue. Passion fruit is known in Hawaii as lilikoi, golden passion fruit in Australia, maracuja peroba in Brazil, and yellow granadilla in South Africa. Passion fruit is a vigorous, climbing vine, which produces an edible round or ovoid fruit and has a tough, smooth, waxy dark purple hued rind with faint, fine white specks. Inside, the fruit is more or less filled with an aromatic mass of double-walled, membranous sacs containing yellow or orange colored pulpy juice and many small, hard, dark brown to black pitted seeds.

COMPOSITION AND USES

Composition

Both yellow and purple passion fruit are good sources of pro-vitamin A, niacin, riboflavin and ascorbic acid; besides, fair amounts of minerals like sodium, potassium, phosphorus and iron (Table 15.1). The yellow passion fruit has a juice yield of 30-33 per cent, while the purple has a yield of 45-50 per cent. The yellow passion fruit has somewhat less ascorbic acid content than the purple but is richer in total acid (mainly citric) and in carotene content. Free amino acids in purple passion fruit juice are: arginine, aspartic acid, glycine, leucine, lysine, proline, threonine, tyrosine and valine. Carotenoids in the purple form constitute 1.16 per cent; in the yellow, 0.06 per cent; flavonoids in the purple, 1.06 per cent; in the yellow, 1.00 per cent; alkaloids in the purple (particularly Harman), 0.012 per cent; in the yellow, 0.70 per cent, and the juice is slightly sedative. Starch content of purple passion fruit juice is 0.74 per cent, while of the yellow is 0.06 per cent.

Table 15.1: Nutritive Value per 100 g of Edible Portion of Passion Fruits (Purple passion fruit, pulp and seeds)

Constituent	Content	Constituent	Content
Calories	90	Phosphorus	64 mg
Moisture	75.1 g	Sodium	28 mg
Carbohydrates	21.2 g	Iron	1.6 mg
Protein	2.2 g	Vitamin A	700 I.U.
Fat	0.7 g	Thiamin	Trace
Ash	0.8 g	Riboflavin	0.13 mg
Calcium	13 mg	Niacin	1.5 mg
Potassium	348 mg	Ascorbic acid	30 mg

Uses

Passion fruit, a popular fruit delicacy, can be cut in halves and the pulp scooped out and eaten fresh or added to fruit salads, ice cream and fruit juices. Other products include tropical fruit cocktail, passion fruit *sherbet* and ice, and jelly and jam combinations. In South Africa, passion fruit juice is blended with milk and an alginate; in Australia the pulp is added to yogurt. After primary juice extraction, some processors employ an enzymatic process to obtain supplementary 'secondary' juice from the double juice sacs surrounding each seed. The high starch content of the juice gives it exceptional viscosity. In North-Eastern state like Manipur, the indigenous people of the Nagas use passion fruit leaves tea and salad. They also make *'Khicheeri'* with rice and passion fruit leaves. In the North-East part of the India, native people use boiled extracts of leaves to treat diarrhea, dysentery, diabetes, hypertension, stomach ailments, and as a liver tonic. Europeans took interest in its sedative, tranquilizer-like property, which is attributed to chemical compound 'passiflorine'. In Madeira, locals drink the juice to promote digestion and as a treatment for gastric cancer. The fruit has been used by the Brazilian tribes as a heart tonic and medicine, and as a favorite drink called *maracuja grande,* which is frequently used to treat asthma, whooping cough, bronchitis and other tough coughs. Several studies have shown health benefits of passion fruit such as peel possessing antidiabetic and anti-inflammatory qualities; presence of antitumor agents with no toxicity in fruits; action against cardiovascular diseases owing to presence of polyphenols; antifungal activity of seeds *etc.* Rind has also been shown to have the ability to treat diabetes, colon cancer, and diseases stemming from diverticulitis. Furthermore, folk medicinal practices include its use as anticonvulsant, antidepressant, astringent, cardiotonic, disinfectant, nervine (balances/calms nerves), neurasthenic (reduces nerve pain) and vermifuge (expels worms) activities. It may have promising and powerful effects on neurological disorders as well. The seeds contain 23 per cent oil similar to sunflower or soybean oil, while the rind residue is used as swine and cattle feed. The Native American Indians, Aztecs and Mayas used *Passiflora* as a remedy for pains and ailments, a tradition which is still followed today. Dried passion flowers are used to brew a pain-killing tea.

ORIGIN, HISTORY AND DISTRIBUTION

The purple passion fruit is native from southern Brazil through Paraguay to northern Argentina. It has been stated that the yellow form is of unknown origin, or perhaps native to the Amazon region of Brazil, or is a hybrid between *P. edulis* and *P. ligularis* (q.v.). Cytological studies have not lent support to the hybrid theory. Speculation as to Australian origin arose through the introduction of seeds from that country into Hawaii and the mainland United States by E.N. Reasoner in 1923. Seeds of a yellow-fruited form were sent from Argentina to the United States Department of Agriculture in 1915 with the explanation that the vine was grown at the Guemes Agricultural Experiment Station from seeds taken from fruits purchased in Covent Garden, London. Some now think the yellow is a chance mutant that occurred in Australia. However, E.P. Killip, in 1938, described *P. edulis* in its natural range as having purple or yellow fruits.

In Australia, the purple passion fruit was flourishing and partially naturalized in coastal areas of Queensland before 1900. Its cultivation, especially on abandoned banana plantations, attained great importance and the crop was considered relatively disease-free and easily managed. Then, about 1943, a widespread invasion of *Fusarium* wilt killed the vines and forced the undertaking of research to find fungus-resistant substitutes. New Zealand, in the early 1930's, had a small but thriving purple passion fruit industry in Auckland Province. In Hawaii, seeds of the purple passion fruit, brought from Australia, were first planted in 1880 and the vine came to be popular in home gardens. It quickly became naturalized in the lower forests and, by 1930, could be found wild on all the islands of the Hawaiian chain. Commercial culture of purple passionfruit was begun in Kenya in 1933 and was expanded in 1960, when the crop was also introduced into Uganda for commercial production. The purple passion fruit was introduced into Israel from Australia early in the 20th Century and is commonly grown in home gardens all around the coastal plain, with small quantities being supplied to processing factories.

Passion fruit vines are found wild and cultivated to some extent in many other parts of the Old World–including the highlands of Java, Sumatra, Malaya, Western Samoa, Norfolk Islands, Cook Islands, Solomon Islands, Guam, the Philippines, the Ivory Coast, Zimbabwe and Taiwan. The yellow passion fruit was introduced into Fiji from Hawaii in 1950, was distributed to farmers in 1960 and became the basis of a small juice-processing industry. Since the introduction of the yellow passion fruit from Brazil into Venezuela in 1954, it has achieved industrial status and national popularity. The purple passion fruit was naturalized in the Blue Mountains of Jamaica by 1913, and both the purple and the yellow are planted to some extent in Puerto Rico. Today, passion fruit is grown nearly everywhere in the tropical belt of South America to Australia, Asia and Africa. South America is currently the largest producer of passion fruit. For years, purple passion fruit was grown at moderate scale in India in the Nilgiris in the south and in various parts of northern India. In many areas, the vine has run wild. The yellow form was unknown in India until just a few decades ago when it was introduced from Ceylon and proved well adapted to low elevations around Madras and Kerala. It was quickly approved as having a more pronounced flavor than the purple and producing within a year of planting heavier and more regular crops. Passion fruit is grown in most of the gardens in Mao-Maram and Poumai Naga dominated areas of Manipur since 1970s. At present, it is being cultivated in Kerala, Tamil Nadu (Nilgiri hills, Shevaroys and Kodai Kanal), Karnataka (Coorg), Kerala (Wynad, Malabar) and North-Eastern states (Manipur, Nagaland, Mizoram, Arunachal Pradesh and Meghalaya) (Table 15.2). Recently, its cultivation has been extended to some areas in Himachal Pradesh and Nasik area (Maharashtra).

Table 15.2: Major Passion Fruit Growing Belts in India

State	Districts
Karnataka	Kodagu (Coorg)
Nagaland	Wokha, Mokokchung, Phek, Dimapur, Kohima, Mon, Tuenchang, Zunheboto, Kiphire, Peren
Mizoram	Aizawl (Saitual), Lunglei (Vanlaiphai), Chhimtuipui (Saiha Tuipang), Champhai
Manipur	Senapati, Churachandpur, Chandel, Ukhrul
Meghalaya	Ri Bhoi, East Khasi Hills, West Khasi Hills, Jaintia Hills, Garo Hills

TAXONOMY AND BOTANICAL DESCRIPTIONS

Taxonomy

Passion fruit (*Passiflora edulis* Sims.) belongs to the family Passifloraceae with the basic number of chromosomes n = 9 (2n = 18). The family Passifloraceae includes 550 species in 12 genera and is represented by more fruiting species than any other plant family. The most important genus *Passiflora* has about 400 species which are mostly native to Tropical America and about 40 species in Asia, Australia and the South Pacific and one in Madagascar. Only single species *Passiflora edulis* Sims. is highly commercialized for its quality fruit production. There are two recognized forms of edible passion fruit; purple passion fruit (*Passiflora edulis* Forma *edulis*

Sims), which grows best under sub-tropical conditions or high altitudes and yellow (*Passiflora edulis* Forma *flavicarpa* Deg.), which is suited to tropical conditions or the plains. Giant granadilla (*Passiflora quadrangularis* L.) is also cultivated to a limited extent for local consumption. It grows best in a hot, moist climate and produces a round or oblong, pale yellow to yellowish-green fruit when ripe, which may reach up to a feet in size.

Table 15.3: Species and Characteristics of Passion Fruit

Sl.No.	Species	Characteristics
1.	Purple passion fruit (*Passiflora edulis* Forma *edulis* Sims)	Requires cooler elevation for cultivation. Less vigorous vines. Both, the purple and yellow passion fruits have trilobed leaves 10-18 cm long with finely-toothed margins and dull white flowers with very deep blue centers. Flowers of the purple passion fruit are normally smaller than yellow forms, approximately 4.5 cm in diameter. Less protandry, anthesis in morning. It bears dark-purple or nearly black, round to ovoid, comparatively smaller (5-8 cm long and 4-8 cm diameter) and less heavy fruits (30-45 g) with tough, waxy smooth rind and orange yellow colored pulpy juice with pleasant and mild-acidic flavored arils. The flavour of the purple type is preferred over that of the yellow type.
2.	Yellow passion fruit/ Golden passion fruit (*Passiflora edulis* Sims Forma *flavicarpa* Degener)	Requires low elevation for cultivation. More vigorous vine than purple passion fruits. Leaves are somewhat larger than purple types. Flowers are larger than purple types *i.e.* about 6 cm in diameter. Stronger protandry, anthesis in afternoon. Round to ovoid fruit with more juice content than purple types, heavier (60-90 g) and larger in size (8-10 cm long and 4-10cm diameter) with smooth, glossy, light and airy thick (3-4mm) rind, seeds (as many as 350) are dark brown rather than black surrounded by yellow to light orange pulp having more acidic juice in comparison to purple types, tolerant to many of the soil borne pests and diseases that affect the purple type, more prolific.
3.	Giant Granadilla (*Passiflora quadran- gularis* Medic.)	The giant granadilla has rounded-oblong leaves 10-20 cm long and its stem is characteristically square in cross section. Flowers of the giant granadilla are quite different; they droop like old-fashioned lampshades and their petals are deep maroon on the inner surface. Oblong-ovoid fruit of very large size (20-30 cm long and 12-15 cm diameter), heavier (225-450 g or more), thick edible rind, greenish-white to pale yellow color, black seeds surrounded by whitish to yellowish sweet-acid arils having mild flavor.

Botany

The passion fruit vine is a shallow-rooted, woody, perennial, climbing by means of tendrils. The plants have a weak tap root and extensive ivory-colored lateral roots. The stem is usually solitary, up to 7 cm in basal diameter, extends 5 to 10 m or more into the crowns of trees, and is covered by a thin, flaky, light brown bark. It can grow 4-6 m per year once established and must have strong support. The stem-wood is light and brittle. The twigs are yellow-green, turning brown, and cling themselves on support by means of tendrils that arise at the leaf axils. It is generally short-lived (5 to 7 years). The alternate, evergreen leaves, deeply 3-lobed when mature, are finely toothed, 7.5-20 cm long, deep-green and glossy above, paler and dull beneath, and, like the young stems and tendrils, tinged with red or purple, especially in the yellow form. A single, fragrant flower, 5-7.5 cm wide, is borne at each node on the new growth. The bloom, clasped by 3 large, green, leaf like bracts, consists of 5 greenish-white sepals, 5 white petals, a fringe like corona of straight, white-tipped rays, rich purple at the base, also 5 stamens with large anthers, the superior ovary borne over the gynoandrophore, and triple-branched style forming a prominent central structure. The flower of the yellow is the showier, with more intense color. The nearly round or ovoid fruit, 4-7.5 cm wide, has a tough rind, smooth, waxy, ranging in hue from dark-purple with faint, fine white specks, to light-yellow or pumpkin-color. It is 3 mm thick, adhering to a 6 mm layer of white pith. Within is a cavity more or less filled with an aromatic mass of double-walled, membranous sacs filled with orange-colored, pulpy juice and as many as 250-350 small, hard, dark-brown or black, pitted seeds. The flavor is appealing, musky, guava-like, subacid to acid. Both passion fruit types take 60-90 days to mature. Growth follows a sigmoid growth curve, reaching maximum size in 21days from anthesis, when sclerification leads to the hardening of the shell. Subsequent fruit mass increases at a slower rate and reaches maximum at about 50 days from anthesis in purple types and 60 days in yellow types. A positive correlation exists between the number of seeds developed and fresh fruit mass and juice content. Growth of giant grandilla also follows single sigmoid curve, reaching its final size in 25 days after anthesis and ripening in 60-65 days.

SOIL AND CLIMATIC REQUIREMENT

Soil

Passion fruit vines grow on diverse nature of soils; however, light to heavy sandy loams with a pH of 6.5 to 7.5 are the most suitable. Soil rich in organic matter (2 per cent) and low in salts are considered to be ideal. Application of lime is pertinent, if the soil is too acidic. The vines are shallow rooted (60 per cent of the roots located within 30 cm of the surface); therefore, moisture stress may be avoided during critical growth phases. However, it can withstand mild drought by defoliating vine. Mulching with a thick layer of organic materials may help conserving moisture in root zone. It should be ensured that there is no impedance to root growth by impermeable, rocky or hardened layers in the top 60 cm. Furthermore, site with water table at less than 2 m should not be selected to avoid the appearance of dry rot. Passion fruit vine is highly susceptible to water logging as such soils favor the occurrence of root diseases or collar rot; thus, provision of drainage is essential. Flat and smoothly undulated lands (gradients less than 8 per cent) are the most suitable as they facilitate various field operations such as crop management, mechanization, harvest and soil cultivation and conservation. On steeper slopes (in the range of 8 to 30 per cent), besides erosion control measures (including levelling to create terraces, *etc.*), irrigation and/or fertigation are more difficult. In very steep areas, passion fruit should be grown individually and the soil constantly be replenished to maintain a natural soil covering.

Climate

The purple passion fruit is subtropical. It grows and produces well between altitudes of 650-1300 m in India. The yellow passion fruit is tropical or near-tropical. Biological processes, such as flowering, fertilization, fruit formation, maturation and fruit quality depend on temperatures. Low temperature (15°/10°C) reduces vegetative growth, associated flowering and yield while high temperatures (30°/25°C) can prevent flower production and inhibit pollen germination. The temperature range between 21 and 25°C is considered as the most favorable for the growth of the plant, being best between 23 and 25°C, but passion fruit is being successfully cultivated in temperatures between 18°C and 35°C. At intermediate temperatures of 23°C to 28°C, the fruit growth period is 60 days, when the temperatures were lower (23°C) and higher (33°C) the period was 75 days. Purple passion fruit thrives and yields well at night temperatures of 4.5-13°C and day temperatures of 18-30°C. Mature vines of the purple passion fruit withstand light frost, though are injured at 1-2°C below freezing.

Relative humidity has a great influence on vegetative development and the phytosanitary state of the passion fruit. Air relative humidity of around 60 per cent is the most favorable for the passion fruit. Relative humidity of greater than 60 per cent, when associated with rains, favors the incidence of disease, like *Citrus* scab and anthracnose (black spot) in the above ground parts of the vine.

The susceptibility of passion fruit to strong winds is also an important factor for this crop. Because passion fruit evolved on the margins of tropical rainforests, they are very sensitive to wind damage. Strong wind causes plants to fall and cold wind causes flowers and new fruits to fall, as well as delays plant growth. In regions prone to high winds the use of windbreaks, like bamboo, grevillea, pine, hibiscus, eucalyptus and grass species is indispensable.

Light is an important factor affecting growth due to its effects on photosynthesis. Inadequate light affects the formation of the flowers and fruit. Flowering of yellow passion fruit is suggested to be photoperioclic. Lower average irradiance in the cool season, during the wet season with cloudy weather and with self-shading reduces vine growth, and the number of floral buds and open flowers. Short periods (1 out of 4 weeks) of heavy shade significantly reduce flowering and potential yield. Regions in which the day length is greater than 11 hours have the best conditions for flowering. In the winter months, the plants do not flower because the days are shorter. The flowers normally open at 12.00 hrs, immediately following the maximum incidence of photosynthetically active radiation and close at 15.00 hrs; however when light intensity is lower, they close at 14.30 hrs. There is an interaction of sunlight with temperature, as no flowering occurs at high temperatures with low irradiance.

Although passion fruit withstands droughts relatively well, prolonged drought damages its vegetative development, causing, in severe cases, leaf fall and the formation of smaller and lighter fruits. Moisture stress may be one of the major environmental factors responsible for seasonal fluctuations in passion fruit yields. Passion fruit develops continuously and so needs a constant supply of water. A well-distributed annual rainfall is necessary for passion fruit production, particularly if supplemental irrigation is not available. In regions where the rains occur in specific periods, resulting in shortage for a few months, irrigation is indispensable to guarantee good production and fruit quality. Rainfall must be minimal during the flowering period, as pollen wetted by free moisture bursts open and becomes non-functional. Furthermore, rain minimizes insect activity and hinders pollination. The yellow passion has been grown quite successfully with rainfall of 800-1750 mm uniformly distributed throughout the year or with supplemental irrigation during dry periods. Yields of around 40 t/ha have been obtained with a total water supply of 1300-1470 mm.

IMPORTANT VARIETIES

Very few well-established cultivars are regularly available in India, so plants are customarily grown from seed or cuttings of vines selected for desirable characteristics. A form of yellow passion fruit that sets fruit abundantly from self-pollination is often grown from cuttings or seeds. Its fruit is usually smaller than that of other cross-pollinated yellow passion fruits. Seeds of large fruited selections of yellow passion fruit were brought from Hawaii and plants from this source are grown. Locally selected purple passion fruit is vegetatively propagated for commercial production. One interspecific hybrid had been developed in India, which has been mentioned below:

Kaveri

It is a hybrid between Purple and Yellow passion fruits developed at Central Horticulture Experimental Station, ICAR-Indian Institute of Horticulture Research, Chettalli, Karnataka. It is a high yielding variety and each fruit weighs 85-110 g. The fruits are purple in colour, fruit quality comparable to that of Purple variety. The variety is reported to have field tolerance to brown leaf spot, collar rot, wilt and nematodes. It is recommended for cultivation in the states of Karnataka, Mizoram, Manipur, Meghalaya.

PLANT PROPAGATION

Passion fruits are easily propagated by seeds, cuttings, air layers or grafting on a selected seedling rootstock. For propagation through sexual means, fruits are collected from superior vines in respect of yield and quality. Seeds can be sown immediately or stored at about 10-13°C for future use. Seeds stored at room temperature for 3 months give better than 85 per cent germination. Seeds germinate in about 2 weeks, although germination can extend over 2-3 months because of seed-coat dormancy. Mechanical scarification by cracking seed coat increases germination. However, this method is not feasible for large quantities of seeds. Giant passion fruit seeds from 45-day-old fruit have about 23 per cent germination, while seeds from 60-day-old fruit have 98-100 per cent germination. The best treatment for optimum germination is to ferment the seeds in their juicy pulp for 72 hours followed by washing and superficial drying out for 3 days. Seeds at room temperature lose viability in 1 month. The seeds are sown in well prepared seed beds during March-April. The seedlings after attaining 4-6 leaves stage are transplanted in 10 cm x 22 cm poly-bags filled with a mixture of soil, compost and sand (2:1:1). The seedlings will be ready for transplanting in the main field in about three months, when they are 20-30 cm tall.

Passion fruit can also be propagated by semi-hardwood or mature wood cuttings of 30-40 cm long, 1-2 cm diameter with 3-4 nodes stuck directly in nursery bags. The cuttings are to be first placed in sand beds/pots for root initiation and then transferred to polybags for better root development. The rooted cuttings are ready for planting in about three months.

Grafting and budding can also be performed. Cleft grafting is more successful with adult scions, while whip grafting works better with juvenile material. When the purple passion fruit or its hybrids are the desired cultivars, plants are propagated by grafting on seedlings of the yellow passion fruit as it is vigorous and has resistance to root and stem rots and nematodes. Grafted vines are more vigorous than their seedling counterparts and have longer life spans. Grafting is done 50-55 cm above the ground to prevent soil contact with the scion. Despite the standardization of different propagation methods, most farmers raise nurseries from the seeds and vegetative propagation is not popular as it is time consuming.

PLANTING AND ORCHARD ESTABLISHMENT

Seedlings at the two- to four-leaf stage are transplanted into individual polybags, grown in partial shade for 1-2 months and then hardened off by gradually exposing them to more sunlight. For grafted vines, the scion portion should have grown until about 25 cm and hardened before field transfer. The orchard of passion fruit can be established in square/rectangle system by maintaining row-to-row and plant-to-plant distance of 3-4m x 3-4m. Planting should be done with the onset of monsoon *i.e.* in the month of June-July. Root-pruning should precede transplanting of seedlings by 2 weeks. Transplanting is best done on a cool, overcast day. Plants are transplanted during the cooler part of the day (early morning or late evening). The soil should be prepared and enriched organically a month in advance if possible. Grafted vines must be planted with the union well above ground, not covered by soil or mulch, otherwise the disease resistance will be lost. Mounding of the rows greatly facilitates fruit collection.

POLLINATION

Purple passion fruits are self-fruitful; thus, bearing more fruits/vine. However, pollination is best under humid conditions. Giant grandillas are, usually, self compatible but sometimes protandrous as anther dehisces before stigma becomes receptive and stigma remains receptive from the time of flower opening to closing. Certain lines can be self sterile, needing the presence of other plants for cross pollination. The flowers of the yellow are perfect but self-incompatible (of sporophytic type) which lead to poor fruit set. The style of passion flower shows rhythmic movement as at anthesis, its style is in upright position, which starts curving in due course of time. The most effective time of pollination is after the styles have completely recurved, with the stigma being receptive only on the day of anthesis. Yellow passion fruit has three types of flowers according to the curvature of the style, sometimes on the same plant: TC (totally curved; constitutes about 71 per cent of total flowers), PC (partially curved; about 20-30 per cent of total flowers), and SC (upright-styled; about 6-10 per cent of total flowers). TC flowers are most prevalent. Flowers with upright styles fail to set fruits with pollens from compatible vines and wither under natural conditions or upon hand pollination. Pollen of flowers from upright style effect good fruit set when deposited on stigmas of TC and PC flowers of different vines, indicating high pollen viability. Some autonomous self-pollination has also been reported in yellow passion fruit in absence of any effective pollinators from Karnataka, India. During anthesis, the sepals, petals, stamens and styles spread out rapidly. The receptive surface of the stigma moves down and the anther filaments spread out. During this movement of the anthers and styles, a part of the pollen-bearing surface of one or two anthers comes in contact with the receptive surface of the stigma located just above the respective anthers and deposits some pollen grains onto the stigmatic surface before the anther moves down completely; thereby, initiating autogamous selfing in an otherwise self incompatible species.

The pollens in passion fruits are heavy and sticky; making wind pollination ineffective thus pollen transfer must occur *via* pollinating insects or manual hand pollination where populations of pollinating insects are insufficient. Since flowers are large, attractive, colourful and fragrant and produce plentiful nectar and pollen, it facilitates insects for cross-pollination. The principal insects visiting passion fruit include *Apis mellifera* (honey bee) and *Xylocopa* sp. (carpenter bees). Carpenter bee is the most effective pollinator as it has large body and its body brushes along the anther and stigma while obtaining nectar. Carpenter bees are the main pollinator of passion fruit in areas like New South Wales, Queensland, Western Australia, the Northern Territory and South Australia, Philippines and in Sao Paulo. However in Australian passion fruit crops, honey bees are the primary agent used in the transfer of pollen with recommended beehive densities of 2-3 hives per hectare. In North-East India, floral biology of purple passion fruit, yellow passion fruit, giant grandilla and *Passiflora foetida* was studied. Purple, giant and *P. foetida* had major bloom during March-April, July-August and September-October. While major bloom in yellow was mainly during May-June and September-October. Purple, giant and *P. foetida* had the maximum duration of bloom of 42.4, 22.5 and 32.6 days, respectively during March-April with the maximum duration of effective bloom of 12.5 8.6 and 10.4 days in purple, giant and *P. foetida*, respectively. Yellow had the maximum duration of bloom for 28.4 days and effective bloom of 10.5 days during May-June. Most of the flowers of purple (54.5 per cent) and giant (58.5 per cent) opened between 6-7 hrs, while the maximum per cent of anthesis in yellow (70 per cent) took place between 12-13 hrs. Pollen dehiscence and pollination in purple and giant mainly occurred between 7-8 hrs, while 13-14 hrs was the major period of pollen dehiscence and pollination in yellow. The earliest anthesis (5-6 hrs), anther dehiscence (6-7 hrs) and pollination (6-7 hrs) were recorded in *P. foetida*. The maximum stigma receptivity was recorded on the day of anthesis in all the passion fruits. The maximum number of bees observed between 7-8 hrs in purple and giant and between 13-14 hrs in yellow. *Apis mellifera, A. cerana* and *Xylocopa* spp. have been reported to be the most common floral visitors.

INTERCULTURAL OPERATIONS

Training and Pruning

For commercial orchard of passion fruit, trellises are required. Trellises contribute most to the cost of production of the crop and should be constructed wisely. Trellises should be constructed in the same direction with the wind wherever possible. On sloping terrain, trellises should be constructed across the slope. Trellis rows should be oriented North-South for maximum exposure to sunlight, and the vines should be allowed to grow together along the trellises to promote cross-pollination. The most economical training method is the two arm kniffin system in which 2.5 m long posts/pillars are erected 3 m apart and four lines of 9 to 11 gauge wire is allowed to run across. Posts, either bamboo or iron, should be fixed at the distance of 3-4 m and on these posts 3-4 wires should be strung by keeping the distance of 30 cm between wires. Once the vines reach the wire, the tips are pinched to facilitate two leaders and these leaders are directed on either side of the wire, which in turn develops laterals. These laterals are intended to bear fruits as passion fruits bear fruits only on current season's growth. Passion fruit can also be trained on pergola system. In this system vines are spread over a criss-cross network of wire (15-20 cm apart), usually 1.8-2.0 m above the ground, supported by bamboo posts or iron posts. The vines are allowed to grow single shoot till it reaches to wire. When vine reaches to wire its tip is pinched to facilitate side shoots on the wire.

Passion fruit vines produce flowers on the current season's growth; therefore, any intervention which encourages new lateral growth increases flowering. Occasional pruning is required; however, their intensity may vary. After laterals have completed fruiting, they should be cut back to developing side shoots closest to the main leader. If no side shoot has formed then the lateral should be cut off at the fourth or sixth node from the main stem. Following laterals have reached the ground after 12-15 months of growth period, they should be cut back to 30 cm of the main leader. Pruning should be performed when bearing is low *i.e.* after harvesting the crop in April and November-December. If left unpruned, laterals begin growing on the ground and the vines tend to form a tangled mass on the wires. Under such circumstances, only the new outside growth bears fruits. Resorting to pruning of old growth can limit the current crop, while removal of new growth restricts future cropping; therefore, following general recommendations are suggested:

1. Light selective pruning, particularly at the end of the annual production cycle, enhances new growth and maintains high yields the following year. This consists of removing all vines about to reach the ground or growing on the ground. Vines are cut near the trellis wires; however, a few nodes away from the main stems. Vines hanging only halfway to the ground are left.

2. Long vines should not be thrown over the trellis, as this only increases entanglement on the trellis and depresses yields.

3. Some vine growth on top of the entangled tops of the trellis may be pruned as fruit produced on these vines is apt to be lost in the maze of vines, especially, when the planting is 2 or 3 years old. Pruned vines on

the trellis should be left to dry in place as attempts to remove the cut vines can damage uncut vines.

Pruning is also recommended for the giant passion fruit. The initial formation pruning consists of eliminating all side shoots to leave a single stem and is similar to that used for purple and yellow passion fruit. Production pruning consists of eliminating shoots that have already produced fruit in order to stimulate new shoots that will flower. At the same time, all damaged or diseased shoots should be removed.

Nutrition Management

The nutrient requirement of passion fruit depends up on the age and stage of growth *viz.*, vegetative growth, initial bearing and full bearing stages. The approximate nutrient requirement of passion fruit is 150 kg N, 100 kg P_2O_5 and 200 kg K_2O per hectare. The nutrient should be applied in splits after fruit harvest. Application of 110, 60 and 100 g N, P_2O_5 and K_2O/per vine/year for cv. Kaveri has been recommended to achieve optimum crop. The roots normally form mycorrhizal associations, which may help acquisition of minerals from soil, particularly phosphorus. Therefore, inoculation of plants with superior strains, at nursery stage, will be desirable.

Water Management

Adequate soil moisture is required to sustain the desired vegetative growth and production. Floral buds do not develop under dry conditions due to reduction in vine growth and extension. Regular watering will keep a vine flowering and fruiting almost continuously. Least flowers develop during the winter season due to short day length. Water requirement is high when fruits are approaching maturity. If soil is dry, fruits may shrivel and fall prematurely.

Weed Management

Weed management can be achieved manually in the rows and mechanically between the rows. During harvest, thorough weeding needs to be done in the rows parallel to the planting lines as the fruits are usually collected from the ground. Mechanical weeding (close to the plant less than one meter) is not recommended in order to prevent damages to the roots, which are largely concentrated within 15 to 45 cm from the stem. Chemical weeding, using selective herbicides eliminates not only the weeds, but reduces operational costs and simplifies the work.

PLANT PROTECTION

Major Insect-Pests and their Management

Several insect-pests infest passion fruit and as result fruits are blemished, plant vigour and productivity is reduced. This leads to severe economic losses, sometimes.

Fruit Fly (*Daucus* sp.)

It is a most serious pest of passion fruit, which causes heavy losses. The pest punctures the immature fruit before the rind hardens. As the fruit enlarges, a corky area develops around the affected portion. If the fruit is undeveloped at the time of puncture, damage may be sufficient to cause it to shrivel and drop off. If the fruit is well-developed, it may grow to maturity. At the time of ripening, the area around the puncture has the appearance of a small woody crater which, while it does not impair juice quality, does disfigure the fruit. Fruit fly adults may be destroyed by application of Malathion 25 per cent wettable powder, sprayed @ 1.5 kg/ha in 500 l of water.

Thrips

Thrips (*Thrips hawaiiensis* Morgan) feeds on buds and developing fruits. Affected fruits are deformed and fruit weight and juice content are reduced. The incidence of this pest is severe in summer crop. Lady bug can be used as a biological control of thrips.

Mites

Mites (*Tetranychus neocaledonicus* Andro) feed on leaves and tender fruits. It leads to defoliation and formation of undersized fruits. Mites can be normally controlled by hosing them with a soap solution for at least three days in succession - make sure you spray every part of the plant, especially the underside of the leaves.

Major Diseases and their Management

The major diseases in passion fruits are collar rot, root rot, brown spot and woodiness virus.

Brown Spot

Brown spot disease is caused by *Alternaria macrospora* Simes. The disease appears as concentric brown spots with greenish margin. Under high humidity, spots norrnally grow larger up to 2 cm in diameter become round and zonate. Spores can form a black thin mass covering the middle of the lesion, being more abundant on the abaxial surface. Abscission of the affected leaves occur rapidly causing intense defoliation. On twigs, dark brown lesions are more elongated and may cause girdling and death of the terminal portion of these organs. Slightly circular spots occur on the mature fruits or when they are half way through their growth process. They are reddish brown, sunken affecting the pulp and damaging the commercial value. Girdling of branches and premature defoliation occurs in severe cases. For its management, affected branches should be pruned and burnt. Application of copper compound based fungicides is able to minimize the incidence of the disease. Further, spray of mancozeb along with iprodione has also been found effective under high humid conditions for the containment of the disease.

Root Rot

Root rot is caused by *Phytophthora nicotianae* var. *parasitica*. The roots are affected and ultimately the plants die. Drenching with 1 per cent Bordeaux mixture helps in checking the disease. The affected plants should be mounded with soil to encourage new root formation.

Wilt or Collar Rot

Wilt or Collar rot is a devastating disease caused by *Fusarium oxysporum/F. passiflorae*. Collar rot is the most common disease of purple passion fruit. Above ground symptom includes mild die back of the plant followed by changing of leaf colour to pale green. Wilting, defoliation and finally plant death occurs resulting from the complete necrotic girdling of the plant collar. Tumescence and fissures in the affected collar bark show purple lesion borders, where reddish structures appear under high relative humidity. For its management, avoid water stagnation in field and ensure proper provisions for drainage. Drenching of copper oxychloride, at the interval of two weeks, reduce the number of plants developing collar disease. The use of a resistant root stock (yellow passion fruit) is an effective way to deal with the problem in the contaminated areas.

Woodiness

Passion fruit woodiness virus (PWV) and cucumber woody virus (CWV) causes woodiness in passion fruit. Symptoms includes a noticeable reduction in the development of plant, severe mosaic of leaves, rugosity and distortion. Affected plants produce woody and detonated fruits. Severe mosaic, epinasty, defoliation and premature death of plants are associated with infection of PWV. Other common symptoms are leaf mottling and ring spot on the younger leaves. Fruits are symptom less or may show mild molting. Chlorotic spots on the leaves and dappled or faded fruits are often found. Viruses are normally transmitted by several species of aphids in a non-persistent, non-circulative way. They can also be transmitted through grafting and experimental mechanical inoculation. Mechanical transmission by knifes, scissors and nails during cultural practices of trimming are observed. None of the viruses are found to be transmitted through seeds. Chemical control of vectors is usually ineffective for the virus because of the non-persistent relationship between the virus and aphid vectors. Usage of virus free seedlings of new plantings, eradication of old and abandoned orchards before starting new crops, care during trimming operations to eliminate mechanical transmission of viruses, avoiding leguminous plants which may harbor the virus near the orchard and rouging of diseased plants by means of systematic inspections during the first five months after transplanting can aid in checking the incidence and spread of potyvirus infection in passion fruit vineyards.

HARVESTING AND YIELD

Yield of passion fruit vary with climate, species and variety of passion fruit, cultivation practices, training system, occurrence of diseases and the abundance of appropriate pollinating agents. Purple passion fruit produces more fruits than yellow and giant granadilla due to self fruitfulness. First fruiting can be noticed from the 10th month, while full bearing reaches by 16-19 months. Although passion fruit continues to flower and fruit throughout the year, there are 2 main periods of fruiting from August to December and March to May. At the latter time, the fruits are somewhat smaller, with less juice. About 60-90 days are required from anthesis to harvest. The ripe fruits fall down from the vine; therefore, fallen fruits should be collected daily to avoid spoilage from soil organisms. Slightly purple coloured fruits along with a small portion of the stem should be picked up rather than being allowed to fall. It has been found that fallen fruits are lower in soluble solids, sugar content, acidity and ascorbic acid content. Purple variety yields 8-10 kg/vine, whereas cv. Kaveri may yield 16-20 kg/vine, the total yield being12-20 tons/ha annually. Passion fruit is a climacteric fruit and; thus, ripening may also take place off the tree. Both purple and yellow passion fruits begin to lose moisture as soon as they ripen and fell and quickly become wrinkled. Fully matured fruits may be harvested and may be kept for ripening. Under-ripe yellow passion fruits can be ripened and stored at 20° C with relative humidity of 85 to 90 per cent. Ripe fruits keep for one week at 2.22°-7.22° C. The fruits can be stored in polyethylene bags at 7-9°C for as long as 3 weeks without loss. Coating with paraffin and storage at 5° to 7° C and relative humidity of 85 to 90 per cent can prevent wrinkling and help retaining quality for 30 days.

VALUE ADDITION

A range of processed products such as nectar, squash, carbonated drinks, juice concentrate *etc.* can be prepared from passion fruit. Such products are in great demand both in domestic as well as export market. Passion fruit juice can be boiled down to a syrup which is used in making sauce, gelatin desserts, candy, ice cream, *sherbet*, cake icing, cake filling, meringue or chiffon pie, cold fruit soup, or in cocktails. Passion fruit is used for making a refreshing slushie by blending ice with fruits like orange juice, lemon, and guava followed by stirring. Alcohol can be added, if desired. Passion fruit icing is prepared by combining passion fruit juice with beaten powdered sugar, margarine, vanilla extract, and soymilk, which may be lathered over citrus-based, coconutty, or vanilla-based sweet breads, cookies and cupcakes. The seeded pulp is made into jelly or is combined with pineapple or tomato in making jam. To avoid impairing of flavor of passion fruit juice by heat preservation agitated or 'spin' pasteurization in the can should be adopted. The frozen juice can be kept without deterioration for 1 year at -17.78 °C and is a very appealing product. The juice can also be 'vacuum-puff' dried or freeze-dried. Swiss processors have marketed a passion fruit-based soft drink called 'Passaia' for a number of years in Western Europe. Costa Rica produces a wine sold as 'Parchita Seco.' A juice drinks 'Pasip', from fruits of passion fruit grown and processed in Manipur, is manufactured by Exotic Juices Ltd. There is an immense potentiality of boosting passion fruit based processing industry in North-Eastern states of India as climate of this region is very conducive for its cultivation.

16

Peach

Peach along with its smooth peeled mutant, the nectarine is a temperate fruit of excellent quality. In India, peach is considered as the most important stone fruit as it grows profitably well both in plains and hills.

COMPOSITION AND USES

Composition

Peaches are considered as a good source of sugars, vitamins and minerals (Table 16.1). Nectarines have a small amount more of vitamin C, provide double the vitamin A, and are a richer source of potassium than peaches.

Table 16.1: Nutritional Value of Peaches per 100 g Edible Portion

Attribute	Contents	Attribute	Contents
Energy	39 kcal	Choline	6.1 mg
Carbohydrates	9.54 g	Vitamin C	6.6 mg
Sugars	8.39 g	Calcium	6 mg
Dietary fiber	1.5 g	Magnesium	9 mg
Protein	0.91 g	Phosphorus	20 mg
β-carotene	162 µg	Potassium	190 mg
Niacin (B₃)	0.806 mg		

Source: USDA Nutrient Database.

Uses

Peaches are highly valued as table fruit, although several value added products like canned peach, dried peach, jams, nectar are also made from it. Yellow fleshed, free from stone; regular producer and relatively free from fuzz peaches are ideal for table purposes. For canning proposes, peaches having yellow flesh, free from stone, small pit, non-splitting, good symmetrical size and evenly mature are preferred. Similarly, for dehydration, white-fleshed, sweet peaches having free stone are preferred. Suitable cultivars for canning are Golden Bush, Foster, Crawford's Early, while for drying, free stone peaches with golden yellow flesh and high sugar content are preferred. Seed kernels are used for the extraction of oil, which is used in the preparation of cosmetics and pharmaceutical industries.

These fruits rank low on the glycemic index, which make them ideal for those needing to control sugar levels. Additionally, peaches are a hydrating summer fruit and provide an ideal combination of sugar and electrolytes. Peach seeds contain cyanogenic glycosides, including amygdalin. More than 80 chemical compounds contribute to the peach aroma. Several esters such as linalyl butyrate or linalyl formate, acids and alcohols, and benzaldehyde contribute to it.

ORIGIN, HISTORY AND DISTRIBUTION

Peach is native to China where it was known to grow as far back as 2000 BC. Commonly found wild species of peach in China are *P. davidiana* in north, *P. mira* in Tibetan plateau and *P. ferganensis* in West China indicating the centre of origin to be China. The primary center of diversity are mountain areas of Tibet and South China and secondary diversity centers are Iran, Central Asia, Italy and Spain.

The history of the nectarine is unclear; the first recorded mention in English is from 1616, but they had probably been grown much earlier within the native range of the peach in Central and Eastern Asia. Nectarines were introduced into the United States in 1906. Furthermore, peacherine, a hybrid between peach and nectarine, is also marketed in Australia and New Zealand.

The peaches are now commercially grown throughout the world in the temperate zone, between 20° to 45° latitude, which indicate that these require lower chilling hours than pome fruits such as apple and pear. With the development of low chilling varieties, the cultivation of peach has been extended to the subtropical zone and even the tropical areas at very high elevations. While peach can be grown in most apple growing areas, it extends for most parts closer to equator because it can tolerant high temperatures and require less winter chilling period to break the dormancy.

The major peach growing countries in the world are the USA, Italy, China, France, Spain and Greece, which account for about 70 per cent of total peach production of world. The peaches are also commercially cultivated in China, Italy, Spain, USA, Turkey, Iran, Chile and France (Table 16.2). However, countries like Japan, Argentina, Australia, Mexico, Korea, Russia, Germany, Portugal, New Zealand, South Africa, Canada, Austria and India also grow peaches commercially. In India, peach is largely cultivated in Jammu and Kashmir, Himachal Pradesh, sub-mountainous tracts of Punjab, Delhi, Haryana, Western Uttar Pradesh and Uttarakhand hills. Limited cultivation of peach is also done in Nilgiri hills and north eastern states. Broadly peaches can be grown between 1,000 to 2,000 m above MSL. But with the development of low chilling peach cultivars, which require chilling duration for less than 500 hours, peaches can now been grown in plains at an elevation of 250- 450 m above MSL also. The low chilling subtropical peach cultivars. such as Pratap, Prabhat, Sharbati, Shan-e-Punjab, Khurmani, Florda Red, Florda Sun, Florda Prince and Early Grande are successfully grown in low hills and plains of Punjab, Rajasthan, Haryana and Uttar Pradesh.

Table 16.2: Top Ten Peach and Nectarine Producer Countries of the World

Country	Production (Metric Tones)	Productivity (MT/ha)	Country	Production (Metric Tones)	Productivity (MT/ha)
China	11.50	15.03	Turkey	0.55	18.6
Italy	1.64	18.5	Iran	0.50	10.5
Spain	1.34	16.4	Chile	0.32	16.6
USA	1.18	20.7	France	0.30	23.4
Greece	0.69	17.3	Argentina	0.28	11

Source: Food and Agriculture Organization-2012.

TAXONOMY AND BOTANICAL DESCRIPTION

Taxonomy

Peach belongs to family Rosaceae, sub-family Prunoideae, Sub genus Amygdalus and section Euamygdalus. All commercial varieties belong to *Prunus persica* (L.) Batch. Some wild species of peach are *Prunus davidiana, P. niora* and *P. ferganensis. Prunus behmi* is considered as a natural hybrid between almomd and peach, which grows extensively in dry temperate zones of H.P. and is primarily, used as a rootstock. The basic chromosome number of peach is 8 and most of the cultivated peach varieties are diploid with somatic number, 2n = 16. Peaches are fuzzy, while its smooth skin single gene mutant, the nectarines are without fuzz.

Peach varieties have been categorized into two groups: *viz.*, Flat peaches (*P. persica* var. *compressa*) and Nectarines (*P. persica* var. *nectarina*). Nectarines are identical to peach in tree and flower characteristics but differ in the absence of fuzz or pubescence on the fruits. Nectariens are usually small in size, having greater aroma, less melting flesh and more prone to thrips attack than peaches. They are also referred to as a 'shaved peach' or 'fuzzless peach', due to its lack of fuzz or short hairs. Though fuzzy peaches and nectarines are regarded commercially as different fruits, with nectarines often erroneously believed to be a cross breed between peaches and plums, or a 'peach with a plum skin'. Nectarines have arisen many times from peach trees, often as bud sports. Furthermore, peacherine, a new entry in the group of peaches, is claimed to be a cross between a peach and a nectarine.

Botany

The peach has small tree, growing 4–10 m tall. solitary pink flowers and deeply pitted stone. The leaves are lanceolate, and pinnately veined. The flowers are produced in early spring before the leaves, and are solitary or paired, pink, with five petals. The fruit has yellow or whitish flesh, a delicate aroma, and a peel that is either velvety (peaches) or smooth (nectarines) in different cultivars. The single, large seed is red-brown, oval shaped, and is surrounded by a wood-like husk. Botanically, the fruit of peach is a drupe. It develops entirely from a superior ovary and consists of skin (exocarp), fleshy mesocarp and hard endocarp, which is called as stone. It bears fruits mostly on one-year-old shoots.

Cultivated peaches are divided into cling stones and free stones, depending on whether the flesh sticks to the stone or not; both can have either white or yellow flesh. Peaches with white flesh typically are very sweet with little acidity, while yellow-fleshed peaches typically have an acidic tang coupled with sweetness, though this also varies greatly. Both colours often have some red on their peel. Low-acid white-fleshed peaches are the most popular in the countries like China, Japan, and neighbouring Asian countries, while Europeans and North Americans favour the acidic, yellow-fleshed kinds.

Like peaches, nectarines can be white or yellow, and cling stone or free stone. On average, nectarines are slightly smaller and sweeter than peaches, but with much overlap. The lack of skin fuzz can make nectarine skins appear more reddish than those of peaches, contributing to the fruit's plum-like appearance. The lack of down on nectarines' skin also means their skin is more easily bruised than peaches.

The fruit of peacherine is intermediate in appearance between a peach and a nectarine, large and brightly colored like a red peach. The flesh of the fruit is usually yellow but white varieties also exist.

SOIL AND CLIMATIC REQUIREMENTS

Soil

Peach can be grown on a variety of soils. However, it thrives best on light, gravely clay loam soils, which are fairly fertile, deep and well-drained. Water logged conditions are unsuitable for peach cultivation as it is highly susceptible to logging condition. Water logging injury is associated with metabolism of the cynogenic glucoside 'Prunasin' in *Prunus* sp. In absence of oxygen, the prunasin get hydrolyzed to hydro cyanic acid, which is auto-toxic. The tolerance of *Prunus* sp. depends upon their ability to hydrolyse prunasin. Very fertile and heavy clay soils are hazardous to peach because such conditions induce heavy vegetative growth susceptible to winter injury. Similarly, acidic and saline soils are unfit for peach cultivation. The pH of the soil should be between 5.8 and 6.8. Sites located in deep valleys are not suitable for peach cultivation as air settles down in these areas and freeze injuries are quite common. Thus, the land, which is not too steep, is the most desirable site for profitable cultivation.

Climate

In comparison to other temperate fruits, peach requires the warmest climate for successful production. It can be successfully grown in areas experiencing 750 to 800 chilling hours during winters. However, the low chilling peaches of Florida group require only 300 to 500 chilling hours to break dormancy. The limiting factors for the peach cultivation are low winter temperature, spring frost, hail storm, high humidity and desiccating winds during summer. However, most limiting factor in temperate region is the lack of flower bud hardiness either due to low dormant temperature or to frost or freeze conditions during spring. In general, peach is very sensitive to low temperature injury after bud break. The swelling buds are injured if the temperature falls below -6.5 °C even for few hours. Use of a water-soluble, exterior grade, white latex paint on stone fruit tree trunks may reduce winter injury. Even good grade lime white washing may also work well. The paint may be applied to the southern and south-western sides of stone fruit tree trunks, including the bases of main branches. The paint helps reflect much of the sunshine falling on the trunk during bright winter days and reduces the amount of heat penetrating the bark. This lowers the possibility of bark splitting and subsequent tree damage. It is particularly important to paint tree trunks and the bases of main limbs up to 8 to 10 years of age. Smaller trunks and main limbs of younger trees respond more to the extreme fluctuations of winter temperatures and may be injured more severely than older trees. A well-distributed rainfall of 80-100 cm per year or assured irrigation is desirable for a good peach crop. Dry climate with low humidity for 2-3 weeks before harvest is ideal for development of a good quality fruit.

IMPORTANT VARIETIES

On basis of separation of pit from pulp, peach varieties have been divided in two, cling stone and free stone.

☆ **Cling stone:** In these varieties, the pulp adheres with the stone even at maturity and is not easily separated. Important varieties of this group are Sun Haven, Red Haven, Shimizu Hakuto and Kanto-5, Sharbati, and Florda Prince.

☆ **Free stone:** In these varieties, the stone separates freely from the pulp at maturity. Important varieties in this group are July Elberta, Rich Haven and Elberta, Flordasun.

On the basis of their use, peach varieties have been classified as table, canning and hedyration varieties as under:

☆ **Table varieties:** The peach varieties for table purpose should be regular bearer, large sized with firm and yellow pulp, relatively free from fuzz, good in taste and attractive in appearance. Free stone varieties are preferred for table purpose. However, most of early ripening and low chilling varieties are cling stone. Important table varieties are Elberta, JH Hale, July Elberta, Babcock, Cardinal, Dixigem, Red top, Red Haven, Red Globe.

☆ **Canning varieties:** The varieties for canning purpose should have good quality, yellow flesh, free stone, small non-splitting pit, firm flesh, symmetrical size and large fruit size. Pit should not leave red colour in the flesh. Important canning varieties are Rich Haven, JH Hale, Golden Bush, Foster, Halford, July Elebrta, Loaded, Crawford's Early and Peak.

☆ **Drying varieties:** White pulped, sweet cultivars having free stone kernels are preferred.

The peach varieties for the Indian hilly conditions are as follows:

☆ **Mid hills**

Early ripening: Word's Earliest, Early White Giant, Red Haven, Sun Haven

Mid-season: Alton, July Elberta, Kanto-5, Shimizu Hakuto

Late ripening: JH Hale, Elberta, Helberta Giant

Low hills and valley areas: Sharbati, Sun Red, Shan-e-Punjab, Totapari, Matchless, Sufeda, Honey, Sweet, Florda Red, Florda Sun, Prabhat and Pratap. The vaieties like Shalin and Yunnan are nematode resistant varieties for plains.

☆ **Low chilling varieties:** Florda Sun, Sun Red, Sun Gold, 16-33, Florda Bell and Florda Red.

☆ **Varieties of nectarines:** The best known cultivars are Annqueen, Charokee, Nectred, Sun Grand, Sunlite, Sunrise and Sunripe.

Chief characteristics of some varieties of peach are as under:

Crest Heaven

Produces top notch free stone fruit with golden yellow skin and flesh. Mid to late season variety, blooms late, fruit lasts well on the tree. Excellent for freezing and canning.

Early Grande

It ripens in the first week of May. Its fruits are large with red blush surface. Flesh yellow, firm with some red color next to pit, semi-free stone when fully ripe. The fruits possess excellent shipping qualities.

Elberta

It is a mid-season cultivar. Its fruits are very large, blushed-red deep golden skin; yellow, slightly bitter flesh; free stone.

Fantasia

Vigorous tree, fruits are large ovate in shape, bright yellow with red blush over the major part of fruit skin and early to mid-season maturity.

Flavorcrest

It is also a mid-season cultivar. Its fruits are large round with red-blush over yellow skin; firm, yellow flesh has excellent flavour and fine texture.

Florda Prince

It is an early self-pollinating low–chill cultivar. Its fruits are large, and uniformly firm. The fruit develops red blush with dark red stripes over a yellow ground colour, with semi-cling stone pits.

Florda Sun

It matures in the last week of April. Fruits are medium to large, roundish and yellow with red blush. Flesh is yellow, juicy and sweet and free stone.

Florda Red

An excellent, mid-season table peach, it matures in the beginning of June. Fruits are large, almost red at maturity, juicy, with white flesh and free stone.

Glo Heaven

A large peach with yellow free stone flesh has mostly red skin with no fuzz, milder flavour, excellent for canning and fresh eating and free stone.

Golden Gem

It is a dwarf and early variety. Its fruits are large with yellow peel; yellow flesh with red pit cavity. Pulp has an excellent flavour; from California.

Golden Glory

It is a dwarf variety which ripens late in the season. Its fruits are very large with golden peel and red blush. Its flesh is yellow, juicy and good flavored.

Golden Jubilee

It is a mid-season variety producing medium to large sized fruits with mottled bright red skin. Its pulp is soft with melting texture and good quality and free stone.

J.H. Hale

It is a midseaon to late variety bearing very large fruits with yellow peel blushed with red and little fuzz. The pulp is yellow with good flavour and aroma and free stone.

July Elberta

It is mid-season cultivar which produces medium round fruits with dull red streaks over greenish yellow peel. Yellow fruit is firm, fine grained and very good flavour having small pit. Good for fresh use or canning.

Khurmani

Its fruits are large, attractive with red coloration, cling stone with white soft juicy flesh.

Paradelux

Trees are medium in vigour, fruits are large in size and oblong flat with prominent beak in shape, yellow skin and flesh and late season maturity.

Prabhat

It is the earliest-maturing peach (mid-April), and fetches good income to the growers. Fruits are medium-sized, roundish and develop attractive red blush. Fully ripe fruit has white- flesh.

Pratap

It matures in the third week of April (one week earlier than Flordasun). The fruits are yellow with red blush. Flesh color is also yellow with red coloration. It has better firmness and keeping quality than Parbhat and Flordasun.

Red June

It is an early-season variety. Its trees are medium in vigour, fruits are large in size, roundish with rounded beak in shape, distinct suture, yellow with red blush on the shoulder, free stone.

Red Haven

The standard and most popular peach variety in the industry. It bears attractive red colour fruits with good fruit quality, although fruit size may be small if it is not properly thinned. The fruit is good for freezing but not for canning., Redhaven is a superior cultivar with good winter hardiness.

Shan-e-Punjab

It matures in the first week of May. Fruits are very large, yellow with red blush, juicy, sweet, excellent in taste, and free stone. Since fruits are firm in texture, they can withstand transportation. Suitable for canning.

Sharbati

Its fruits are large, greenish-yellow with rosy patches, very juicy with excellent taste and flavour. Fruits ripen during June-end to first week of July.

Snow Queen

Trees are spreading, vigourous, fruits are small to medium in size, bright red colour on cream white background having smooth surface, flesh white, cling stone, maturity during mid June.

Suncrest

It is a mid-season variety which bears large, round fruits with bright red over yellow peel. The pulp is firm, yellow having melting but good texture and flavour and free stone. Suitable for fresh use or canning.

Sunhaven

It is an early to mid-season cultivar which bears bright red over gold coloured fruits. Fruits are juicy, with fine-grained yellow pulp with sweet flavour, and semi-cling stone.

Sun Red

Trees are low in vigour, fruits are small to medium in size with bright red skin, semi-free stone and early season maturity.

ROOTSTOCKS AND PROPAGATION

Propagation

Peach is commonly propagated by grafting or budding on seedlings of peach, plum or apricots. However, wild peach or wild peach x almond hybrid seedling are most preferred rootstocks. Peach seeds require the stratification, characterized by need for "after-ripening" of seeds before they can germinate. This is achieved by keeping seeds at about 5°C for 70 to 90 days before sowing under moist conditions, for proper and uniform seed germination. In cooler climate, direct sowing in nursery beds during November-December meets the stratification requirement. The stratification period can be reduced by treatment of seeds with growth regulators like gibberellins or by chemicals like thiourea. Peaches are usually propagated by T-budding or ring budding from May- June to September. Peach scion varieties can also be propagated by tongue grafting during February-March with more than 90 per cent success.

Rootstocks

Clonal rootstocks for peach could not be popularized with success to a great extent. Peach seedlings are still the principal rootstock source for commercial peach trees. Seedlings can be divided into three classes: wild types, commercial cultivars (Halford and Lovell), and seedlings developed for use as rootstocks (Siberian C, Bailey, Rutgers Red Leaf). Peach x almond hybrids, GF 556 and GF 677 are popular in Europe, which are suitable for alkaline soil conditions. Likewise, Citation, a peach-apricot-plum hybrid developed in California is reported to be tolerant to wet soils. St. Julian is compatible with peaches and plums and supposedly reduces scion vigor by 10 to 15 per cent. Damas (GF 1869) is a plum hybrid compatible with peaches and plums but not nectarines. Amandier (GF 677) is a vigorous rootstock fully compatible with peaches, plums, and nectarines. Rootstocks Krymsk 1 (VVA-1) and Krymsk 2 (VSA-1) were developed in Russia along the Black Sea for their dwarfing ability and cold-hardiness. Both are compatible with peach and plum. Hiawatha is another rootstock that is available commercially. It provides about 10 per cent dwarfing capabilities. Two size-controlling rootstocks *viz.*, Controller 5 and Controller 9 for peaches and nectarines are also being developed at The University of California at Davis. *Prunus besseyi* and *P. tomentosa* are the dwarfing rootstocks but express poor compatibility with peach varieties. Peach is severely affected by nematodes. Some of the root-knot nematode resistant rootstocks for peach are Nemagaurd, Nemared, Shalin, Okinawa, Higama, Yunnan and S-37. Behmi (*Prunus mira*) is used as rootstock in some parts of Himachal Pradesh.

ORCHARD ESTABLISHMENT

Peach is usually planted in December-January when the plants are dormant. In general two-year-old saplings are planted to achieve good success. In flat areas, square, rectangle or hexagonal system of planting can be followed. In hills, contour planting or terrace planting is most desirable. The plants on vigorous seedling stocks can be planted at 5 or 6 meter distance either way in well prepared pits. In high density plantation, the distance can be reduced to 3m x3m. In Tatura trellis and Meadow system, peach is planted at a distance of 5m x 1m (2,000 plants/ha) and 2 m x1 m (5,000 plants/ha) respectively. The size of the pit should be at least 1 x 1 x 1 m. If there is any hard pan or the rock in subsoil, it should be removed while digging pits for proper root development. The pits should be dug during September–October and refilled with fertile top soil mixed with 40kg well-rotten farmyard manure and solution of chlorpyriphos (1 ml/liter) to each pit to avoid any damage from insects. The grafting point should be kept at least 15-20 cm above the ground level at the time of planting.

FLOWERING, POLLINATION AND FRUIT SET

Peaches start bearing at the age of 2-4 years. Flowering commences in the month of February and terminates in March. Peach trees produce several thousand flower buds, but usually only 20 to 50 per cent of the flowers set fruit. However, 10-15 per cent fruit set gives sufficient crop. Practically, all the commercial varieties of peach are self-pollinated. Thus, most peach plantations do not require the provisions for cross-pollination for assured crop. However, a few male sterile/self infertile varieties like JH Hale, June Elberta, Halberta and Chinese Cling and require pollinizers for fruitful production. For such varieties, every third row should be of a pollinizer like July Elberta. Honeybees accomplish the pollination in peach. Thus, it is advisable to place 4-5 beehives in the orchards of such varieties to ensure adequate pollination, and to have good crop.

Fruit Thinning

Peach trees have the tendency to bear heavily than expectation. When there are numerous fruits on the tree, there is competition for nutrition and water among themselves and as such, the fruits are small in size with poor colour and taste. Moreover, due to heavy crop load, the branches may also break. To prevent limb breakage and ensure good fruit quality, excess fruits must be removed or "thinned." From bloom until about 50 days after bloom (Stage I of fruit growth) fruit growth is the result of cell division. Stage II of fruit growth begins when cell division ceases and is characterized by a lack of visible fruit growth. Stage III of fruit growth begins about six weeks before harvest and is characterized by very rapid fruit growth as the cells fill with water and expand. Fruit thinning anytime before harvest will reduce fruit to fruit competition for water and carbohydrates, but thinning early in the season will result in the greatest increase in fruit size. Bloom thinning with fingers or stiff brushes will remove about 60 per cent of the flowers. Additional follow up thinning is usually required at about 45 days after bloom. Because some flowers may be killed by spring frost, many growers prefer to only partially thin at bloom. At 45 to 50 days after bloom, there are usually two or three different sizes of fruits on the tree. The smaller fruits likely were not fertilized and will fall on their own. The largest fruits at thinning time will be the largest fruits at harvest. A general rule of thumb is to thin fruits so the average distance between fruits is 6 to 8 inches along the shoot, but no two fruits should be closer than 2.5 inches.

Chemical thinning is nowadays preferred over hand thinning which is slow and costly. Thus, a large number of chemicals like DNOC, 3CPA, NAA, GA_3 and thiourea have been used, but ethephon is the most widely used chemical for fruit thinning. Thiourea (40-100 ppm) and 2,4-D and 2,4,5-T (40-100 ppm) are also good fruit thinner for peaches.

INTERCULTURAL OPERATIONS

Training and Pruning

Training: Training and pruning is considered as vital operations to provide strong framework and facilitate easy management practices, penetration of light and aeration. It is primarily decided by the bearing habit of the trees. Peach bears on one-year-old wood. The terminal and lateral shoots are most important for fruiting. The regular pruning not only assures the production of new vegetative growth for regularity of fruiting but also maintains proper fruit size and quality. Peach usually requires moderate to heavy pruning annually. The pruning in initial years of plantation is done to shape the framework of plants, which is termed as training.

Usually, the modified central leader system or the open centre system of training is adopted in peach. Modified central leader system can be followed in areas, which experience plenty of sunshine or in areas of heavy snowfall where risk of limb breakage is more. However, vase or open centre system is recommended when sunlight exposure is a limiting factor. In recent years, open centre system has become more popular in peach because of its high yield efficiency. In this system, 3 or 4 branches arising in opposite direction with wide-angle crouches are selected and headed back and unwanted branches are thinned out. In second and third year, 5 to 7 secondary limbs on main scaffolds are selected and headed back. Diseased, weak and dry shoots should be pruned out on the secondary scaffolds. Another method of training peaches is 'V' shaped *Tatura trellis*, which is becoming popular because of high yield efficiency. In this method the plants are spaced in rows 6 m apart. The plant-to-plant distance within the rows is kept 1.5 to 2.0 m. Two limbs of the tree are trained across the inter row space at an angle of 60° from the horizontal forming 'V' shaped canopy. The permanent wire trellis in 'V' shape supported by steel or concrete posts supports the 'V' shaped scaffolds. The fruiting area is developed on the 2 'V' shaped scaffold branches. The other training systems, which have been found suitable for peaches are Pillar, Belgium Fence, The Kearny Agricultural Center Perpendicular "V" (KAC-V) and Quad "V".

Pruning: Young peach plants are pruned for developing strong framework and afterwards, trees should be pruned every year to produce new growth for regular and good fruiting. Peach trees bear fruits laterally on one-year-old shoots. The best quality fruits are produced on upper 1/3 of the tree canopy. Fruit buds of peach are plump and roundish whereas the leaf buds are small narrow and pointed. There may be 1, 2 or 3 buds at a node depending on variety and vigour. Usually the middle bud is vegetative. In vigorous shoots, more than 30 per cent lateral buds are vegetative particularly in lower portion. All the shoots should be headed back to at least by one half to two third to maintain the balance in fruiting and vegetative growth. Well-grown bearing tree seldom fails to develop enough fruit buds for a heavy crop. Hence, heavy pruning of bearing trees is desirable for not only retaining average crop load but also limiting the fruiting canopy closer to main trunk and primary scaffolds.

Avoidance of Winter Injury

Mid-winter temperatures of -10 to -15°F often kill peach flower buds, and shoots and branches may be injured or killed at -20 to -25°F. There is no control over winter temperatures, but there are several ways to minimize low temperature injury as described below:

1. **Planting of hardy varieties:** Varieties such as Red Haven, Crest Haven, Encore, Harken, and Harcrest are quite winter hardy, but Loring, Topaz, O'Henry, and Rich Lady are marginally hardy.

2. **Pruning in late winter:** Pruning temporarily reduces tree hardiness. Therefore, delay pruning until two or three weeks before bloom.

3. **Painting trunks white:** During winter, the low sun angle increases the trunk temperature. The rapid alternating of heating and cooling during the day and night can cause bark splitting, especially on the south side of the tree. White paint on the trunk reflects the light and heat and minimizes such injury. In November, paint trunks and lower branches with white latex paint. However, use of oil-base paint should be avoided.

Nutrition Management

Among stone fruits, peach is considered as a heavy feeder and thus requires more nitrogenous fertilizers. The nitrogen application should be based on previous years growth and fruiting of the tree. Young trees should make 30-45 cm annual extension growth, whereas, 30 cm of growth is sufficient to maintain a mature bearing tree is good vigour and optimal production. The need for fertilizers or nutrients is affected by several factors, and it is best guided by tissue and soil analysis. The optimal leaf nutrient standards are N: 2.5-3.3 per cent, K: 1.25-3.0 per cent, Mg: 0.25-0.54 per cent and Ca: 1.2-2.5 per cent.

In general, the mature tree of over 6 years requires 60 kg FYM, 500 g N, 250 g P_2O_5 and 700 g K_2O annually, which varies from state to state (Table 16.3).

Table 16.3: Manurial and Fertilizer Recommendations for Peach in India

State	Age of Tree	Farmyard Manure (kg/tree)	N	P_2O_5 (g/tree)	K_2O
Himachal Pradesh	6	60	500	250	700
Uttar Pradesh	10	40	300	500	300
Tamil Nadu	–	20	200	1,000	1,000
Arunachal Pradesh	7	50	350	210	210

Whole quantity of farmyard manure along with P and K is given during December-January. Half of N should be given in spring before flowering and the remaining half a month later if irrigation facilities are available. Under rainfed conditions, N fertilizers should be applied in one lot 15 days before bud break. The manures and nitrogenous fertilizer should always be applied by broadcasting evenly in the tree basins, which should be sufficiently large and should encompass the entire canopy of the tree. It should be sufficiently large and should encompass the entire canopy of the tree. It should be thoroughly mixed in soil by gentle raking. Phosphatic and potassic fertilizers should be applied in trenches of 20-25 cm width and 10-15cm deep made beneath the tree canopy at a distance of 1-2m from the main trunk. The trees should be irrigated lightly immediately after the application of manures and fertilizers.

Peach is very susceptible to Fe deficiency, which can be controlled by foliar application of 0.5-1.0 per cent ferrous sulphate or by soil application of 50-250 g chelated Fe (Fe-EDDTA) at 20-30 spots around the tree in small holes. Trunk injection of 1 per cent ferrous sulphate or ferric citrate is also beneficial in extreme cases.

WATER MANAGEMENT

To get optimum-sized and quality peaches, irrigation is very much essential. Irrigation of peaches is particularly necessary during hot and dry summer (April-June). This season coincides with the rapid vegetative growth and fruit size also increases at a faster rate in peaches culminating into a high demand for water. Moisture stress at this stage may lead to fruit drop, reduced fruit size and poor quality. Therefore, it is essential to irrigate peach orchards at 3-4 days intervals in summer in light soils and after a week in heavy soils. Irrigation should be stopped a week or so before harvesting a particular variety. The critical moisture requirements period for different cultivars is different (Table 16.4). Due to scarcity of water in hills, drip irrigation is recommended.

Table 16.4: Critical Periods of Moisture Requirements in Low-chill Peach Cultivars

Cultivar	Critical period
Prabhat and Partap	March-end –first fortnight of April
Florda Sun	April
Shan-e-Punjab	Mid-April or beginning of May
Florda Red	Beginning of May-mid-June
Khurmani and Sharbati	June

Weed Control

Atrazine (2.9 kg/ha), Turbacil (0.8kg/ha) as pre-emergent and Glyhosate (4.32kg/ha) as post-emergent herbicides are quite effective to control weeds without any phytotoxic effect. In nursery, Oxyflurofen (0.5kg/ha) and Diuron (2kg/ha) are good to control weeds.

PLANT PROTECTION

Major Insect-Pests and their Management

The major insect-pests of peach and their control measures are given below.

Peach Leaf Curl Aphid (*Brachycaudus helichrysi*)

It is the most destructive pest of peach, which causes damage by sucking sap from the the buds and foliage. As a result of extensive feeding, the leaves and buds become distorted, curl upward and drop. It can be effectively checked by spraying methyl demeton 255EC (0.025 per cent) or dimethoate 30 per cent EC (0.03 per cent) or monocrotophos 36$SL (0.04 per cent) 7-10 days before flowering (pink bud stage). Spray of imidacloprid 17.8 per cent EC 0.006 per cent is also very effective. In addition, remove the alternate host plants like golden rod and ageratum from the vicinity of peach orchards.

Green Peach Aphid (*Myzus persicae*)

It is also a destructive pest of peach and nectarine, which feed primarily on the underside of leaves, which causes them to curl, become distorted and yellow, and drop prematurely. Feeding may also occur on flowers and fruit resulting in distortion and drop. When abundant, aphid feeding results in excretion of large amounts of honeydew, which supports the growth of a black sooty fungus that causes spotting of leaves and fruit. This aphid may also serve as a vector for several viral diseases of stone fruits. Green peach aphid is one of several aphids that can transmit plum pox virus. Spray methyl demeton 255EC (0.025 per cent) or dimethoate 30 per cent EC (0.03 per cent) or monocrotophos 36 per cent EC (0.04 per cent) 7-10 days before flowering (pink bud stage) for its control. Spray of imidacloprid 17.8 per cent EC 0.006 per cent is also very effective for its control.

Peach Fruit Fly (*Dacus dorsalis*)

Fruit fly causes considerable damage to peaches. The females lay eggs inside fruits and maggots feed on pulp. As a result, fruit becomes soft, ferments and drops. The incidence reduces yield and quality of the fruit. Neem oil/horticulture tree oil spray can be followed during April-May when flies appear on leaves. Spray of bait consisting of malathion (0.1 per cent)+ sugar/gur (1 per cent) is also very effective for its management. Provision for bait station (25g gur + 10 ml malathion + water) can also be made to attract and kill the adult flies.

Peach Tree Borer (*Synanthedon exitiosa*)

Borers feed on the inner bark of trees, where they may kill the tree by girdling or cause the bark to peel away, exposing the tree to other pests and diseases, thrusting the metal wire into holes to kills the insects, spray of *Bt* preparations *e.g.* dipel, thuricide and Javeline during bloom can be done. Spray with lime-sulphur or with 3 per cent oil emulsion at dormancy can also be adopted, alternatively.

Nematodes

Nematodes affect peaches by adversely affecting the establishment, growth, productivity and longevity of the trees. Several nematodes like root knot nematode (*Meloidogyne* sp.), root lesion nematode (*Pratylenchus* sp.), dagger nematode (*Xiphinema* sp.) are associated with peach. Besides influencing the growth and productivity of trees, nematodes act as vector for transmitting several viral diseases like tobacco ring spot and peach rosette. Several practices are in vogue for the control of nematodes, which include the use of resistant rootstocks (Nemagaurd, Shalin, Nemared), nematicides (furadon, 100-300 g/tree) and soil fumigation *etc.*

Major Diseases and their Management

Important diseases of peach and nectarines are bacterial gummosis, peach leaf curl and brown rot, which are briefly discussed below:

Peach Leaf Curl

Peach leaf curl, caused by a fungus, *Taphrina deformans*, is a common disease of peach and nectarine, which occurs primarily in wet springs. Infected leaves become misshapen, deformed, and necrotic, resulting in premature defoliation with subsequent re-sprouting of new leaves. Some "tolerance" has been observed in varieties like Red Haven, Candor, Clayton, and Frost. Spray with Bordeaux mixture/lime sulphur at leaf fall and bud swell for the management of the disease. In addition, peach curl aphid, which is known vector of this disease, should be kept under control by spraying recommended insecticides.

Bacterial Gummosis

Like other stone fruits, it is the most serious disease of peach orchards. It is caused by a bacterium, *Pseudomonas syringae.* The affected portions like bark, outer sap wood and fruits develop circular to elongated soaked gumming lesions. For its management, spray streptocycline @ 10g/100 L water before the onset of rainy season or alternatively spray Copper oxychloride/Bordeaux mixture @ 0.3 per cent after leaf fall. In addition, cut and burn the affected plant parts immediately to restrict the further spread of the disease.

Brown Rot

The brown rot fungus (*Monilinia fructicola*) causes blossom blight, fruit rot, twig blight, and branch canker. Brown rot of ripening fruit is very common, and it generally occurs as the fruit approaches maturity. The first evidence of fruit infection is the appearance of a small brown spot, frequently originating in a slight wound caused by insect feeding or egg-laying activities. The rotted area rapidly expands and eventually becomes covered with tan-gray fungal fruiting tufts. Fruit rotted by brown rot usually retain their form and usually remain attached to the tree for some time after being completely rotted. Later they may drop off the trees. For its management, remove all rotted fruits after harvest to reduce the amount of fungus over-wintering in orchards. Adequate pruning will increase air circulation, allowing faster drying and fewer fruit infections. Apply fungicide sprays during bloom and as fruit ripens or as suggested for brown rot of plums. Fungicides such as Indar or

Orbit alternated with Topsin-M, Captan or Ziram are fairly effective for brown rot management if applied at frequent intervals.

Bacterial Spot

Bacterial spot is a sporadic, but potentially devastating disease, caused by a bacterium, *Xanthomonas arboricola* (*X. campestris* pv. *pruni*). Its symptoms first appear on leaves as small water-soaked lesions on the underside of leaves, especially along the leaf mid-vein, tip or margins. Lesions become brown to black and generally angular in outline. Often the centers of spots fall out giving the leaf a 'shot-hole' appearance. Lesion margins have a reddish coloration. Severely infected leaves turn yellow and drop prematurely. Infected fruit develop brown to black lesions. Fruit lesions may coalesce and cause the fruit to become pitted and cracked. For its control, grow resistant or tolerant varieties such as Candor, Cresthaven, Earliglo, Encore, Harbelle, Harbinger, Harken, Norman, Ranger, Redskin *etc.* Applying a dormant application of fixed copper or antibiotics such as oxytetracycline can provide some control. Similarly avoid excessive use of nitrogenous fertilizers.

Anthracnose

Anthracnose in peach can be caused by two related fungi, *Colletotrichum gloeosporioides* and *C. acutatum*. Similar species of *Colletotrichum* also causes bitter rot of apple and anthracnose of strawberry. Anthracnose begins as small chlorotic spots on the fruit surface. The spots gradually enlarge and become noticeable as circular, sunken, tan lesions on ripening fruit. The sunken lesions have a glistening or slimy surface, unlike brown rot disease with which it could be confused. The decay continues to enlarge and can be quite extensive in fruits on the tree. When the decay becomes quite advanced, the fruit surface eventually takes on a grayish black colour. Leaf and twig symptoms are generally not seen in anthracnose. For its control, remove the infected fruits from the orchard, remove wild *Prunus* species growing near the orchard and spray fungicides recommended for the control of brown rot.

Phytoplasma Rot

The characteristic symptoms of the disease include yellowing of leaves along with rotting, red spotting, occasional tattering and stunting. Fruits on severely infected trees are shriveled, undersized and drop prematurely. Trees may be partially infected and the symptoms may be exhibiting symptoms on a few branches. Use disease-free bud wood for raising healthy nursery plants to avoid this disease.

Peach Scab

Peach scab is caused by a fungus, *Cladosporium carpophilum*. Its symptoms first appear as small, round, green to black spots on the fruits about six or seven weeks after petal fall. These circular spots later become black and velvet, primarily on the stem end of half-grown to mature fruit. When the disease is severe, the lesions often run together, resulting in fruit cracking or abnormal fruit development. Although the most symptoms of peach scab occur on the fruit, the disease can also affect twigs and leaves. For its control, prune the trees to increase air circulation and facilitate the drying of fruit and foliage. Using fungicides such as captan, benlate, bravo, topsin-M, pristine, and ziram is quite effective.

Rhizopus Rot

Rhizopus rot is most common diseases on peaches and nectarines, which is caused by a fungus, *Rhizopus stolonifer*. This rot appears as large masses of black-grey fungus extending outward from the fruit in a whisker-like effect. Symptoms rarely develop in the orchard unless fruit is left to tree ripen. Because this fungus is a wound parasite, breaks in the skin favour disease development. Fruit picked and shipped to market becomes vulnerable to decay as the fruit ripens and sugar concentrations increase, especially if temperature during shipping is allowed to go above 10°C. One diseased peach in a container can infect many other fruits in a few days. Pristine is registered fungicide for control of *Rhizopus* rot. Apply it the day before harvest for best results.

Postharvest Diseases

In addition to brown rot, and *Rhizopus* rot, several other postharvest diseases such as blue mold rot (*Penicillium expansum*), grey mold (*Botrytis cinerea*), black mold rot (*Aspergillus niger*) and green mold (*Alternaria alternata*) also infest peaches during storage. These diseases occur either due to contamination and infection occurs during fruit growing, harvesting, handling, packing and transportation of the fruits. Hence, it is always advisable to reduce latent infections, bruises and avoid injuries during postharvest handling. In addition, hot water treatment (50°C) for 3 minutes and refrigerated storage of the fruits should be done in time.

Physiological Disorders and their Management

Sun Scald

It is the most destructive disorder of peaches, which causes severe damage to the exposed trunk and main scaffold branches. Painting of exposed trunk and branches with lime paste and shading by wrapping with straw or hay reduces its incidence considerably.

Skin Discoloration

Skin discoloration is characterized as brown and black spots or stripes, which are restricted to the skin. Abrasion damage in combination with heavy metal contamination is pre-disposing factors for skin discolouration. The damaged skin cells, where the anthocyanin/phenolic pigments are located, collapse and their contents react with heavy metals turning their color dark brown/black. Iron, copper and aluminum are the most deleterious contaminants. Foliar nutrient, fungicide and insecticide pre-harvest sprays, which contain the above-mentioned metals in combination with abrasion damage, have the capacity to induce this disorder on peach and nectarine fruit when sprayed close to harvest. For it management, make proper care to reduce fruit abrasion damage, treat fruit gently, avoid long hauling and spray of foliar nutrients containing Fe, Cu, or Al during fruit maturation.

Internal Breakdown

Internal breakdown is distinguished by flesh browning, flesh without juiciness due to leatheriness or mealiness, black pit cavity, flesh translucency, red pigment accumulation, and loss of flavor. Symptoms normally appear after placing fruit at room temperature, while some ripening is occurring, following cold storage. Early season cultivars are least susceptible and late-season cultivars are most susceptible. Storage below 0°C but above the freezing point is recommended to delay occurrence of internal breakdown and extend market life of produce.

Split and Shattered Pits

The split pits/shattered pits are more common in early ripening cultivars such as 'June Gold' and 'Spring Gold'. The reason being that during the pit-hardening phase, the pit gradually loses flexibility and becomes very rigid while the flesh of the fruit is still tightly adhering to the pit. Consequently, the expansion of the fruit flesh creates internal forces pulling out on the pit. If great enough, this force will cause the pit to break in the weakest spot, which is along the suture. Such fruits become unfit for human consumption and fetch low price in the market. This disorder primarily occurs in fruits if rains occur after a long dry spell.

Fruit Splitting

Splitting of fruits generally occurs at dorsal and ventral sides mostly at the time of pit hardening stage. Sometimes gum exudes from the fruit making it unfit for consumption. Splitting and gumming are accentuated during heavy rains after a long dry spell. The exact cause of this problem is still to be determined.

Double Fruit/Cleft Sutures

Double fruit or twin fruits is a problem of minor importance and can be eliminated by hand thinning. Heat and water stress during the summer months increases the development of doubled fruit. Thus to minimize this problem, it is essential to irrigate the orchard adequately during the hot, dry summer months. Another related disorder is the cleft suture that is formed when a double fruit develops in such a way in which one of the fruits is very small. At the stem end, the suture crease is deepened and lengthened forming the cleft suture or a spurred fruit with a cleft suture. This disorder is seen more often on 'Rio Grande', 'June Prince' and 'June Gold'.

MATURITY, HARVESTING AND YIELD

Maturity

Peaches are one of the earliest fruits in the season to reach the market and hence growers have the tendency to pick immature fruits and send them to the market. Such fruits do not fetch good price in the market. Thus, to get premium price and reduce the losses during packaging and transporting, peaches should be harvested at optimum stage of maturity. A large number of maturity indices such as days from full bloom to maturity, calendar date, fruit size, firmness, sense of touch, pit discoloration, freeness of pit, taste, ground colour, sugar, acidity, starch, sugar:

acid ratio *etc.*, have been assessed on different cultivars. The days from full bloom to maturity in different cultivars in India are Alexander: 86 days; July Elberta: 101 days; Babcock: 122 days and Elberta: 127 days. However, the main index is the change of ground colour of the fruits. Thus, the best time for harvesting peaches is when ground colour starts changing from green to pale. The fruit should be firm at the time of harvest in order to stand transportation.

Harvesting

All peach fruits do not mature simultaneously. It is, therefore, advisable to make 3-4 pickings starting with the largest and best coloured fruits. Fruits are harvested by twisting with hand. The peak harvesting period for different peach cultivars in hills of India is mid-May (Shan-e-Punjab) to mid-July (July Elberta and Shimizu Hakuto).

Yield

The peach comes into bearing after two years of planting in the field. The plants bear commercially for about 20 years. Although the yield depends on several factors but full bearing peach tree in conventional plantations amy produce 7-10 tonnes fruits/ha and in Tatura Trellis about 22-25 tonnes fruits/ha.

POSTHARVEST HANDLING

Grading and Packing

Peaches are highly perishable fruits and must be handled carefully to avoid cuts and bruising injury. The picking containers should be padded or lined with cushion materials. Fruits should be immediately pre-cooled before grading and packing. Fruits should be graded into A, B or C grades according to size and quality, and then packed in wooden boxed lined with newspapers or some locally available material. Usually three layer packs containing about 10kg fruits are preferred in the Indian market. Peaches can also be packed in two kg ventilated cardboard boxes of 28cm x 20 cm x 5cm size. These boxes packed in ventilated wooden boxed can be sent to market.

Storage

Peaches have a shorter storage life than most other temperate fruits. The recommended cold storage conditions are 0°-0.3°C and 85-90 per cent relative humidity. In these conditions, free stone peaches and nectarines can be kept for two weeks and cling stone for four weeks. Pre-cooled peaches can be stored for 28-36 days. Peaches are frozen in cold storage at 0.9°C. In controlled atmosphere storage containing 5 per cent CO_2 + 1-2 per cent O_2 at 0°C, peaches can be stored up to 42 days.

17

Pear

The pear is the 2nd most important temperate fruit, which is known for good taste and eating quality. Due to higher productivity and wider adaptability under different agro-climatic conditions, it is grown in temperate and subtropical climatic conditions profitably. However, due to lack of attractive shape and colour and perishable nature of fruits, its area under cultivation is not increasing at a faster rate in our country.

COMPOSITION AND USES

Pears are a good source of dietary fiber and a good source of vitamin C. Most of the vitamin C, as well as the dietary fiber, is contained within the skin of the fruit. Most of the fiber is insoluble, making pears a good laxative. Pears are a concentrated source of phenolic phytonutrients, flavanols (catechin, epicatechin), anthocyanins and carotenoids (β-carotene, lutein, zeaxanthin). Due to richness in antioxidants, pears are also rated as healthy food.

Table 17.1: Nutritive Value of Ripe or Near Fruits per 100 g of Edible Portion

Attribute	Contents	Attribute	Contents
Energy	57 kcal	Vitamin C	4.3 mg
Carbohydrates	15.23 g	Calcium	9 mg
Sugars	9.75 g	Magnesium	7 mg
Dietary fiber	3.1 g	Phosphorus	12 mg
Protein	0.36 g	Potassium	116 mg

Source: USDA Nutrient Database.

Pears are usually eaten raw, however, some processed products such as jam, jelly and cider are also prepared from pears. Pears can also be used as salads and desserts, and making juice.

ORIGIN, HISTORY AND DISTRIBUTION

The region including Asia Minor, Caucasus, Central Asia and Western Himalayas is considered as the primary centre of origin of pear. Some Pomologists believe that it originated from Western China. More than half of the *Pyrus* species are found in Europe, North America and Asia Minor around the Mediterranean Sea, whereas others are native to Asia. The European pear has been selected and improved since prehistoric times, and was cultivated in Europe in 1000 BC. Pears probably came to the new world with the first settlers on the East coast, and spread westward with pioneers. Asian pears were domesticated in China about the same time European pears were in Europe, 3000 years ago. *P. pyrifolia* is native to central and southern China, and probably the first to be domesticated. Almost all the commercial cultivars of the European pear predominate throughout the world except in Japan and China, where oriental pears are grown commercially.

The pear production is generally confined to the temperate zone and is cultivated in all the continents of the world. The exact estimate of area under pear is not known. Major pear producing countries are China, Italy, USA, Argentina, Spain, Canada, Italy, Turkey, South Africa, Japan and India (Table 17.1). It is also grown in countries like Mexico, Brazil, Chile, Russia, Germany, France, Switzerland, Austria, Poland, UK, Australia, New Zealand, and Korea.

The pears are not popular in India, yet its cultivation has picked up in the recent years. In spite of very favourable conditions for pear growing in northern India, its extensive cultivation is rather limited due to poor shelf life and lack of proper transportation facilities. In India, pears are cultivated mainly in Jammu and Kashmir, Himachal Pradesh, Punjab, Arunachal Pradesh, Meghalaya, Mizoram, and Uttarakhand. Low chilling pears are grown in the plains of Punjab, Haryana, Uttar Pradesh and Nilgiri hills of South India.

Table 17.2: Top Ten Pear Producer Countries of the World

Country	Production (Tonnes)	Country	Production (MT)
China	15,945,013	Turkey	386,382
Italy	926,542	South Africa	350,527
United States	876,086	Netherlands	336,000
Argentina	691,270	India	334,774
Spain	502,234	Japan	312,800

TAXONOMY AND BOTANICAL DESCRIPTIONS

Taxonomy

The cultivated pear (*Pyrus communis* L.) belongs to order Rosales, family Rosaceae and sub family Pomoideae with basic chromosome number (N) = 17.

The genus *Pyrus* is composed of about 22 species, which are found in Asia, Europe, and northern Africa. Two species of *Pyrus* are commercially cultivated: European pear (*Pyrus communis* L.) and Asian pear (*Pyrus pyrifolia* L.). *P. communis* is possibly derived from the cross between *P. caucasia* and *P. nivalis* (snow pear). This is the major pear of commerce. Asian pear: *P. pyrifolia* (Burm. f.) Nak. [syn. *P. serotina* L., also called as Japanese or Oriental pear, or 'Nashi'. The commercially important oriental pear cultivars have been derived from the species of *P. pyrifolia*, *P. ussurensis*, and their natural hybrids. *P. pyrifolia* is resistant to fire blight and drought, while the *P. ussuriensis* is considered as winter hardy. Fruit of wild *P. ussuriensis* is astringent, small, and course-textured, so that it was probably hybridized with *P. pyrifolia* prior to domestication. Chinese writings dating from 200-1000 BC describe pear propagation and culture. Asian pears moved from China to Japan, Korea, and Taiwan, where they are cultivated commercially today. Almost all the commercial cultivars of the European pear are descendent of *P. communis*. The ancestor of this species is considered to be *P. communi* var. *pyrester*, *P. communis* var. *caucasia* and *P. nivalis*. *P. nivalis* (snow pear) is mainly utilized for making pear cider in Europe.

Many species are used as rootstocks for European and Asian pears and as ornamental trees. The Manchurian or Ussurian Pear, *Pyrus ussuriensis* (which produces unpalatable fruit) has been crossed with *Pyrus communis* to breed hardier pear cultivars. The Bradford pear (*Pyrus calleryana*) in particular has become widespread in several parts as an ornamental tree, and a blight-resistant rootstock for *Pyrus communis*. The Willow-leaved pear (*Pyrus salicifolia*) is grown for its attractive, slender, densely silvery-hairy leaves

Most of pear varieties are diploids with somatic number 34, and triploids (51) or tetraploids (68) are rare. Major contributors to the present day varieties are *P. communis*, *P. nivalis* and *P. serotina* whereas *P. syriaca*, *P. ussuriensis* and *P. longipes* are minor donors.

Botany

Pear is a medium-sized semi-spreading tree, reaching nearly 15–17 metres height. Some species are shrubby as well. The leaves are alternately arranged, simple, glossy green on some species, densely silvery-hairy in others. Most pears are deciduous, but one or two species in South-East Asia are evergreen. Most are cold-hardy, withstanding temperatures between –25 °C and –40 °C in winter, except for the evergreen species, which only tolerate temperatures down to about –15 °C. The flowers are white, rarely tinted yellow or pink. Like apple, the pear fruit is a pome. The fruit is composed of the receptacle or upper end of the flower-stalk is greatly dilated.

SOIL AND CLIMATIC REQUIREMENTS

Soil

Pear thrives well on a variety of soils but deep, heavy loam, fertile, medium-textured, moisture retentive and well drained soils are the best for its profitable cultivation. Pears are more tolerant to wet conditions but more susceptible to drought. Thus, pear can do better in areas of high water table, poorly aerated or heavy textured soil as compared to any other deciduous fruit. As compared to apple, pear is less tolerant to drought but more tolerant to wet soils than apple. A soil depth of about 180 cm is ideal for proper root growth and fruit production. Pear prefers neutral soil (pH 6.0 to 7.5) because in alkaline soils, iron deficiency may appear. There should not be hard rock or pan within two meters of soil depth, which might restrict proper root growth. Highly fertile soils rich in nitrogen are not suitable for commercial cultivation of pear as such conditions may favour the incidence of pear psylla and fire blight.

Climate

Pear is amongst few fruit crops, which is adaptable to a wide range of climatic conditions. Pears can be grown in a climate ranging from very cold temperate to humid subtropical, and can tolerate as low as −26°C temperature during dormancy and as high as 45°C during growing period. Most of the European varieties require 1,000-1,500 chilling hours during winter to complete their chilling requirement for satisfactory flowering and fruiting. In contrast, the Oriental pears require only 200-500 chilling hours below 7 °C during winters.

Spring frost causes extensive damage to the blossoms, which are killed below 3.3 °C. Hence, lowlands should be avoided for its planting. Similarly, hail-prone areas are also unsuitable for its profitable cultivation because it damages both plants and fruits.

A well-distributed annual precipitation of 100-125 cm is desirable for successful cultivation of pear in rainfed areas. Winter precipitation in the form of snow is important for cultivation of high quality European varieties. In hilly areas, the northern aspect is most suitable for pear cultivation as it provides adequate moisture for plant growth, and fruit development and sufficient winter chilling required for breaking rest period. In North Indian conditions, pear can be successfully grown from foothills to high hills (600 to 2,700 m above mean sea level).

IMPORTANT VARIETIES

The cultivated varieties of pears are categorized in two groups: European pears and oriental pears. European pear (*Pyrus communis*) is native to Europe and its varieties have high chilling requirement. The varieties of this group have fine quality, and are juicy, pyriform in shape, usually pear flavoured, very soft and melting in texture. These varieties have few grit cells in the pulp, and are primarily used for table purposes and not for processing. The Oriental pears (*Pyrus serotina*) are also called Sino-Japanese, which are native of Japan, China and Korea. Oriental pear varieties have low chilling requirement and can be cultivated in warmer areas. These varieties are crisp, juicy, oval shape, having apple like flavour, heavy bearer, hard, grittily and poor in quality. These varieties are more suitable for processing.

The leading varieties of pear, which are commercially grown throughout the world are William Bartlett, Anjou Bosc, Hardy, Comice, Winter Nelis, Seckel, Kieffer and Clapp's Favourite. Other important varieties being grown in different parts of the world are Dr Jule Guyot, Flemish Beauty, and Conference.

The European group of pear varieties cultivated in high hills of India having cooler climate can be categorized into different maturity groups (Table 17.3). Similarly, pear varieties have also been classified on the basis of colour of fruits at the time of maturity (Table 17.4).

Table 17.3: Classification of Pear Varieties on the Basis of Maturity Period

Sl.No.	Group	Maturity Period	Varieties
1.	Extremely early	1st -15th June	Early China
2.	Early	16th –30th June	Basuiodose Favourite
3.	Mid early	1st -15th July	Doyenne Du Commice, Max Red Bartlett, Red Bartlett, Starkrimson
4.	Mid-season	16th-31st July	Baggugosha, Bartlett, Flemish Beauty, Laxton's Superb, Starking Delicious
5.	Late mid-season	1st –15th August	Jargonelle, Monarch, Thumb pear, Smart pear
6.	Late season	16th –31st August	Beurre Hardy, Common pear, Country pear, Kashmir pear, Le Conte
7.	Very late	1st-15th September	Awal number, New pear, Weldon, Shinsui
8.	Extremely late	16th September onwards	Conference

Table 17.4: Classification of Pear Varieties on the Basis of Fruit Colour

Sl.No.	Colour of Fruit	Varieties
1.	Green	Awal Number, Nepolean, Conference
2.	Yellow green	Baggugosa, Bartlett, Beurre Hardy, Seckel
3.	Pale Yellow	Beurre Bosc, Common Pear, Country Pear, Shinsui, Starking Delicious
4.	Green with red blush	Early China, Flemish Beauty, Leconte, Monarch
5.	Yellow green with red blush	Beurre Dial, Jargonelle, Laxton's Superb, Victoria
6.	Yellow with red blush	Doyenne-Du-Comice, Red Bartlett
7.	Red	Max red Bartlett, Starkrimson
8.	Pale brown russetted	Thumb Pear, Yakumo

The varieties for low hills and valley areas with warmer climate belong to Oriental group. These varieties have low winter chilling requirement. The important varieties for these areas are Sand pear (*Pathar Nakh*), Kieffer, China, Gola, Le Conte, Smith, Punjab Beauty, Punjab Nector, Punjab Gola. William Bartlett is the leading variety of pear in Indian conditions. The coloured strains of Bartlertt, *viz.*, Max Red Bartlett, Red Bartlett and Starking Delicious are gaining popularity due to fancy red colour of fruits.

The chief characteristics of important varieties of pear are described briefly hereunder:

☆ **Bartlett:** It is the best known variety of pear throughout the world. The Bartlett pear is unique in that its colour turns from bright green to golden yellow as it ripens. Its creamy, sweet and aromatic flesh is perfect for eating fresh, as well as for canning or adding to salads or desserts. This is commonly referred to as *Williams* pears.

☆ **Red Bartlett:** Red Bartlett pears turn a gorgeous bright red as they ripen, and have a smooth, sweet, and juicy flesh. These delicious pears are used as salads and desserts, and making juice.

☆ **Max Red Bartlett:** Trees are medium in vigour, fruits are large in size, typical pyriform obovate in shape, red blushed skin and mid-season maturity.

☆ **Conference:** Trees are medium to moderately vigourous, fruits are medium in size, pyriform with long neck, brown rusted colour over light green ground and late season maturity.

☆ **Flemish Beauty:** Trees are vigourous, fruits are large in size, obovate to obtuse pyriform in shape, creamy yellow colour skin with slightly red blushed, mid-season maturity and acts as a good pollinizer.

☆ **Laxton's Superb:** Trees are medium in vigour, fruits are medium in size, yellowish green with slight red blush and maturity in mid-season.

☆ **Fertility:** Trees are vigourous and upright, fruits are small to medium in size, pyriform in shape, brown russeted in colour and mid-season maturity.

☆ **Keiffer:**Trees are vigourous, fruits are large in size, pyriform in shape, golden yellow in colour and early season maturity.

☆ **Bosc**: Its fruits are cinnamon brown coloured and have long tapered necks. They have a dense, fragrant, honey-sweet flesh with a smooth texture. It is an excellent variety for eating fresh as well as for cooking.

☆ **Comice:** Its fruits are round in shape with a short neck and stem. They are most often green and sometimes have a red blush in spots. Fruits are succulent and juicy and have a mellow sweetness that makes it an elegant dessert pear which is also delicious when paired with cheese.

☆ **Concorde:** The Concorde pear is known for its tall, elongated necked firm fruits and dense pulp. Its peel is golden-green and often times has golden yellow russeting in spots. Its vanilla-sweet flavour and firm texture is considered good for processing.

☆ **Red Anjou:** These are compact and short-necked pears. Fruits may be green or red coloured, having refreshingly sweet flavour and moist texture similar to their green counterparts.

☆ **Green Anjou:** It is an all-purpose pear whose dense flesh makes it excellent for snacking, cooking, and slicing fresh into salads. Its peel color remains green as it ripens.

☆ **Seckel:** It is the smallest of the commonly eaten pears. Its fruits are olive-green skin with a maroon blush. These pears are known for their crunchy flesh and ultra-sweet flavor, especially used for children's snacks, and pickling.

☆ **Kashmiri Nakh:** Trees are medium in vigour, fruits are small to medium in size, obovate to slightly conical in shape, skin light to dark green which turns to light yellow and mid-season maturity.

☆ **Jargonellae:** Trees are upright with spreading branches, fruits are obtuse pyriform to oblong pyriform in shape, lemon yellow over green base with faded red blush on skin and mid-season maturity.

☆ **Starkrimson:** Its fruits are bright crimson red in colour which brighten as they ripen and more narrow-necked that Anjou pears. It has a smooth flesh, sweet flavour, and a subtle floral aroma, making it perfect choice for snacking, salads, or any fresh use.

☆ **Forelle:** It is an uncommon variety known for its smaller size and its unique yellow-green fruits which are naturally decorated with crimson freckles. It has a crisp texture even when ripe, and is perfect for snacking, cooking, and pairing with wine and cheese.

☆ **Pathernakh:** It is also called sand pear. It is heavy bearing and good keeping quality. Fruits are round in shape and green with prominent dots. The flesh is crisp and juicy. This fruits are tough and firm and can stand transportation very well for long distance without any spoilage. Highly suitable for low hills and plains of North India.

☆ **Baggugosha:** Trees are upright and vigorous and irregular bearing habit. It has a small green, yellow fruit with tapering stem-end. The fruits are sweet and somewhat gritty. It ripens an August.

FLOWERING, POLLINATION AND FRUIT SET

Pear starts flowering at the age of 5-8 years, depending upon the variety, soil and climatic conditions, cultural practices and altitude. Pear bears mostly on spurs and terminal shoots. Flower bud differentiation takes place in July and flowering occurs in the following spring. Usually pear flowers in March-April and fruit setting takes place in April. In general, pear varieties are self-fruitful and do not require pollinizers. However, most of the varieties having higher chilling requirement are cross-pollinated, Hence, provision of pollinizers in the orchard increases their productivity. Thus, by-and-large, cross-pollination is essential for good crop. Most of the pear varieties are cross compatible. China, Bartlett, Comic, Beurre Hardy, Winter Nelis, Flemish Beauty and Kieffer are very good pollinizers for other pear varieties of commerce. Beurre Hardy and Winter Nelis is a good pollinizer for William Bartlett and should be planted in the ratio of 16:1. The pears should be planted in combination of 2-3 varieties for proper cross-pollination. Magness and Waite are pollen sterile varieties for which adequate pollinizers must be planted in the commercial orchards.

PROPAGATION AND ROOTSTOCKS

Pear varieties are usually propagated by grafting them on the seedling rootstocks. Tongue grafting and 'T' budding are most successful propagation methods. Seedling rootstocks of *P. communis* are compatible with all pear cultivars. The seedling rootstocks of *P. pashia* (Kainth or Mahal) and *P. serotina* (Shiara) and root suckers of *Pyrus pyrifolia* are used in India. Several strains Quince (*Cydonia oblonga*) are used as a dwarfing rootstocks in pear. Quince A, Quince B and Quince C are the three rootstocks not fully compatible with pear varieties. Quince A is relatively compatible semi-dwarf clonal rootstock whereas Quince C and BA 29C are dwarf and incompatible with commercial varieties. Therefore, inter-stock of Old Home or Beurre Hardy must be used with clonal rootstocks of quince to overcome the problem of graft incompatibility.

Old Home x Farmingdale (OH x F) series of clonal rootstocks are graft compatible, resistant to fire blight, cold hardy and have high yield potential. Apart from OH x F series, other clonal rootstock series are Oregon series (USA), BP series (South Africa) and D series (Australia). However, clonal rootstocks have not gained popularity as in case of apple. Clonal and seedling rootstocks commonly used throughout the world are given in Table 17.5.

Table 17.5: Common Rootstocks of Pears and their Attributes

Rootstock	Pear Decline	Fire Blight	Cold Hardiness	Tree size (Per cent of Old Home clonal)
Old Home clonal	Resistant	Resistant	Good	100
OH x F	Resistant	Resistant	Good	60-100
Quince	Moderately resistant	Susceptible	Poor	50-60
P. communis seedlings	Moderately resistant	Susceptible	Good	90
P. calleryana seedling	Moderately resistant	Resistant	Poor	90

There are several other rootstocks as well, which are also tolerant to environmental and edaphic stresses. For example, Ore-211, Ore-249, Ore-260, Ore-261 and Ore-264 are tolerant to high temperature, drought, high and low soil pH and are resistant to nematodes.

ORCHARD ESTABLISHMENT

Before planting, it is necessary to prepare the land by carrying out preliminary operations before planning the trees which largely depend on the condition of the land. In general, the land should be leveled properly giving a gentle slope for water drainage. A planting plan is prepared adopting a particular layout system before actual planting. The layout system depends on plant density to be adopted and topography of land. In general, square or rectangular system of planting is followed. However, in hilly areas, contour system is followed. In this system, first row is drawn at the highest elevation and all the trees in a row come at the same elevation. The distance between rows depends on the slope, being closer on the steeper slope. The planting distance depends upon soil fertility, cultivar, rootstock, training system and climate of the region. In hilly areas, the trees on seedling rootstock are planted at a distance of 5 x 5 m but for clonal rootstocks, the distance can be reduced to 3 x 3 m. However, for Baggugosha and Patharnakh, the planting distance may be increased to 6 x 6 m and s x 8 m, respectively. The planting of trees can be done anytime from December to mid February in plains. However, in hills, late fall or early spring are the common planting periods. In regions where winter is mild and soil has enough moisture, late fall plantings is desirable but in the contradictory conditions early spring planting is the best. A pit of 1m x 1m x1m size is dug at such places and filled with a mixture of soil and well rotten farmyard manure or compost and 30g chloropyriphos dust. Irrigation is given after filling pits to settle down the mixture. At the time of planting, a small hole is dug just big enough to accommodate all the roots. Very long roots can be shortened and plant should remain straight in its position when roots are being covered with soil firmly.

CULTURAL OPERATIONS

Training and Pruning

Proper training and pruning of pear trees is essential for the development of strong framework, to maintain vigour and growth, spread the fruiting area uniformly, secure fruits of good size and quality, encourage regular bearing and to provide convenience of pruning, spraying and harvesting. Like other fruit plants of temperate region, pear trees also need proper training and pruning for the development of proper framework and production of good quality fruits. Most of pear varieties are upright in growth habit and need a training system, which would induce spreading habit. Modified central leader system is most appropriate for training the standard trees. Usually 2 to 3 scaffold branches are selected around the trunk every year for 3 to 4 years in a spiral stair case fashion. The central leader is headed back every year during winters and a terminal side branch is allowed to develop as modified central leader in order to check the upright growth and develop spreading habit.

Pear plants moderately pruned during winters every year. Both heading back and thinning out of branches are followed. All the shoots need to be headed back and about ¼ to ½ length should be removed to encourage spreading habit. Crowding branches should be selectively thinned out. Dead, broken or diseased branches should also be removed every year. Since pear is spur bearer, renewal of spur is done after 8 to 10 years in order to encourage new healthy spurs.

Nutrition Management

Pear trees remove heavy amount of nutrients from soil, and application of manures and fertilizers should be done from the beginning itself. However, doses of manures and fertilizers depend on several factors, like soil fertility, soil type, age of the tree, climatic conditions, previous fertilizer use and cultural practices. Thus, the fertilizer schedule should be based on soil and leaf analysis. The optimal leaf nutrient levels in pear should be N 2.0-2.8 per cent, P 0.1-0.2 per cent, K 1.0-2.0 per cent, Mg 0.3-0.5 per cent, Ca 1.5-3.5 per cent, S 125-300 ppm, Fe 100-250 ppm, Mn 20-75 ppm, Zn 15-40 ppm, B 20-50 ppm and Cu 4-10 ppm. Usually one year old plant requires 10 kg FYM, 70 g N, 35 g P_2O_5 and 70 g K_2O which can be increased annually till 10th year and the levels of fertilizers maintained thereafter. Thus, 10-year-old mature tree requires 60-100 kg FYM, 700 g N, 350 g P_2O_5 and 700 g K_2O. FYM, P_2O_5

and K_2O should be applied during December at the time of basin preparation. N is applied usually in February-March in single or in split doses. However, this schedule is different in different states of India (Table 17.6).

Table 17.6: Recommended Doses of Manures and Fertilizers for Full Grown Pear Trees in different States of India

State	FYM (Kg/tree)	Fertilizer (g/tree)		
		N	P	K
Himachal Pradesh	100	700	350	700
Jammu and Kashmir	–	600	130	750
Punjab	50	500	320	900
Uttrakhand	–	500	–	325
Haryana	40	750	600	1500
Arunachal Pradesh	50	350	210	210
Tamil Nadu	40	600	150	300

Pears are sensitive to boron deficiency. Cracking of young fruits and pitting of older fruits are common symptoms of boron deficiency. Two sprays of 0.1 per cent boric acid during April-March at 15 days interval can overcome B deficiency. Similarly, the deficiency of Zn and Fe on young foliage can easily be controlled by spraying 0.4 -0.5 per cent zinc sulphate and ferrous sulphate, respectively during April.

Water Management

Although, pears are grown on rainfed conditions in India, yet for commercial and profitable cultivation, arrangement of irrigation is mandatory. Immediately after planting, a light irrigation is mandatory. Second irrigation is applied after 2-3 days. Subsequent irrigations should be given as and when required. Moisture excess and stress affects colour, composition and keeping quality of fruits. After harvesting in July-August, the trees should be irrigated at 20days intervals up to the end of October. Afterwards, no irrigation is required up to January except when the manures and fertilizers are applied in December.

Intercropping

In the initial years of planting, intercropping is useful in pear orchards. Green gram, *urd, toria* and sunflower can be grown in summer, while wheat, peas, gram and *senji* in winter season may be intercropped in young orchards. However, an additional dose of fertilizers should be given to intercrops. Peach can also be planted as fillers in pear plantations.

Weed Management

Like apple, several weeds grow in pear orchards. For fruitful production, the basins of pear trees should be kept weed free. In the inter basin spaces, permanent sod may be allowed to develop. In areas where moisture conservation is important, plant basins can be mulched with hay, at least 10 cm in thickness, after spring rains and retained throughout the summers. The mulch should be removed before onset of monsoon in order to avoid excessive soil moisture and root suffocation. In cooler areas black polyethylene mulch is better as it conserves moisture and helps in weed control.

Weed Management before Planting

It is easier and cheaper to control perennial weeds before planting the orchard than after, because there is a better selection of treatment options available when the ground is fallow. Established weeds can be controlled either chemically or mechanically. If the weeds are annuals, control them before they set seed by mowing, discing, or using herbicides. Perennial weeds can be mechanically controlled by repeated discing in summer, control them with herbicides, or with a combination of two techniques. Apply glyphosate when the grasses are actively growing and then cultivate them 2 weeks after the herbicide is applied. Many other weeds, like nut sedges, can be effectively controlled by cultivating them with a soil-inverting plough. Such plough buries the underground tubers or nutlets of the weeds deep into the soil profile where they desiccate or rot.

Weed Management in Newly Planted Orchards

In orchards that have received an herbicide treatment, disturb the soil as little as possible once the trees are planted. Glyphosate can be used to suppress nut sedges and perennial broadleaf weeds. Avoid spraying pear foliage or trunks with glyphosate. Regular pre-emergent and post-emergent treatments during the establishment years remove much of the competition from weeds and facilitate irrigation and other cultural practices. If mechanical control is used, additional control measures (hand hoeing or spot treatment with herbicides) will be needed for weeds growing adjacent to the trees that are not controlled with tillage operations.

Weed Management in Established Orchards

If vegetation (either resident vegetation or cover crop) has been maintained in the orchards, it can either be mechanically managed by mowing or chemically managed by applying low rates of a post-emergent herbicide. Within the tree row, pre-emergent and post-emergent application of herbicides is common management tools. For best results, most pre-emergent herbicides need to be sprayed onto the soil just before an irrigation or rainfall so that the water carries the chemical into the soil where the weed seeds are located. Post-emergent herbicides are used on established weeds. They act either by contact or by translocation throughout the plant.

PHYSIOLOGICAL DISORDERS

Hard End

This malady is quite common in oriental pear (*P. cerotina*). Fruits approaching maturity become hard and black over the blossom-end. It may be attributed to unfavourable water relationship between the other plant parts and fruits. The use of European pear as rootstock helps to minimize this problem.

Pink End

In this disorder, pre-mature ripening of the fruit begins with pink colouration near the blossom end. Consequently, core break down (brown heart) and softening occur in affected fruits, which do not ripen properly. This disorder is caused by abnormally cool growing seasons (7.1°C during night and 21°C in day time) preceding harvest. The crop should be harvested as soon as the initial symptoms of this disorder start appearing on the fruit.

Pear Scald

During storage, core breakdown takes place particularly if over mature fruits are stored. The breakdown of tissues takes place in the core area, which may spread to entire fruit. At first, the pulp becomes soft and watery, which later turns brownish and the peel also gets discoloured. On prolonged storage, the fruit surface becomes russeted, brownish and discoloured. This malady is quite common in Conference variety.

PLANT PROTECTION

Major Insect-Pests and their Management

About 70 insect-pests attack pear in different parts of the world. However, the most common pests of pear are as under:

☆ **Pear psylla** (*Pyrylla pyricola*): Psylla is considered as the most serious pest of pear. It causes damage to foliage and fruits by sucking cell sap. In addition, it is known agent for spread of pear decline. These insects secrete honeydew on which sooty mould develops, which causes blackening of the fruits and foliage. As a result of severe attack, the affected foliage and fruit may fall prematurely. This pest can be controlled by restricting vegetative growth of the trees, summer pruning of water sprouts, use of overhead tree sprinklers to wash out the honeydew and spraying fenvelrate or fenoxycarb (0.05 per cent) before flowering initiation.

☆ **Bark eating caterpillar** (*Indarbela quadrinotata*): It is a serious pest in old and neglected orchards. Its larval stage is only damaging, which bore holes into stem and branches. Its presence can be noticed from the excretory pellets coming out from the holes on the trunk or branches. Insertion of insecticide into the holes is very effective for its control, but it is a tedious and cumbersome. Hence, spraying monocrotophos (0.01 per cent) in the affected parts during October is equally effective.

☆ **Pear aphid** (*Toxoptera* sp.): Grayish-brown aphids live and feed on the folds of young leaves and cause extensive damage by sucking sap. The affected leaves curl along the margins and their cupping takes place. It can be controlled by repeated sprays of rogor or metasystox (0.01 per cent), starting from March onwards.

☆ **Fruit fly** (*Bactocera dorsalis*): Nowadays, fruit fly is becoming a serious pest of pear. Adult flies lay egg on the developing fruits. After hatching, maggots bore the fruits and feed on pulp. Such fruits are unfit for human consumption. The menace of fruit fly can be reduced by collecting and destroying the affected fruits, exposing the soil and killing the pupae and spraying malathion (0.05 per cent) or keeping a mixture of malathion + jaggery and eugnol in open containers at 4-5 places in the orchard to kill the adult flies.

Some other pests like codling moth, mites, hairy caterpillar, leaf roller and fruit moth also attack pear, which can be controlled by the insecticides used for the control of other insect-pests suggested above.

Major Diseases and their Management

Several diseases like powdery mildew, pink disease, flyspeck, stem brown, *etc.*, are common to both apple and pear. However, fire blight, leaf spot and pear decline are specific to pear only, which are described briefly hereunder:

☆ **Fire blight:** This is the most serious disease of pear throughout the world, which is caused by a bacterium, *Erwinia amylovora*. The symptoms of this disease first appear on blossoms as small spots, which later spread to shoots, developing fruits and in severe cases to all plant parts. Infected blossoms may turn brown or black and remain attached to the tree, thereby giving a burnt appearance. Its attack is severe if the environmental temperature is above 19°C and relative humidity remains between 80-100 per cent for at least 48 hours. Preventive as well as chemical control is required for its effective management. It is essential to keep the orchard clean and to destroy the affected plant parts. Remove water sprouts regularly and disinfect pruning tools before and after their use. Similarly, avoid Laxton's Superb and Bartlett varieties, which are susceptible and boost to grow resistant varieties like Keiffer, Honeysweet *etc.* In addition, spray, Bordeaux mixture (8:8:100 along with 2-3 per cent tree oil at green tip stage and during dormant season or streptomycin (100 ppm) after rains, throughout the spring and early summer.

☆ **Blossom blight:** This disease is caused by *Pseudomonas syringae*, and is called as *Pseudomonas* blight, bacterial canker or false fir blight as its initial symptoms are almost similar to fire blight. It primarily affects flowers and not other plant parts. Initially black lesions appear on the floral parts, which enlarge and form necrotic lesions. The infected blossoms may wilt soon and drop down. It develops rapidly at lower temperature. Bordeaux mixture spray (4:4:50) at the onset of leaf fall and later at bud burst is quite effective in controlling it.

☆ **Pear scab:** Pear scab is an economically important disease throughout the world and can cause serious losses in susceptible cultivars. It is caused by a fungus, *Venturia pirima*. Although, it is similar to apple scab, however unlike apple scab its pathogen overwinters in imperfect stage on infected twigs in addition to fallen leaves. In addition, it frequently appears on twigs. Most suitable conditions for its spread are a combination of temperature (23.9°C) and continuous wet period for 5 hours or more.

It is better to remove and destroy the infected plant parts and clean the orchard regularly. Follow spray schedule recommended for apple.

☆ **Stony pit**: Stony pit of pear is presumed to be caused by a destructive virus, but the virus has not been isolated. Affected fruits are unsightly and unmarketable. This disease is sometimes referred to as 'dimpling' because of the symptoms observed on fruit.

☆ **Shoot-fruit blight**: It is caused by *Phoma glomerata*. The symptoms appear small, circular and brown spots on bud scars, wounds, twig stubs or in crotches. As these spots enlarge, their centers become sunken with edges raised above. To keep this disease under control, remove the affected parts, especially bark alongwith some healthy part during dormant period and destroy them. Apply Bordeaux paste to affected parts. Spray Bordeaux mixture (2:2:250) or copper oxychloride (0.15 per cent) during January, March, and June.

☆ **Root rot and sap wood rot**: The fungi like, *Polyporus paolustris* and *Ganoderma lucidum* are associated with these diseases. In the initial stage, gummosis of main trunk and limbs occur, and later reddish brown bodies appear at the base of the tree in rainy season. The roots become soft, spongy and whitish. The affected trees may wilt and die soon. Mix bavistin (10 g) + vitavax (5 g) in 50 litres of water and irrigate the trees during April-May and September-October. Avoid injury and piling up of soil near tree trunks and growing intercrops, which require more irrigation during winter.

MATURITY, HARVESTING AND YIELD

Pears should be harvested at right time of maturity. It is a climacteric fruit, which ripens on the tree as well as after harvesting. Fruit size, weight, fruit and pulp colour, TSS, acidity, pressure, days after full bloom (DFFB) *etc.*, are the maturity indices used in harvesting the pear, which varies from variety to variety. Flesh firmness is the most reliable indicator of pear maturity. Firmness in the range of 10-15 lbs as measured by a pressure tester is desirable for most cultivars. Fully mature pear fruits are harvested while still firm and green for distant marketing and processing as the ripe pears are delicate and do not withstand long transportation and have poor shelf life. However, the picking can be delayed till green colour of peel starts turning pale for domestic or short distance markets. European pears are harvested when 'firm mature'.

Yield

Average yield of pears in proper management conditions is about 30-40 tons per hectare.

Harvesting

Pear fruits are picked individually be giving a gentle twist rather than direct pull. Harvesting should be done in 2-3 pickings at 3-4 days intervals rather than single picking.

POSTHARVEST HANDLING

Precooling

Since respiration and breakdown processes in pear fruit are rapid, it is very important to remove the field heat as quickly as possible after harvest. Otherwise, it is difficult to retain fruit quality in storage. Pears after harvesting are cooled to a core temperature of – 0.6 to -1.6°C to remove field heat and arrest ripening. Pre-cooling is not necessary if fruits are to be consumed within a few weeks of harvesting. Hydro-cooling reduces the incidence or shriveling and brown core without affecting weight loss or incidence of rot.

Grading and Packing

Grading of pears is very important to get better returns. Pear fruits are graded as extra large (8 cm diameter), large (7 cm diameter), medium (6.5 cm diameter) and small (5 cm diameter). These grades are also known as extra class, class1, class 11 and class 111. The misshapen, damaged, blemished and scared fruits should be excluded while grading. The wooden, plastic or cardboard boxes are generally used for packing pears. The fruits should be packed in layers. The bottom and top of the containers are properly cushioned with newspaper or dry grass for avoiding compaction and bruises to fruits. The fruits can also be wrapped individually in 10 micron HDPE bags before packing which maintains freshness and improves fruit quality compared with unwrapped fruits. Labeling of boxes indicating grade, cultivar and name of the orchard should be pasted, printed or stamped on the container.

Storage

Pears can be stored at –1 to 0 °C at 80-85 per cent relative humidity for 30 to 45 days. In controlled atmospheric storage, pears can be stored at 2 per cent O_2 and 1 per cent CO_2 level and –1 to 0 °C for 2-8 months depending on variety and controlled atmosphere storage conditions (Table 17.7).

Table 17.7: Recommended Controlled Atmosphere Storage Conditions for some Pear Varieties

Variety	Temperature (°C)	Oxygen (per cent)	Carbon Dioxide (per cent)	Storage Period (months)
Anjou	1	1.5	1.0	8
Bartlett	−0.5	1.5	1.5	2
Blanquilla	0.5	3.0	5.0	8
Comice	0	5.0	5.0	
Conference	−0.5	2-3	0.5	6
Kaiser	0-1.0	4-5	2-3	5
Packman's Triumph	0	5.0	3.0	7

18

Persimmon

INTRODUCTION

Persimmons are the edible fruit of a number of species of trees in the genus *Diospyros*. *Diospyros* is in the family Ebenaceae. The word Diospyros comes from the ancient Greek words 'Dios' and 'pyros'. this means more or less 'divine fruit', though its literal meaning is closer to 'Wheat of Zeus'. It is, however, sufficiently confusing to have given rise to some curious interpretations, such as 'God's pear' and 'Jove's fire'. The most widely cultivated species is the Asian persimmon, *Diospyros kaki*.

COMPOSITION AND USES

Composition

Persimmons contain higher levels of dietary fiber, sodium, potassium, magnesium, calcium, iron and manganese than apples, but lower levels of copper and zinc. They also contain vitamin C and β-carotene (Table 18.1). Persimmon fruits contain phytochemicals, such as catechin and gallocatechin, as well as compounds under preliminary research for potential anti-cancer activity, such as betulinic acid. In a study, a diet supplemented with dried, powdered Triumph persimmons improved lipid metabolism in laboratory rats. Un-ripened persimmons contain the soluble tannin 'shibuol', which, upon contact with a weak acid, polymerizes in the stomach and forms a gluey coagulum, a 'foodball' or phytobezoar that can affix with other stomach matter. These phytobezoars are often very hard and almost woody in consistency and may cause ingestion of un-ripened persimmons.

Table 18.1: Nutritional Value per 100 g of Raw Japanese Persimmon Fruit

Attribute	Contents	Attribute	Contents	Attribute	Contents
Energy	70 kcal	Thiamine (vit. B$_1$)	0.03 mg	Calcium	8 mg
Carbohydrates	18.6 g	Riboflavin (vit. B$_2$)	0.02 mg	Magnesium	9 mg
Dietary fiber	3.6 g	Folate (vit. B$_9$)	8 µg	Phosphorus	17 mg
Fat	0.19 g	Vitamin C	7.5 mg	Potassium	161 mg
Protein	0.58 g	Vitamin E	0.73 mg	Iron	0.15 mg
Sugars	12.53 g	Vitamin K	2.6 µg	Manganese	0.355 mg
β-carotene	253 µg	Choline	7.6 mg	Sodium	1 mg

Source: USDA Nutrient Database.

Uses

Persimmons are eaten fresh, dried, raw, or cooked.When eaten fresh they are usually eaten whole like an apple or cut into quarters, though with some varieties it is best to peel the skin first. One way to consume very ripe persimmons, which can have the texture of pudding, is to remove the top leaf with a paring knife and scoop out the flesh with a spoon. Riper persimmons can also be eaten by removing the top leaf, breaking the fruit in half and eating from the inside out. The flesh ranges from firm to mushy, and the texture is unique. The flesh is very sweet and when firm due to being unripe, possesses an apple-like crunch. American persimmons and *Diospyros digyna* are completely inedible until they are fully ripe. Interestingly, in some countries such as China, Korea, Japan, and Vietnam after harvesting, 'Hachiya' persimmons are prepared using traditional hand-drying techniques, outdoors for two to three weeks. The fruit is then further dried by exposure to heat over several days before being shipped to market. Dried persimmons are eaten as a snack or dessert and used for other culinary purposes. In Taiwan, fruits of astringent varieties are sealed in jars filled with limewater to get rid of bitterness. Slightly hardened in the process, they are sold under the name 'crisp persimmon'.

ORIGIN, HISTORY AND DISTRIBUTION

Japanese persimmon is native to China or North-Eastern Asia extending upto Japan. Other species are not cultivated commercially but have their origin from different countries. Of several species of *Diospyros*, Japanese persimmon is widely grown in China, Korea, Japan, Brazil, Spain and Italy. Among different countries, China is the largest producer of Japanese persimmons in the world (Table 18.2). In India, it is successfully grown at the elevation between 900 to 1500 m above mean sea level in Himalayan ranges.

TAXONOMY AND BOTANICAL DESCRIPTIONS

Taxonomy

Japanese persimmon is botanically *Diospyros kaki* and is a member of family Ebenaceae. The basic chromosome number is 15 and the cultivated varieties are hexaploids with somatic number, 6X= 90. Wild ancestor of persimmon is *D. roxburghii* which grows in the forests from Assam to Indo-China region. *D. lotus* is a temperate Asiatic species cultivated to a small extent and popularly called as Amlook. While there are many species of *Diospyros* that bear fruit inedible to humans. However, some species such as *Diospyros kaki, D. lotus, D. virginiana, D. texana, D. digyna* and *D. discolor* produce edible fruits (Table 18.3). Of these species, *D. lotus, D. virginiana* are used as rootstocks commercially for *Diospyros kaki*.

Table 18.2: Top Ten Persimmon Producing Countries of the World (2011)

Country	Production (MT)	Country	Production (MT)
China	3,259,334	Spain	70,000
Korea	390,820	Italy	50,236
Japan	207,500	Pakistan	2,526
Brazil	154,625	Iran	2,123
Azerbaijan	146,084	Australia	642

Source: FAO Production Year Book-2012

Botany

Persimmon is a dioecious, long-lived deciduous tree growing to 25-30 ft. It has ovate or obovate leaves, that are shiny on top and pubescent beneath. The leaves are borne on pubescent branchlets. The persimmon has ovoid winter bud with 3 outer scales and terminal buds in the shoot is lacking. The fruiting is axillary on the shoots. Persimmon flowers are yellowish white and 0.75 inch long. Male and female flowers are usually borne on separate trees; sometimes perfect or female flowers are found on male trees, and occasionally male flowers on female trees. Male flowers, in groups of 3 in the leaf axils, have 18- 24 stamens in 2 rows. Female flowers, solitary, have a large leaflike calyx, and pale-yellow corolla, 8 undeveloped stamens and oblate or rounded ovary bearing the style and stigma. Perfect flowers are intermediate between the two. Fruits are usually set in clusters which hang on the branches during winter. The fruit is usually capped by the persistent calyx. The colour of ripe fruit ranges from light yellow-orange to dark red-orange depending on the species and variety. In some species, it is black as well. Fruits vary in size from 1.5 to 9 cm in diameter, and in shape. They may be spherical, acorn, or pumpkin-shaped. The calyx becomes easy to remove once the fruit is ripe. Generally, the pulp is bitter and astringent until fully ripe, when it becomes soft, sweet and pleasant, but dark-fleshed types may be non-astringent, crisp, sweet and edible even before full ripening. Like the tomato, persimmons are not popularly considered to be berries, but in terms of botanical morphology the fruit is in fact a berry.

SOIL AND CLIMATIC REQUIREMENTS

Soil

Persimmons tolerate a variety of soils, but like so many plants, they perform best in a deep, well-drained one. Light, sandy soils are not suitable, but it will grow

Table 18.3: Some Species of Diospyros and their Chief Characteristics

Common Name	Scientific Name	Origin	Chief Characteristics
Asian or Japanese persimmon	*Diospyros kaki*	China	Most widely cultivated species, deciduous, with broad, stiff leaves, fruits are sweet, and slightly tangy with a soft but fibrous texture. Numerous cultivars have been selected, but te Japanese cultivar 'Hachiya' is widely grown. It is edible in its crisp firm state, but has its best flavor when allowed to rest and soften slightly after harvest. Triumph, an Israeli variety is also popular because it is seedless, sweet, and can be eaten whole.
Date-plum/Amlook	*Diospyros lotus*	South-West Asia and South-East Europe	Known to the ancient Greeks as 'the fruit of the gods', or often referred to as 'nature's candy'. 'date-plum', referring to the taste of this fruit which is reminiscent of both plums and dates. In India, it is commonly called as Amlook
American persimmon	*Diospyros virginiana*	Native to the eastern United States	Fruits contain higher levels of vitamin C, calcium, iron and potassium than the Japanese Persimmon, traditionally eaten in a special steamed pudding in the Midwest, valuable source of food for white tail deer, because the fruit ripens late and hang on the tree throughout the winters.
Black persimmon	*Diospyros digyna*	Native to Mexico	Its fruit has green skin and white flesh, which turns black when ripe.
Mabolo or Velvet-apple	*Diospyros discolor*	Native to the Philippines/China	Its fruits develop bright red colour on ripening, known as shizi in China and sometimes as Korean mango.
Indian persimmon	*Diospyros peregrina*	Native to coastal West Bengal	A slow growing tree. The fruit is green and turns yellow when ripe. It is relatively small and has an unremarkable flavor and is better known for its medicinal than its culinary uses.
Texas persimmon	*Diospyros texana*	Native to central and west Texas (US), and North-Eastern Mexico.	Fruit is black on the outside and ripens in August. Fruit becomes edible when it turns dark purple or black. Can be eaten fresh or made into pudding or custard.

on many other soil types and is tolerant of heavy clay soils if drainage is not severely impeded. Yield is reduced on heavy alluvial soils due to increased fruit drop. The soil pH for optimum growth is 6.0-6.8.

Climate

Persimmon can be grown in a wide range of subtropical and warm temperate climate. In India, persimmon is grown commercially at an elevation of 900 m to 1500 m. It is sometimes grown as a home garden fruit in cool locations at lower elevations. The trees are deciduous and enter a rest period and complete their dormancy by mid-February in India. Temperature 8-11°C for about 900 h is enough to complete dormancy. The trees, when dormant, can tolerate fairly low minimum temperature. Annual rainfall of at least 30 inches is required for commercial cultivation. It is susceptible to wind damage, and trees should be protected from strong winds. In the spring, the young foliage is easily damaged. In the fall, premature defoliation by wind affects fruit quality and the next year's production. Branches with heavy crop loads may be broken during windy weather. Shading by windbreak trees should be avoided. If persimmon does not receive full sun, weak growth and fruit drop may result.

IMPORTANT VARIETIES

Most of the persimmon varieties produce fruits parthenocarpically. The persimmon flowers are also cross pollinated by wind. The parthenocarpic fruits are more astringent than pollinated fruits. Hachiya and Hyakume are important astringent varieties. Fuyu is non-astringent variety, which can be eaten raw. The astringent varieties are superior, soft and sweet but can be used only when fully ripe. Details of varietal attributes of some important cultivars have been mentioned below:

A. Astringent Varieties

Eureka

Its tree is small but vigorous. It is a drought and frost resistant, precocious and heavy-bearing variety. It produces medium to large oblate fruit, which is puckered at the calyx. Its peel is orange-red and has good fruit quality. It ripens late in the season.

Hachiya

Its trees are vigorous, and upright-spreading. Fruits are large, oblong-conical with glossy and deep orange peel. Flesh is dark yellow, sweet and rich in carbohydrates. Good for drying. It is a mid- season to late variety.

Honan Red

It is a late variety and has tall, upright, and moderately vigorous tree. It is good cropper and produces small, roundish oblate fruit with thin peel. Peel develops distinct orange-red colour. Pulp is very sweet and rich in carbohydrates. It is an excellent variety for fresh eating and drying.

Saijo

It is a cold-hardy and regular bearer variety. Its fruits are small, and elongated. Its fruits have dull-yellow colour on maturity. Flavour is sweet. An excellent variety for drying.

Tamopan

Large, somewhat four-sided fruit, broad-oblate and indented around the middle. Fruit peel is thick, and orange-red in colour. Flesh is light orange, sweet and rich when fully ripe.

Tanenashi

It is an alternate bearer but an early variety. Tree is vigorous, with round canopy, and prolific bearer. It produces medium-sized round-conical fruits. Peel is light yellow or orange in colour, turning orange-red, thick on ripening. Flesh is yellow and sweet.

Triumph

Sold as Sharon fruit after astringency has been chemically removed. Medium-sized, oblate fruits. Ripens in October.

B. Non-astringent Varieties

Fuyu

It is a late ripening variety. Its tree is vigorous, spreading, and productive. It produces medium-large oblate fruit, faintly four-sided. Its peel is deep orange, and pulp is light orange, sweet and mild. Keeps well and an excellent packer and shipper. Most popular non-astringent cultivar in Japan.

Giant Fuyu (Gosho)

It is regular bearer, dwarf variety but bears light crops in some seasons. Its fruits are large, and roundish-oblate in shape. Its peel is reddish orange in colour and very attractive. When fully ripe, fruits have deepest red colour. Flesh quality good, sweeter than Fuyu. It ripens in late-October. It is prone to premature shedding of fruit.

Imoto

It is almost similar to Jiro in plant characteristics as it is considered as a bud sport of Jiro. Its peel is reddish brown. Occasional male flowers and seeds. Ripens in late October to early-November.

Jiro

It is similar to Fuyu. Its fruits are large and have orange-red peel. Flavour and quality excellent. It ripens in late-October to early-November. Usually sold as Fuyu. Most popular non-astringent variety in California (USA).

Izu

It is dwarf variety which ripens early, from the end of September to mid-October. It produces medium-sized fruits of orange colour. Pulp is soft, with a good amount of syrup, of fine texture. Flavor is very good. Not reliably non-astringent. It bears only female flowers.

Maekawajiro

It is bud mutant of Jiro. Medium-sized, rounded fruit, smoother and less indented than Jiro. Peel has rich orange colour. Pulp is sweet and of good quality. It ripens in mid-season. Must be planted with a suitable pollinator to ensure good fruit yield.

Okugosho

Its trees are medium-sized, vigorous, and spreading. It produces medium-sized, round fruits of orange to deep-red colour. Pulp is sweet, of good texture, and flavour. It ripens in early-November. It is a very good pollinizer for other varieties.

Suruga

It is a late ripening cultivar. It produces large fruit of orange-red colour. Pulp is very sweet, and of excellent quality. Difficult to soften on tree (fruit becomes spongy rather than soft). Trees are almost free from alternate bearing. It has been recommended for growing in warmer climates.

C. Pollination Variant Varieties (Astringent when seedless)

Chocolate

Small to medium-sized, oblong-conical fruit. Skin reddish orange. Flesh brown-streaked when pollinated, must be soft-ripe before eating. Ripens late October to early November. Tree large, vigorous, producing many male blossoms. Recommended as a pollinator for pollination variant cultivars such as Hyakume and Zenji Maru.

Gailey

Fruit small, roundish to conical with a rounded apex. Skin dull red, pebbled. Flesh dark, firm, juicy, of fair flavor. Tree small to medium. Bears many male flowers regularly and is an excellent cultivar to plant for cross-pollination. It has attractive autumn foliage and ornamental value.

Hyakume

Fruit large, roundish oblong to roundish oblate. Skin buff-yellow to light orange, marked with rings and veins near the apex. Flesh dark cinnamon when seeded, juicy, of firm texture, nonmelting. Flavor spicy, very good. Non-astringent even while the fruit is still hard. Ripens in mid-season, stores and ships well.

Maru

Small to medium-sized fruit, rounded at the apex. Skin brilliant orange-red, attractive. Flesh dark cinnamon, juicy, sweet and rich, quality excellent. Stores and ships especially well. Tree vigorous and productive.

Nishimura Wase

Fruit medium, round conical to oblate. Orange colour. Mediocre flavour, ripens in September. Bears male flowers.

FLOWERING, POLLINATION AND FRUIT SET

In persimmon, flowering takes place in spring. Persimmons varieties have been divided into two main pollination groups:

- ☆ **Pollination constants:** In these varieties, no change in flesh colour occurs after pollination and seed formation (Fuyu) and,

- ☆ **Pollination variants**: The flesh is light coloured when seedless, and dark reddish brown as a result of pollination and seed formation.

Cultivars are further subdivided according to the type of flower they bear. Most Japanese cultivars are pistillate constants (female only) with sterile stamens (male part) while staminate constants produce mainly male flowers and include the pollinators, Gailey is considered as the best pollinator for Fuyu. Some pistillate constant cultivars, such as Fuyu, can set fruit parthenocarpically (without seed) and still produce good yields regardless of pollination. In Japan, Fuyu has a low set and requires cross-pollination, often by hand, to improve yields. Pollination is desirable to help reduce natural fruit drop and improve fruit quality, especially in non-astringent cultivars. A recommended layout for pollinators is every third tree in every third row. As bees mainly pollinate persimmon flowers, 4 to 5 bee-hives per hectare should be maintained in the orchard during flowering.

ROOTSTOCKS AND PROPAGATION

Persimmon is usually propagated by grafting the scion on seedling rootstocks, using the whip-graft for smaller diameter stocks and the cleft and veneer grafting on the larger stock. Veneer grafting is generally more successful than budding and should be carried out in September when plants have active sap flow. Tongue grafting is also done with a success rate of 60-65 per cent. Green grafting in summer has also been successful in New Zealand. T-budding may be done in spring or early-autumn.

Three rootstock species have been used for propagation of Japanese persimmon. These are *D. kaki*, *D. lotus* and *D. virginiana*. *D. lotus* is more susceptible to crown gall. *D. virginiana* is used as rootstock for Japanese persimmon in Israel and USA. It is particularly well adapted to damp heavy soils and very cold hardy conditions. However, *D. virginiana* often suckers badly and trees on this rootstock are not always uniform in size and vigour. It is also highly susceptible to *Cephalosporium* wilt. For these reasons, *D. kaki* is preferred to *D. lotus* and *D. virginiana* as rootstock for persimmon. In Japan, Saijo produces very uniform seedlings and is extensively used as rootstock. In India, *Diospyros lotus* is used as the rootstock. The fruits of *D. lotus* ripen during late-October. On ripening, the fruits become soft and should be pulped and fermented for about a week. After all the flesh is fermented from the seed, the seeds are washed thoroughly and the floaters are discarded. Improved germination *of D.kaki, D.lotus* and *D. virginiana* seeds has been reported after stratification for 60-90days. In Australia, seed germination of *D. kaki* has been reported to be high when extracted from soft but non-decomposed fruit and planted immediately. The best seed germination is obtained at 28°C. The seeds at this temperature take about 2-3 weeks to germinate. Young seedlings usually take a year to be of a suitable size for grafting.

PLANTING AND ORCHARD ESTABLISHMENT

The plants are planted in autumn in well prepared pits. The planting density for persimmon depends on cultivar, rootstock and soil type. Dwarfing cultivars (Jiro) can be closely planted at 5m × 2.5m (800 trees/ha), semi-dwarf cultivar, Fuyu at 5m x 3m (660 trees/ha) and vigorous cultivars at 6m x 4.5 m (370 trees/ha). Generally, wider spacing is used on deeper and more fertile soils. The plants should be headed back to a well developed lateral bud before planting to encourage the shoot growth as the terminal bud is lacking. Care is necessary when transplanting to the field, because persimmon roots are fragile and easily damaged by drying or rough handling.

INTERCULTURAL OPERATIONS

Training and Pruning

The persimmon trees should be trained to form a low head by heading back at 60 cm above ground level. 4 to 5 shoots with good crotch angle are developed on the stem. Persimmons require only light pruning in initial years of growth and grown up trees do not require pruning except removal of broken or diseased branches. Pruning mature plants is done during the dormant winter months to remove cross-over, diseased, or broken branches. Pruning is also done to remove weak, shaded branches, open the canopy to prevent self-shading, reduce excessively vigorous shoot growth, and regulate crop load.

Persimmon fruit is borne on the current season's branch growth. After three to five years, bracing may be needed to prevent the weight of the fruit from breaking branches. Pruning secondary branches so that bearing shoots are kept close to the main branches may help to avoid a drooping habit and reduce the need for bracing. Varieties which bear heavily (*e.g.*, 'Fuyu') fruit clusters are usually thinned to increase fruit size.

Nutrition Management

Persimmon does not require high fertilizer doses. However, relatively high amount of K and rather low amount of P are required. A large amount of K is transferred to the fruit during its growth; and if K supply is low, fruit growth is reduced. On the other hand, excessively high K content leads to rough skin and lower fruit quality. Rates of fertilizes need to be adjusted according to cultivar, crop load and leaf nutrient levels. Young tees up to three years of age require an application of complete fertilizer (11:4:15) before bud –break and 3-4 lighter follow-up applications of urea at monthly intervals at the peak growing period. Once trees begin cropping, and application of complete fertilizer is applied 4-6 weeks before harvesting. Leaf nutrient levels and sampling procedures for persimmons have also been established. In Japan, leaf samples of persimmon are collected about two months prior to harvest. The youngest fully expended leaves are selected from non-fruiting laterals. Commercial cultivation requires a dose of 10-20-20 N-P-K fertilizer, applied in February or March when new shoots emerge. Excessive nitrogen fertilization will force vegetative growth, so moderate fertilizer applications are desirable.

Magnesium deficiency causes leaf tissues between veins to become gradually pale and change to yellow, and leaf drop ensure. In order to prevent Mg deficiency (soil having a strong acidic reaction), Ca should be added; foliar strays of 2-3 per cent $MgSO_4$ are also very effective. As a consequence of Mn deficiency, leaves show abnormal patterns, assume yellow color and drop early. Foliar stray of $MnSO_4$ (0.3-0.5 per cent) with addition of Ca is very effective.

Furthermore, several persimmon cultivars are susceptible to fruit drop and calyx cavity if N levels are too high when fruits are sizing. Nitrogen levels should be reduced if these problems are evident.

Water Management

Irrigation is considered essential for the successful production of persimmon. Dry periods during fruit growth reduce the size, quality and number of fruits carried to maturity. The early summer period is most important in determining the yield and quality of fruits. High levels of soil moisture are required for better leaf growth and flowering. Moisture deficiency during early summer may increase fruit drop. The water requirements of plants start decreasing during fruit maturity, although good soil water status is still necessary. Moisture stress during this period can cause premature leaf drop, reduction in sugar levels in maturing fruits and increasing susceptibility to sunburn. Line cracking of fruits may also develop due to water stress at this stage. Irrigation on a regular schedule should be applied to maintain uniform levels of moisture in the soil.

Persimmon like other deciduous plants sheds its leaves during winter and enters in dormancy period. Creation of water stress to achieve dormancy is not necessary. Once the harvesting is over, irrigation can be reduced. Before the trees enter into dormancy, irrigation is essential to keep trees healthy during dormancy period. The young plants at the time of planting require watering at 8-10 days intervals.

PLANT PROTECTION

Major Insect-Pests and their Management

Queensland Fruit Fly (*Bactrocera tryroni*)

It is the most serious pest of persimmons. Damage appears as small black spots. In astringent cultivars, the larvae often fail to develop. Early cultivars are particularly prone to attack.

Mealybugs (*Pseudococcus* spp.)

The females cause damage by sucking sap from the calyx of fruit. The presence of this pest can restrict movement of fruit to interstate or overseas markets. Biological control is available using natural predators such as ladybirds and lacewing larvae.

Fruit Spotting Bug (*Amblypelta nitida*)

Due to its damage, there is appearance of black, sunken spots between the calyx and the shoulders of the fruit. The flesh below the spot is dark and severely damaged and fruit often drop.

Vertebrate Pests

Birds and flying foxes can cause serious crop losses to ripening fruit in coastal orchards in some seasons. Permanent netting, with a mesh size of 20–45 mm, has proven to be the most effective long-term control method.

Major Diseases and their Management

Cercospora Leaf Spot (*Cercospora kaki*)

Symptoms appear a small dark brown angular spots on leaves in late summer and autumn. Severe infections can cause early defoliation.

Circular Leaf Spot (*Mycosphaerella* spp.)

Small to large circular spots surrounded by a watermark like halo. Spray of mixture of Carbendazim (0.03 per cent)+Mancozeb (0.25 per cent) for the management of leaf spots.

Crown Gall (*Agrobacterium radiobacter* var. *tumefaciens*)

In this disease, large galls (swellings) develop around the crown with smaller marble-size galls on larger roots. Persimmons are very susceptible to crown gall and it is essential to treat seedlings and nursery trees with a registered inoculant before planting.

Major Physiological Disorders and their Control

Fruit Drop

Fruit drop is one of the most important physiological problems, which may be related to a number of causes including excessive fruit rot, lack of pollination, water stress, excessive nitrogen application and insect damage. The first wave of drop occurs in early-June just after petal fall and continues up to late July. Thereafter, no fruit drop occurs in most of the varieties. But in some varieties, late drop is also noted which is not equivalent to pre-harvest drop of apples and seems to be a unique feature of persimmon. The late drop is affected by the nutrition conditions of trees. Ringing, blossom thinning, and nitrogenous fertilizer applications reduce fruit drop. Moreover, a high negative relation exists between leaf area per fruit and fruit drop. All fruits drop immediately after defoliation under ringed conditions. This shows that fruit drop in persimmon is closely related to the nutrient status of tree.

Calyx Cavity

Calyx cavity is also known as calyx-separation or calyx dehiscence. It can be a serious problem in persimmon. The symptoms of this disorder are a sparse space or cavity that occurs directly beneath the calyx of the fruit. This cavity becomes a habitat for mealy bugs and fungal growth. Some cultivars are more susceptible than others. The incidence of calyx cavity appears to be less on trees which have heavier crop loads and where fruits have been pollinated. Control measures include the avoidance of excessive N and K fertilizers, especially in later spring/summer and close to harvest; thinning early in the season to enhance calyx growth and optimizing pollination to produce more than three seeds/fruit. Sites with deep, fertile and poor-draining soils are likely to encourage this disorder as also areas with high autumn rainfall.

Skin Russeting (Rings)

The symptoms of this disorder are concentric shallow rings around the fruit. It appears damage from thrips during flowering, excessive nitrogen, irregular irrigation or high relative humidity during fruit ripening are possible causes of this disorder. Skin russeting is more prevalent on conical oblate shaped astringent cultivars such as Hachiya.

MATURITY, HARVESTING AND YIELD

Maturity

Fruit needs to be well developed and show the characteristic colour for the cultivar before being harvested. Firm fruit that remains on the tree until it develops a good colour will develop a higher sugar content and have good flavour and consistency after harvest. Immature fruit does not soften evenly after harvest and may remain partly astringent and generally lacking flavour. Sugar levels and colour are a good indication of fruit maturity. Persimmons are considered ready for harvest when they have reached a full orange to orange-red colour with no visible green background (colour maturity charts are useful here) and a sugar level of 14–15 per cent soluble solids (14–15° Brix) as measured using a refractometer. For the cultivar Fuyu a soluble solid of 15° Brix is recommended at harvest. In Japan, the Fuyu cultivar can attain soluble solids of up to 18° Brix. Japanese growers also use colour charts frequently to determine when each cultivar is ready for harvesting.

Harvesting

The best way to harvest persimmons is to clip the fruit from the tree with small secateurs, leaving the calyx and a short stem attached to the fruit. It is possible to snap the fruit from the tree but this requires skill and may injure the fruit.

Yield

The persimmon trees start bearing 4-5 years after planting. However, dwarf and semi-dwarf cultivars start bearing 2-3 years after planting. Mature trees of Fuyu are capable of producing 50kg fruit/plant. Jiro cultivar has recorded over 80 kg plant, whereas in the Hachiya, the yield is over 100kg/plant.

POSTHARVEST HANDLING

Handling

Field heat in fruit should be removed as soon as possible by a cooling system, such as fan-forced or hydro-cooling system, to bring pulp temperature below 20°C before packing. Persimmons are graded according to size, colour and freedom from skin blemish. Fruit showing blemishes from insect damage, wind abrasion, skin russeting or skin puncture damage caused by birds or flying foxes should be sorted out. Any fruit that is poorly or unevenly coloured should be rejected for quality local or export packs. Dipping of fruits in 500 ppm ethephon solution ripens the fruit in 2-3 days. The fruits do not stand long storage for more than one week after ripening. Astringency can also be removed by giving CO_2 treatment to the fruits.

All fruit packed needs to have the calyx and stem intact and be fresh in appearance. Fruit can easily be bruised and is graded and packed mostly by hand. However, fruit graders can be used if soft brushes and sponge rollers are used in them. In some advanced countries, waxing is also done in persimmons. The most popular package for persimmons is a single layer tray 90 mm deep with a plastic insert liner as commonly used for stone fruit.

Storage

Persimmons can be successfully cool stored for up to 3 months at 0°C and 90–95 per cent relative humidity. During storage, the skin colour intensifies and turns darker and, if stored with other fruits with ethylene gas present flesh firmness may decrease. The cultivar Fuyu can be stored for up to 5 to 6 months using controlled atmosphere storage with 5–8 per cent CO_2 and 2–3 per cent O_2 at 0°C. Good results can also be obtained by placing fruit in a 0.06 mm thick low-density polyethylene bag and storing at 0°C. The postharvest life of the fruit can be improved if the fruit has been pollinated.

Drying

Persimmons are dried in Japan and Brazil, and are commonly used in the Oriental diet. Fruits used for drying are harvested when ripe and firm. After being peeled and sun dried for 30-40 days, they are stored at about 65° F and 50 to 60 per cent R.H. Kneading every 4–5 days is necessary to give uniform texture and improve their flavour. During storage or slow drying, a surface covering of sugar crystals gradually appears, and this improves the appearance of the product. Dried persimmons contain a large amount of dextrose and are comparable to dried peaches in food value.

Removing Astringency

When persimmons are picked ripe but still firm, they are sometimes slow to soften and lose astringency. This process may take 2 to 3 weeks at 70° F but can be speeded up by placing the fruit in a freezer for about 24 hours. When the persimmons are removed and thawed, they are both soft and free of astringency, and may be eaten fresh immediately or cooked. Firm, ripe persimmons may be placed with an apple in a plastic bag or fruit ripening bowl. Ethylene gas released by the apple, speeds up the process of softening, and astringency loss in persimmons. Astringency is also removed by chemicals, and such fruits are sold as 'Sharon fruit' in some countries.

19

Plum

INTRODUCTION

Plums are important temperate deciduous fruit crops and ranks next in importance to peaches. This is because of diverse and wider adaptability in different conditions of soil. These are considered as widely adapted stone fruits, domesticated in Europe, Asia and North America. The plum trees are considered hardy and a large number of its species and hybrids have been domesticated world over.

COMPOSITION AND USES

Composition

The plums are rich source of carbohydrates, carotene, vitamin C and minerals such as magnesium, and potassium (Table 19.1).

Table 19.1: Nutritional Value of Raw Plums per 100 g Fruit

Attribute	Contents	Attribute	Contents
Energy	46 kcal	Vitamin C	9.5 mg
Carbohydrates	11.42 g	Vitamin K	6.4 µg
Sugars	9.92 g	Calcium	6 mg
Dietary fiber	1.4 g	Phosphorus	16 mg
Protein	0.7 g	Potassium	157 mg
β-carotene	190 µg	Magnesium	7 mg

Source: USDA Nutrient Database.

Uses

The taste of the plum fruit ranges from sweet to tart; the skin itself may be particularly tart. It is juicy and can be eaten fresh or in dried form (prune) or used in jam-making or other recipes. Plum juice can be fermented into plum wine. In central England, a cider-like alcoholic beverage known as plum jerkum is made from plums. Dried plums (or prunes) are also sweet and juicy and contain several antioxidants. Plums and prunes are known for their laxative effect. This effect has been attributed to various compounds present in the fruits, such as dietary fiber, sorbitol, and isatin. Prunes and prune juice are often used to help regulate the functioning of the digestive system. Dried, salted plums are used as a snack, sometimes known as *saladito* or *salao*. Various flavours of dried plum are available at Chinese grocers and specialty stores worldwide. Pickled plums are another type of preserve available in Asia and international specialty stores.

Like other members of the rose family, plum seeds contain cyanogenic glycosides, including amygdalin, which decomposes into a sugar molecule and hydrogen cyanide gas, which may be hazardous to human health. Prune kernel oil is made from the fleshy inner part of the pit of the plum.

ORIGIN, HISTORY AND DISTRIBUTION

The five different centres of origin for plums have been identified. These are, Europe for European plums (*P. domestica*), Western Asia for Damson plums (*P. insitia*), Western and Central Asia for cherry plum (*P. cerasifera*), China for Japanese plums (*P. salicina*) and North America for American plums (*P. americana*).

It is believed that some varieties of Japanese plum were introduced in Japan during 1500 AD. Kelsey variety was introduced in the the USA in 1870 and then to other parts of Europe. In India, plum was first introduced at Mashobra (Shimla) by Alexander Couth and later European settlers and Missionaries started its cultivation during 1870 in Kullu (H.P.) from where it spread to other parts of India.

Table 19.2: Top 10 Plum Producing Countries of the World (2012)

Country	Production (MT)	Country	Production (MT)
China	5,873,656	United States	281,499
Serbia	581,874	Turkey	268,696
Romania	573,596	Spain	230,877
Chile	293,205	India	199,241
Iran	288,205	Italy	191,989

Source: UN Food and Agriculture Organization.

Europe is the largest producer of plums where it ranks second in importance after apple among temperate fruits. In USA, plum ranks fourth after apple, peach and pear. At present, China is the largest producer of plums in the world. Other plum producing countries of the world are Serbia, Romania, Chile, Iran, USA Turkey, Spain, India, and Italy (Table 19.2). In addition, sizeable amount of plums is also produced by Germany, Russia, Bulgaria, Hungary, France, Poland, Austria, Japan, Mexico, Argentina, Australia, New Zealand, Afghanistan, and Pakistan. In India, the plums are grown in Jammu and Kashmir, Himachal Pradesh, Uttarakhand, Nilgiris, Eastern Himalayan ranges and sub-mountainous Punjab, Mizoram, Tamil Nadu, and Arunachal Pradesh.

TAXONOMY AND BOTANICAL DESCRIPTION

Taxonomy

Plums belong to family Rosaceae, subfamily Prunoideae, genus *Prunus* and sub-genus Prunophora. Most of the plum species belong to section Euprunus, which includes *Punus domestica* (European plums) (6n), *P. insititia* (Damson plums) (6n), *P. cerasifera* (cherry plum, myrobalan plum) (2n, 3n, 4n, 6n), *P. spinosa* (Black thorn sloe) (4n) and *P. salicina* (Japanese plum) (2n, 4n) and *P. ussurensis* (Ussurian plum). However, the American plums (*P. americana*) (2n) belong to section Prunocerasus. The basic chromosome number of plums is 8. The varieties of plums may be diploids, tetraploids and hexaploids with somatic chromosome number of 16, 32 and 48, respectively.

Botany

The plum trees are of medium in size in the plains but in hills these are of small size, attaining a height of 5-8 m or more. The bark of twigs and trunk is of blackish-brown in colour, usually shredding from the old scaffolds. Leaves are thick with glossy dark green above and pale green underneath, oblong in shape and sharply pointed with serrated margins. It flowers in first fortnight of February. The flowers are bronze on small spurs with very close rings. Thus, flowers occur closely and inflorescence is termed as cymose. Plum flowers are either solitary or in umbel and white in colour. Fruit is a drupe (stone) with thin edible exocarp and fleshy mesocarp. Mature plum fruit may have a dusty-white coating that gives them a glaucous appearance. This is an epicuticular wax coating and is known as 'wax bloom'. Dried plum fruits are called dried plums or prunes, although prunes are a distinct type of plum, and may have antedated the fruits now commonly known as plums. Plums come in a wide variety of colors and sizes. Some are much firmer-fleshed than others, and some have yellow, white, green or red flesh, with equally varying skin color. Plum cultivars in use today include:

- ☆ **Damson:** These plums have purple or black skin, green flesh, cling stone, and are astringent.
- ☆ **Green Gage:** Such plums have firm, green flesh and skin even when ripe.
- ☆ **Mirabelle:** This group of plums has dark yellow peel, and are predominantly grown in northeast France.
- ☆ **Satsuma plum:** This group of plums has firm red flesh with a red peel.
- ☆ **Victoria:** Such plums have yellow flesh with a red or mottled peel.
- ☆ **Yellow Gage or Golden plum:** Such plums are similar to greengage, but have yellow peel.

SOIL AND CLIMATIC REQUIREMENTS

Soil

Plums can be grown on a variety of soils. However, for successful and profitable cultivation, it requires deep, fertile and well-drained sandy loam soils. Although European plums prefer heavy and rich clay loam soils, but Japanese plums can even be grown profitably on inferior soils having shallow water table and high pH. For prunes, sandy soils rich in potassium should be preferred. Usually, plums perform well in slightly acidic soils with pH 6.5 to 7, and alkaline or saline soils are undesirable for its profitable cultivation.

Climate

Due to wider variability between species and varieties, plum can be grown successfully cultivated from temperate to subtropical zone. It can thrive well in areas with cold winters as well as hot summers. Plum can thrive in very high rainfall as well as dry areas. Most of its varieties require 800 to 1,000 chilling hours to break winter rest. The European plums require higher winter chilling (800-1,000 hours) as compared to Japanese plums (500-800 hours) for satisfactory bud burst in the spring. Most of the Japanese plum cultivars are resistant to winter cold and thus can be successfully grown in plains of North India. Plums do not require high humid conditions. As plums bloom early in the season, these are prone to spring frost injury. About 50 to 100 cm well distributed annual rainfall is enough for quality plum production; however, it can withstand drought spells better than other stone fruits. Hence, selection of relatively higher sites with good air drainage is always desirable for profitable cultivation of plums.

IMPORTANT VARIETIES

About 2000 varieties of plums belonging to different species are grown world over. However, only about two dozen have become popular. Keeping in view the origin, adaptation and domestication, the plum varieties have been categorized into 3 groups *viz.*, European plum, Japanese plum and American plum.

A. European Plum

This group is primarily cultivated in the USA and Europe. The trees of this group are moderately vigorous with thick leaves, which are glossy dark green above and pale green with pubescence beneath. The leaf margins are saw toothed. Fruits are born on spurs and are of variable size, colour and shape, which may be cling stone or free stone. European plum, varieties are further grouped in following 5 sub groups:

☆ **Prune:** This is a group of plums distinguished from others, as this is the only group, which can be dried with pit intact. This includes high chilling requiring blue-purple free stone varieties with high TSS and sugars. The pulp is also firm and thick. Prunes are very high in iron and vitamins. The popular varieties in this group are French, Sugar, Italian, German, Imperial and Stanley.

☆ **Reine Claude** (Green Gage). This group is considered as a hybrid between *P. domestica* and *P. insititia*. The fruits are more or less round having slight suture, green, yellow or slight red in colour. Pulp is sweet, tender and juicy. Important varieties of this group are Reine Claude, Jefferson, Washington, Imperial Gage and Green Gage.

☆ **Yellow Egg:** It is a small and relatively less important group. Highly suitable for canning as the fruits are small, round and yellow in colour. Yellow Egg and Golden Drop are important varieties of this group.

☆ **Imperatrice:** It is a large group, which includes all blue plums. The fruits are blue in colour having thick bloom, medium size, oval in shape, firm pulp, thick peel and fair in quality. Important varieties of this group are Grand Duke, Diamond, Tragedy and President.

☆ **Lombard:** This group resembles the Imperatrice group except that fruits are purplish-red in colour, usually small in size and low in quality. Important varieties of the group are Lombard, Victoria, Bradshaw and Pond.

B. Japanese Plum

This group is next in importance to large fruited European plums, which is believed to have originated in China. This group requires less winter chilling than European plums. The trees are early blooming and susceptible to spring frost. However, many cultivars of this group are as cold hardy as peach and can be grown in a wide range of conditions. The trees are spreading types but are upright growing in few cultivars. The bark is rough, peach like as compared to smooth grey bark in European plums bear on many budded spurs as well as on one-year-old shoots. The leaves are medium sized and shape pointed. The fruits of this group are quite variable but easily distinguished from other types by large size, oblate to heart shape and bright yellow, red or purplish but never blue colour. The pulp is yellow, amber or red, and firm. The fruit quality is fair to excellent. Under Indian conditions, the Japanese group varieties are predominantly cultivated. Important varieties of this group are Santa Rosa, Meriposa, Beauty, Methley, Burmosa, Red Ace (Florida), Formosa, Kelsey, Red Heart, Elephants Heart, Burbank and Frontier. This group is suitable for both table as well as processing purposes.

C. American Plum

This group is native of North America and includes several species. The fruits do not have commercial values. Hence, this group is primarily used as rootstock or for culinary purposes. The important species of this group are *P. americana* (cold resistant), *P. hortulana* (vigorous, resistant to brown rot, good for processing), *P. munsoniana* (resistant to spring frost and fruit brown rot), *P. besseyi* (dwarf rootstock for stone fruits), *P. maritime* (suitable for beach cultivation, processing) and *P. subcordata* (used for preserves and sauce).

D. Other Plum Types

Besides above-mentioned types of plums, there are some species, which are mainly used in hybridization or a s rootstocks. These species are myrobalan plum

(*P. cerasifera*), mainly used as roostock; simon plum (*P. simony*), used mainly in hybridization, and damson plum (*P. insititia*), which is mainly used for culinary purposes.

Important varieties of plums for different elevations are:

☆ **High hills**

Early: Sweet Early, Methley, Kelsey, Early Transparent Gage

Mid-season: Santa Rosa, Starkign Delicious, Satsuma, Burbank, Elephant's Heart

Late: Meriposa, Frontier, Prunes

☆ **Mid hills**: Beauty, Santa Rosa, Meriposa, Frontier

☆ **Low hills and valley areas**: Alucha Purple, Titron, Alucha Black, Alubukhara, Kala Amritsari

☆ **Dry temperate zone**: Prunes, Local Mansons

☆ **Sub-tropical plains**: Titron, Golden Zardalu, Alubukhara, Sharbati, Labli, Kala Amritsari, Kabul Green Gage, Katraruchak, Zardalu Yellow and Alucha *etc*.

The low chilling varieties, which can be grown up to tropical regions are FLA-85-1, FLA-86-1, FLA-1-2. They are gradually becoming popular in such areas.

Chief characteristics of some of the important varieties are hereunder:

Santa Rosa

Trees are large, vigourous and upright growth, fruits are large in size, round to oblong conic shape with a slight beak like tapering at the base, purplish crimson skin colour, acts as pollinizer and early to mid-season maturity.

Frontier

Trees are upright and vigourous, fruits are large in size, rounded to slightly heart shape at base, red purple skin and yellowish flesh colour, free stone and mid-season maturity.

Red Beauty

Trees are medium to vigourous, fruits are medium in size with globose in shape, bright red skin colour and yellow flesh and very early season maturity.

Kala Amritsari

It is self-fruitful high yielding local cultivar. The trees are vigorous with profuse branching. This is the most preferred cultivar of plains. Its fruits are medium sized, round oblate and depressed at both ends. The peel turns dark purple on ripening. Its pulp is yellow, juicy and little acidic. Its fruits are preferred for making jam and squash.

Methley

Trees are medium to vigourous, fruits are small to medium in size with roundish heart shape, reddish purple maroon skin with dark red flesh, good pollinizer and very early season maturity.

Burbank

Trees are low in vigour and somewhat drooping, fruits are medium in size, bright red mottled skin colour and deep yellow flesh, and early to mid-season maturity.

Satsuma

Trees are upright, medium to vigorous, fruits are medium to large in size with roundish cordate in shape, dark reddish skin, semi free stone and mid-season maturity.

Titron

It is a self-fruitful cultivar but its yield increases if Alucha Early Round is used as a pollinizer. Its trees are smaller than Kala Amritsari. Its fruits are also smaller than Satluj Purple and Kala Amritsari. It is good for table purpose as well as for jam making.

Mariposa

Trees are upright vigourous, large heart in shape, skin mottled maroon over green base, almost free stone and mid to late season maturity.

Beauty

Trees are upright and medium in vigour, fruits are small to medium in size with round shape, skin translucent red over yellow base and early season maturity.

Satluj Purple

It is a self-unfruitful cultivar and requires Kala Amritsari as pollinizer which should be planted in alternate rows for achieving good fruit set. Tree is medium in vigour with upright habit of growth. Fruits are of medium in size, developing crimson colour on ripening. It is an early ripening low yielding cultivar.

Kataruchak

The cultivar originated in village Kataruchak of Gurdaspur district of Punjab. The cultivar is partially self-fruitful. The yield per tree increases if Kataruchak and Kala Amritsari are inter-planted. The fruit sell at a premium price due to the presence of waxy bloom on the epicarp of fruits. Trees are as vigorous and bear bigger sized, heart shaped fruits of purplish colour. It ripens just after Kala Amritsari. Fruits are good for table purpose, and for making jam and squash.

Alubukhara

It is self-unfruitful and need 'Howe' as pollinizer. The tree is upright and spreading. The fruits are of larger size than all other cultivars. Its yield is less than

Kala Amritsari. The eqicarp yellow in colour with reddish spots. Pulp is juicy and sweet.

FLOWERING, POLLINATION AND FRUIT SET

Fruit setting is problem in plums and all commercial cultivars require cross pollination for fruitful production. In European plums, both self-fruitful and self-unfruitful varieties are there. In self-fruitful varieties, usually 30 per cent fruit set occurs, which is considered quite enough for a good crop. However, in self-unfruitful varieties, fruit set is less than 1.5 per cent without the provision of pollinizers. Thus, 10 to 20 per cent inter plantation of pollinizing varieties is required. The self-fruitful varieties can pollinate the self-unfruitful varieties. Important self-fruitful varieties are California Blue, French Damson, German Prune, Giant, Stanlley, Prune, Victoria and Yellow Egg. The self unfruitful varieties are Belgian Purple, Diamond, Grand Duke, Hall, Italian Prune, Jefferson, Imperial Gage, President, Reine Claude, River's Early, Sultan, Tragedy, Transparent and Washington.

Most of the Japanese plum varieties are self-fruitful and do not require cross-pollination, but some varieties are self-unfruitful as well. However, they behave differently under different sets of climatic conditions. For example, Santa Rosa is self-fruitful in some parts of India and the USA but it requires cross-pollination for effective production. Thus, for fruitful production it is advisable to grow pollinizers in both group of plums. Red Roy, Red Rose (Late Santa Rosa), Santa Rosa, Climax, Beauty and Methyl are self-fruitful varieties, and Meriposa, Inca, Eldorado, Formosa, Kelsey, Burbank and Satsuma are self-unfruitful varieties. The varieties like Wickson, Larodo, Santa Rosa, Red Heart, Elephant Heart and Beauty are considered as good pollinizers for both self-fruitful as well as self-unfruitful varieties.

It is interesting to note that European plum may pollinate Japanese plum, although not very effective, but the reciprocal combinations are unfruitful. Among the American plums, *P. americana, P. hortulana* and *P. munsoniana* are self-unfruitful. Surprise and *P. besseyi* can be used as pollinizers for American plum. Similarly, majority of plum varieties grown in subtropics are self-unfruitful and hence require pollination for fruit set, and provision of pollinizers is important for fruitful production.

ROOTSTOCKS AND PROPAGATION

Seedlings of plums, apricots or peach can be used as rootstock for plum varieties. In Indian conditions, wild apricot and wild peach seedlings are considered good rootstock for plums. Some rootstocks selections from different species are Brompton, St. Julien A, Damson (*P. insititia*), Myrobalan (*P. cerasifera*) and Mariana (*P. cerasifera* x *P. munsoniana*). Of several rootstocks, myrobalan (*P. cerasifera*) is most widely used rootstock for European plum throughout the plum growing countries of the world. However, for utilization special character, several other rootstocks are also used (Tables 19.3 and 19.4).

Table 19.3: Important rootstocks for plums with their important characteristics

Sl.No.	Chief Character	Rootstock (s)
1.	Dwarfing	Pixy, St. Julien K, *P. besseyi*
2.	Cold hardiness	Brompton, St. Julien, Mariana 2624, Micronette
3.	Resistance to nematodes	Nemagaurd, Myrobalan 29-C, Mariana 2624 and GF 8-1
4.	Resistance to Bacterial canker	Pixy, Myrobalan B
5.	Resistance to collar rot	Myrobalan 29 C, Mariana 2624, GF 43
6.	Drought tolerance	Mariana 4001, Myrobalan 2-7, GF-677 and GF 557
7.	Resistance to Water logging	Mariana D 1251, Mariana GF 8/1, Damas GF 1869

Chief characteristics of some of the most commonly used plum rootstocks are hereunder:

Table 19.4: Chief Characteristics of some Commonly Used Plum Rootstocks

Nemaguard	Vigorous, resistant to root-knot nematode. Excellent rootstock for well-drained soils.
Lovell	More tolerant of wet soils than Nemaguard. Also more cold hardy. Susceptible to nematodes in sandy soils.
Atlas™*	Extremely vigorous, nematode resistant, productive, increases fruit size. Not suitable for wet soil conditions, delays fruit maturity in some varieties.
Viking™*	Vigorous, promotes precocity in bearing, nematode resistant, productive, increases fruit size, can be used in wet soil conditions.
Titan	Hybrid between almond x Nemaguard, extremely vigorous resistant to root-knot nematode, provide good anchorage to scion cultivar, and tolerant of calcareous soil conditions. However, due to excessive vigourness, it may delay fruit maturity and not suitable wet soil conditions.
Marianna 26-24	Dwarf plant type, shallow root system, much more tolerant to wet soils conditions, resistant to oak-root fungus, and root-knot nematodes.
Myrobalan 29C	Shallow but vigorous root system. Tolerant to wet soils. Immune to root-knot nematodes, and some resistance to oak-root fungus.
Citation	Induces dwarfness in the scion cultivar, highly tolerant of wet soil conditions, highly winter hardy and resistant to root-knot nematodes.
St. Julian "A"	Semi-dwarf, cold hardy, and resist fluctuating spring temperatures.
Hansen 536	Very vigorous, with excellent anchorage and few root suckers. Not suitable for water-logged conditions, highly susceptible to bacterial canker, phytopthora and oak root fungus.

Plum is usually propagated through tongue grafting, which is done at bud break in spring or 'T' budding at the onset of rainy season. The grafting/budding success is more than 90 per cent in plums. The clonal rootstocks of plums are propagated by normal layering or hardwood cuttings.

PLANTING AND ORCHARD ESTABLISHMENT

The plums are usually planted in December-January when the plants are dormant. In general, two-year-old saplings are planted to achieve good success. In flat areas, square, rectangle or hexagonal system of planting can be followed. In hills, contour planting or terrace planting is most desirable. The plants on vigorous seedling stocks can be planted at 5 or 6 meter distance either way in well prepared pits. Size of the pit should be at least 1 x 1 x 1 m. If there is any hard pan or the rock in subsoil, it should be removed while digging pits for proper root development. The grafting point should be kept at least 15-20 cm above the ground level at the time of planting.

INTERCULTURAL OPERATIONS

Training and Pruning

Like other stone fruits, training and pruning is considered as vital operations to provide strong framework and facilitate easy management practices, penetration of light and aeration. In general, modified central leader system of training is followed in plums as in case of apple. However, in mid hills and valley areas open centre system can also be followed. Burbank and spreading type Japanese plum varieties should be trained in open centre system, whereas upright growing varieties like Santa Rosa, Stanley and Wickson can be trained in modified central leader system.

Plums vary greatly in growth and bearing habit and thus trees of even same variety require training and pruning differently. However, in general, plums produce many lateral shoots and water sprouts, which should be removed from time-to-time. In upright growing varieties, heading back is preferred at the desirable points to induce spreading habit.

In general, plums require moderate pruning, but relatively more than apple and less than peach. It is important to encourage 45 to 60 cm of average extension growth in pre-bearing age and 25 to 30 cm in bearing trees, which can be regulated by pruning. Heavy heading back encourages vigorous growth. The undesirable, diseased, broken limbs and water sprouts should be thinned out. The Japanese varieties with the tendency of over bearing should be pruned harder to improve fruit size and quality.

Nutrition Management

Plum trees require fairly good amount of manures and fertilizers for proper growth and production and thus must be manured and fertilized judiciously. The need for fertilizers or nutrients is affected by several factors, and it is best guided by tissue analysis. However, in general, one year old plum plant requires about 50 g N, 25 g P_2O_5 and K_2O which can be given in the form of calcium ammonium nitrate, single super phosphate and muriate of potash which can be increased annually upon the age of 10 years when the plants became mature. The fully grown fruiting tree should be given 300 g N, 400 g P_2O_5 and 200 g K_2O along with about 40-50 kg well rotten FYM (Table 19.5) should be applied during winter along with P_2O_5 and K_2O at the time of basin preparation. Half dose of N fertilizer should be applied in spring before flowering and the rest half dose a month later. Foliar application of nitrogen in the form of urea at 1 per cent is quite effective in plums and 2 to 3 sprays during fruit development stage can improve the fruit size as well as the plant growth.

Table 19.5: Recommended Doses of Manures and Fertilizer for Plum

Age of the Plant (Years)	Farm Yard Manure (kg)	Doses Per Year (g)		
		Urea	Single Super Phosphate	Muriate of Potash
1-2	5-10	50-100	60-120	25-50
3-4	15-20	150-200	180-240	75-100
5-6	25-30	250-300	300-400	125-150
7 and above	40	300	400	200

Water Management

Plums are shallow rooted and fast growing plants hence need adequate supply of moisture during growing period. Interval of irrigation may depend upon many factors such as soil type, climate and age of trees *etc*. Moisture stress during April-May affects the yield and quality of fruits. Hence, irrigations at weekly interval are required during April, May and June. However, no irrigation should be given at full bloom stage and the ripening stage be given to avoid flower and fruit drop. During rainy season, no irrigation is required. The interval may increase to 20 days in September, October and November. No irrigation should be given during December and January months.

Intercropping

In solid block of plums, intercrops can be grown for the first five years. However, if the plum is planted as filler tree, then there is little space left for growing of crops. In such circumstances, intercrops can only be grown for 1-2 years. The legumes like peas, grams, *moong* or vegetables such as cabbage, cauliflower can be grown. Water needs of inter-crops should be such that also favour plum growth and fruiting.

Weed Management

In plum orchard, several monocot and dicot weeds compete with main crop for water and nutrients; In addition, weeds also act as a source of many insect-pests and diseases. Hence, weeds must be controlled well in time. Weeds can be controlled manually or through the use of weedicides. The manual weed control in plum is quite laborious and expensive. Atrazine at the rate of 6 kg/ha at pre-emergence stage followed by grammaxone at the rate of 2 L/ha at post emergence stage of weed growth can effectively control the weeds for 4 to 5 month in plum orchards. Care should be taken that foliage of the tree should not come in contact with the herbicides.

Fruit Thinning

Many plum cultivars, particularly of Japanese plums, have the tendency of over bearing. Thus, removal of some crop helps in improving the size and quality of remaining fruits. Fruit thinning can be done manually, mechanically or by the use of chemicals. Foliar spray of DNOC (0.04-0.08 per cent), ethephon (200 ppm) or carbaryl (1,000 ppm) at full bloom is quite useful in reducing crop load to a satisfactory level.

PLANT PROTECTION

Major Insect-Pests and their Management

Main insect-pests of plum are sanjose scale, blossom thrips, European red mite, plum weevil, plum fruit moth and nematodes. The damage caused by these pests and management practices for their control are described briefly hereunder.

☆ **Sanjose scale:** This is a serious pest of plums causing considerable loss. The nymphs and adults of this pest feed on bark of trees and develop grayish specks. In heavy infestation, entire surface of the tree is covered. It can be controlled by a spraying methyl demeton 255EC (0.025 per cent) or dimethoate 30 per cent EC (0.03 per cent) or monocrotophos 36$SL (0.04 per cent) about 7-10 days before flowering (pink bud stage).

☆ **European red mite** (*Panonychus ulmi*): European red mites feed by sucking the contents out of leaf cells. Such leaf damage reduces tree vitality and can adversely affect fruit size. Leaf injury caused by European red mite begins as a mottling and browning of leaves. Unless populations are very heavy, European red mite does not cause defoliation. Spray of acaricides like dicofol (0.02 per cent is effective to control mites. Several predaceous species feed on European red mite, including lacewings (*Chrysoperla* spp.) and lady beetles (*Hippodamia convergens*), and can be used in the orchards.

☆ **Blossom thrips:** These are small tiny insects, causes direct damage to plants by laying eggs in flower buds and nymphs and adults scrape tissues there. As a result, there is no fruit-setting. For its management, spray methyl demeton 255EC (0.025 per cent) or dimethoate 30 per cent EC (0.03 per cent) or monocrotophos 36$SL (0.04 per cent) 7-10 days before flowering (pink bud stage).

☆ **Plum weevil:** Adult weevils feed on the epidermis of fresh and tender leaves leaving behind only the network of veins, as a result, affected leaves dry up and fall. Higher incidence may even inflict a serious damage in young plantations. Collect infected fruits and destroy them so as to reduce weevil population in the orchard. Also spray dimethoate 30 per cent EC (0.03 per cent) as and when the first attack of insect becomes visible.

☆ **Plum fruit moth** (*Grapholita* (*Cydia*) *funebrana*): The larvae of this insect bore into developing fruits and feed on the pulp. This ultimately leads to fruit drop and rot. Infested fruits are often slightly mis-shapen and ripen early. This should not be confused with pocket plum, a fungal disease affecting plum fruits. The infected trees have many light brown excrement pellets near the plum stone where the caterpillar feeds. Infested fruits tend to ripen first, fruits that ripen later on the tree often have a much lower infestation rate. Plum moth caterpillars can only be controlled on plum with insecticides before they enter the fruits. On trees small enough to be sprayed, the newly-hatched caterpillars can be killed by using deltamethrin or carbaryl (0.1 per cent). Use one of these sprays in mid-June. As a precautionary measure, the fallen fruits should be collected and destroyed.

☆ **Plum scale:** The females of the unarmoured scale look like brown small nodules and are noticed on the current season's growth of the tree. They suck the sap from the plants and excrete honey dew, to which a large number of flies and ants get attracted. Adult females suck the sap from shoots while nymphs thrive mainly on leaves. Heavy attack of the pest may result in poor fruit set or undersized fruits. Miscible spray oil treatment during the month of January is effective in controlling the plum scale. Spray of monocrotophos (0.036 per cent) or quinalphos (0.05 per cent) after fruit harvest is also quite effective.

☆ **Plum sawfly (*Hoplocampa flava*):** Plum sawfly is notorious pest of plum in some parts of the world. Sawfly emerges from the soil during spring and lays eggs on the blossom of plum trees. When the plums develop the little caterpillars eat their way into the centre of the plum and feed off it as the plum develops. Plum sawfly is difficult to control. However, raking of soil around the tree in February-March for exposing the pupae of pest to birds, destruction of fallen fruits, use of a pheromone trap and spray of deltamethrin are some effective ways to control this pest.

In addition, peach leaf curl aphid, defoliating beetles and bark eating caterpillars also cause considerable damage to plums. These pests can be controlled by the measures suggested for other stone fruits like peach and cherry.

Major Diseases and their Management

The diseases infecting plums and causing economic losses are bacterial spot, bacterial canker, brown rot, powdery mildew and gummosis. Some of viral diseases affecting plums are plum pox (sharka), plum line pattern, prune dwarf and stanley plum decline. Few important diseases and their control measures are described briefly hereunder.

☆ **Bacterial canker:** It is a most serious diseases of plums caused by a bacteria, *Xanthomonas pruni* and *Pseudomonas syringae*. There is development of water soaked gumming lesions on twigs and the main stem. Infected trees become chlorotic and wilted. The bacteria enter through leaf stomata, which are particularly susceptible during rains. Clean the wounds and smear with Mashobra paste during dormancy. Before rainy season, spray streptocyclin (0.03 per cent) and after leaf fall spray blitox (0.03 per cent).

☆ **Powdery mildew:** Powdery mildew, caused by the fungus *Podosphaera oxyacanthae*, also affects apricot, sour cherry, and almond. Infected leaves are covered with a powdery white growth, and when severely affected, the leaves are distorted and curled upward. Later, tiny black dots form on the powdery surface. Shoots are also attacked and may be stunted and distorted. The powdery mildew fungus survives the winter on the buds. Infected shoots produce airborne spores that spread the fungus in humid weather. Powdery mildew can be controlled by 2-3 sprays of wettable sulphur (0.3 per cent).

☆ **Bacterial spot:** Two similar diseases *viz.*, bacterial spot and "shot hole," caused by the bacteria *Xanthomonas pruni* and *Pseudomonas syringae* are problems on apricot and plum and may also affect sour cherry and almond. Leaf spots first appear as watersoaked spots on the under-surface of leaves. These spots are somewhat angular and later turn brown to black. The centers of many spots fall out, leaving red margins around the holes. Leaves with many leaf spots turn yellow and drop, causing premature defoliation that reduces fruit size weakens the tree. Spots on fruits are dark brown or reddish-brown; and if infection occurs early, the spots are sunk-en. Twig infections are not usually noticed. Cultural practices such as avoiding planting young susceptible trees near old ones can help reducing spread of disease. Use a balanced fertilizer and avoid excess nitrogen, as this promotes disease development. Copper fungicides (*e.g.* Bordeaux mixture) have been found reducing the disease incidence.

☆ **Brown rot:** It is caused by a fungus, *Monilinia fructicola*, which also attacks other stone fruits as well. There is development of small, circular, brown spots on leaves, flowers and fruits.However, it becomes apparent on fruits during storage. The varieties like Santa Rosa and Wickson are highly susceptible to it. Spray difolatan (0.02 per cent) or captan (0.02 per cent) 3 weeks before harvest.

☆ **Black knot:** Black knot is caused by the fungus, *Apiosporina morbosa*. Black knot is a common and often serious disease of plum and prune trees in the USA. Once established, the disease becomes progressively more severe each year unless control measures are taken. The disease is characterized by elongated, rough black swellings or knots that develop on the woody portions of infected trees. These knots are most common on small twigs and branches but may be found on main scaffold limbs and even the trunk in heavily infected orchards. Knots often start to form near the point of leaf attachment. They are initially green and soft but then turn brown, harden, and finally become black as they expand and age. Mature knots eventually encircle the infected branch. Old knots are sometimes partially covered with a powdery pink or white fungus growth and are often invaded by insect borers. Numerous infections cause trees to lose vigor, bloom poorly and become increasingly unproductive and susceptible to winter injury. The entire tree may gradually weaken and die if the severity of the disease increases and effective control measures are not taken.

Most plum varieties are susceptible to this disease, however Early Italian, Santa Rosa, and Formosa are much less susceptible; and President is apparently resistant to black knot, and must be grown. Avoid new plant-ings near orchards having black knot problem. Similarly, remove all wild plum and cherry trees from nearby areas. Prune affected parts and burn them.

☆ **Plum pocket:** Plum Pocket is becoming increasingly common in some parts of plum growing areas. The disease is caused by the fungus *Taphrina pruni* and affects plum and damson trees. The symptoms are quite unusual and hard to mistake for any other pest or disease. Young fruit begin to appear longer than normal and slightly larger, which normally becomes visible around mid June. The next symptoms are white marks on the skin of affected plums. Thereafter, plums begin to wither and die. Normally only around 50 per cent of the plums are affected. If the fruits are cut open at any stage, no stone is found, just an empty 'pocket' of white flesh. The affected plums wither, turn brown and fall off the tree. For its control, prune all the branches and twigs which look diseased, remove and burn all affected fruits including those which have fallen to the ground and spray dfenoconazole, once in November and again in March. Spray of Bordeaux mixture two times during the year is also effective alternative.

☆ **Silver leaf:** Silver leaf is a fungal disease caused by *Chondrostereum purpureum*. Leaves develop a silvery sheen.Following the appearance of the silvery sheen, affected branches die. It affects cherries, apples and rhododendron. Since the fungus produces most of its infectious spores in autumn and winter, prune susceptible plants in summer. Paint the pruning cuts with suitable wound paint.

Integrated Plant Protection

The integrated pesticides and fungicide spray schedule can effectively control most of plum pests and fungal diseases. And for achieving this, all available prophylactic (indirect) plant protection measures must be applied before direct control measures are used. The decision for the application of direct control measures must be based on economic thresholds, risk assessments and forecasts including those provided by official forecasting services. Priority must be given to natural, cultural, biological, genetic and biotechnological methods of pest, disease and weed control, and the use of agrochemicals must be minimised.

Plant protection products may only be used when justified and the most selective, least toxic, least persistent product, which is as safe as possible to humans and the environment selected. Populations of the main natural enemies of fruit pests must be preserved. At least two main natural enemies (*e.g.* parasites of scales or coccinellids and syrphid predators of aphids) can be identified and preserved. This means plant protection products toxic to them may not be used. *Bacillus thuringiensis* can be used for control of leaf roller and noctuid caterpillars where effective. Phytoseiid predatory mites must be conserved and utilised in integrated mite management. The cultural practice of removal of sources of infestation or infection (*e.g.* scab, canker, brown rot) is required for effective control. The risk of viral disease must be minimised by timely removal of infection sources from orchards and their surroundings. Populations of pests, diseases and weeds must be regularly monitored and recorded. Predominant weed species present, their growth stage, distribution and extent should also be recorded. Wherever an additional control measure is deemed necessary, a biological, genetic or biotechnological control method (*e.g. Bacillus thuringiensis* or pheromone mating disruption for tortricids/ plum fruit moth) should be used if available and effective. *Cydia funebrana* must be monitored using pheromone traps and control measures should only be applied where necessary. The use of selective insecticides such as insect growth regulators or *Bacillus thuringiensis* should be preferred. Alcohol-baited traps must be used for mass-trapping to control *Xyleborus dispar* where necessary.

MATURITY, HARVESTING AND YIELD

Maturity

Plums are very perishable and hence must be harvested at appropriate maturity to get maximum benefits. Although, plums develop best dessert quality on the tree but for long distance marketing, plums should be picked a few days in advance when the fruits are still hard but have attained proper colour in at least 50 per cent of fruit surface. In general, change in fruit surface colour is good maturity index. Similarly, the fruits with 10-12 psi pressure at fruit maturity are fit for harvesting. The use of ground colour, although an imperfect index, is considered the most practical and reliable method for determining minimum maturity as fruit firmness is an excellent indicator for harvesting the fruits at optimum maturity. However, a combination of ground colour and fruit firmness may be better than a single index for fruit maturity. Fruit soluble solids content (SSC) varies significantly among orchards, as well as from tree to tree and, therefore, is not reliable indicator of fruit maturity. Usually Santa Rosa takes about 104 days after full bloom to maturity, whereas Beauty requires only 84 days. With the development of new technologies such as near infrared (NIR), magnetic resonance (MR), light transmittance (LT) and sound detection, it is hoped that an ideal, non-destructive, reliable maturity index for plums will be formulated soon.

Harvesting

For local marketing fruits should be harvested when ripe and firm. For distant markets, fruits are picked when firm but have developed 50 per cent colour on the skin. Plum should be harvested along with pedicels avoiding any injury to the fruit.

Yield

Plum trees come in to bearing 3-5 years after planting and remain productive up to 30-35 years depending upon the management. However, a full bearing plum tree may produce about 35-50 kg fruits.

POSTHARVET HANDLING

Plum fruit is very perishable in nature, hence should be handled with care. The small baskets should be padded with rice trash or grass at the bottom and sides.

Freshly harvested fruits are transferred in these baskets and covered with paper and tied in gunny cloth.

Grading and Packing

The fruits should be graded before packing in basket or wooden boxes. Several pickings are made as the entire fruit on a tree do not ripen at one time. Fruit is borne on spurs also so care should be taken to the save the spurs from breakage, during harvesting. Different grades of fruits are packed in different boxes and labelled. Three standard sizes as per 'Ag-Mark' System of grading plum (Table 19.6) are as follows:

Table 19.6: Grading and Packing Standards for Plum

Grade	Fruit Size (Diameter, cm)	Inner Box Size (cm)	Number of Layers	Number of Fruits per Layer
Special	4.2 and above	36 x 16 x 16	3	28-32
Grade-I	3.6-4.2	36 x 16 x 16	4	38-42
Grade-II	Below 3.6	36 x 16 x 16	4	50-56

For distant markets wooden boxes are preferred over baskets. To save fruits from injury, each layer of fruit is covered with paper strips and newspaper sheet. Finally the lid of the box is nailed. Now-a-days, small sized CFB boxes (3-5 kg) are in vogue in hilly areas. Mother dairy (Delhi) uses plastic punnets for display of plums.

20
Strawberry

INTRODUCTION

The modern cultivated strawberry (*Fragaria x ananassa* Duch.) is one of the most delicious, refreshing and soft fruits of the world. Worldwide it is also the most widely distributed fruit-crop due to its genotypic diversity, highly heterozygous nature and broad range of environmental adaptation. Its plant is cherished in gardens and in commercial fields for its beautiful red fruit that has a tentalizing aroma. Being a rich source of vitamins and minerals coupled with delicate flavour, strawberry has now become an important table-fruit of millions of people around the globe. It is amongst the few fruit crops, which gives quicker and very high returns per unit area on capital investment, as the crop is ready for harvesting within 6 months of planting. However, due to several technological advances in its cultivation like, introduction of day-neutral cultivars, and protected cultivation, strawberries remain available as fresh fruit throughout the year.

At present, strawberry is grown in wide climatic zones, extending to temperate, Mediterranean, sub-tropical and taiga zones. Its cultivation is influenced by the specific regional adaptations due to critical photoperiods and temperature requirements and thus its cultural systems are highly variable. Due to the constant efforts of strawberry breeders, the world-wide interest for strawberry cultivation has boosted its production tremendously, which has resulted in widespread popularity of strawberry in the last 50 years.

COMPOSITION AND USES

Composition

The fresh-ripe fruits of strawberry are a rich source of vitamins and minerals (Table 20.1). Among vitamins, it is a fairly good source of vitamin A (60 IU/100 g of edible portion) and vitamin C (30-120 mg/100 g of edible portion).

Table 20.1: Nutritive Value of Strawberry Fruit*

Nutrient	Share	Nutrient	Share	Nutrient	Share
Water (per cent)	87-90	Organic acids (mg/100 g of edible portion)		Minerals (mg/100 g edible portion)	
Protein (g)	0.25-0.7	Citric acid	420-1240	Potassium	164
Carbohydrates (g)	8.5-9.2	Malic acid	90-675	Calcium	21
Fats (g)	0.2-0.5	Succinic	100	Phosphorus	21
Fibre (per cent)	1.1	Vitamins	Share	Sugars	Share
Total phenolics	60-120 mg	Vit-A	60 IU	Fructose	1.7-3.5
Anthocyanins	50-100 mg	Ascorbic acid	30-120 mg	Glucose	1.4-3.0
Energy	44 calories			Sucrose	0.5-2.5

Uses

Strawberry is also a rich source of pectin (0.55 per cent), available in the form of calcium pectate, which serves as an excellent ingredient for jelly-making. Besides, it is also a rich source of minerals like, potassium, calcium and phosphorus. The total soluble solid (T.S.S.) content of fruit comprises sugars, acids and other substances dissolved in cell-sap. Water constitutes about 90 per cent of fruit. The mature soft fruit contains about 5.0 per cent total sugars and 0.90 per cent to 1.85 per cent acids. Fructose and glucose are the major sugars found in strawberry, with a small proportion of sucrose. Citric acid is the most abundantly found organic acid, followed by malic, succinic and oxalic acids. These acids determine pH, colour stability and inhibit enzyme activity of strawberry-fruit. Strawberry-fruit contains polyphenols like, chlorogenic acid, catechin and coumaric acid. The contents of the phenolic compounds, however decrease as fruit ripens due to the synthesis of anthocyanins.

The fresh-ripe fruits of strawberry are a rich source of vitamins and minerals. Strawberry is also a rich source of pectin (0.55 per cent), available in the form of calcium pectate, which serves as an excellent ingredient for jelly-making citric acid is the most abundantly found organic acid, followed by malic, succinic and oxalic acids. These acids determine pH, colour stability and inhibit enzyme activity of strawberry-fruit. Strawberry-fruit contains polyphenols like, chlorogenic acid, catechin and coumaric acid. Strawberries are known for their characteristic aroma, which is attributed to the presence of ethyl hexanoate, methyl hexanoate, ethyl heptanoate, ethyl propionate, ethyl butanoate, methyl butanoate, furanone and linalool. The red colour of the fruit is mainly due to the presence of an anthocyanin, pelarogonidin 3-onoglucoside, and traces of cyanidin. Fruits are mostly eaten fresh, and are consumed not for the food value but for the flavour. Besides dessert purpose, strawberries are processed into various value-added products, canned strawberry, jam, jelly, ice-cream, conserves, freeze strawberry, wine and other soft drinks. The strawberry jam is famous the world over.

ORIGIN, HISTORY AND DISTRIBUTION

Fragaria is a genus of perennial creeping herbs, found growing in wild in different climatic zones of the world. The cultivated strawberry (*Fragaria x ananassa* Duch.) is a monoecious octoploid hybrid of two largely dioecious octoploid species, *Fragaria chiloensis* L.) Duch. and *Fragaria virginiana* Duch. It is assumed that hybridization between *F. chiloensis* and *F. virginiana* took place spontaneously in Europe in the seventeen century when female plants of *Fragaria chiloensis* of Chilean origin were grown in proximity to male *Fragaria virginiana* plants of North American origin. The history of the strawberry goes back as far as the Romans and perhaps even the Greeks, but because the fruit had never been a staple of agriculture, it is difficult to find its ancient references. However, it is evident from literature that strawberry was in cultivation in Europe by the thirteen century, and then French began to transplant wood strawberry (*F. vesca*) from wilderness to garden. But the plant was considered more for ornamental flowers than for its fruits, although it was grown to some extent for table-purpose.

The first reference of strawberry cultivation dates back to 1368, when a gardener, Jean Dudoy of King Charles V had planted 1,200 plants of strawberry in royal gardens of the Louvre in Paris. In 1375, Chateau de Couvres had planted 4 blocks of strawberry in his garden in Duchess of Burgundy. Doubtless, its culture was crude, but it was highly appreciated by Duchess of Burgundy. England too was an early admirer of strawberry. The successive modifications of the name from "Streowberige:, "Streowberge", "Streberewyse" and "Strawberry" show a long familiarity with it. There are many theories in England about the derivation of the name strawberry. Anglo-Saxons called strawberry as "hayberry" because it ripened at the time, the hay was mown. Another guess is that the name has been derived from the way children strung the berries on straws or grass or hay to sell, a cushion, still practised at present also in different parts of Ireland. A more likely explanation is that the Anglo Saxons used the name strawberry to describe the way the runners strew or strang away from mother plant to find place in which to grow. Commercial strawberry cultivation in the USA had started at the beginning of 18th century, and within 25 years it had gained significant momentum. Most of the European countries had started growing strawberries commercially since 18th century, and of the total production (4,594,539 tones), today nearly 60 per cent of the world's production is only from Europe (Table 20.2). In Canada, the main growing regions are Ontario, British Columbia and Quebec. Strawberry is also grown in Israel, Japan, Turkey, Australia and Newzealand. Poland produces strawberries only for processing. In India, strawberries were first introduced by the IARI regional Horticultural Research Station, Shimla (Himachal Pradesh) in early sixties. But the early efforts to popularize its cultivation in Himachal Pradesh and Uttar Pradesh had received set back on account of the poor adaptability of cultivars, low returns per unit area and lack of technical know-how. However, after the introduction of cultivars like, Tioga, Torrey, Elista, Chandler, Shasta, Douglas, Fairfax, Senga Sengana, and standardization of different agro-techniques, its cultivation has got a boost during the last decade, and now its area and productivity is increasing slowly. At present, it is being grown in Shimla, Solan, Bilaspur, Kangra, Kullu, Palampur (Himachal Pradesh); Dehradun, Saharanpur, Muzaffarnagar, Ghaziabad (Uttar Pradesh); Hoshiarpur, Ludhiana; Jalandhar, Patiala (Punjab); Gurgaon, Hisar, Karnal (Haryana); Bangalore, Coorg (Karnataka); Kodaikanal, Palani hills and Servoy hills (Tamil Nadu); Pune, Mahabalweshwar (Maharashtra) on a small scale.

Table 20.2: Top Ten Strawberry Producing Countries of the World

Country	Production (MT)	Country	Production (MT)
USA	1,312,960	Russia	184,000
Turkey	302,416	Japan	177,300
Spain	262,730	South Korea	171,519
Egypt	240,284	Poland	166,159
Mexico	228,900	Germany	154,418

Source: FAO Production Yearbook-2012.

TAXONOMICAL ANMD BOTANICAL DESCRIPTION

Taxonomy

Fragaria species belongs to family Rosaceae with a basic chromosome number of x = 7. The cultivated strawberry, *Fragaria x ananassa*, is an octoploid, having chromosome number (2n) of 56. In addition to *F. x ananassa*, the genus *Fragaria* includes at least 11 other species, including diploids, tetraploids, octoploids and a hexaploid. Many of these species have been cultivated at one time or the other, and some are still being grown on a limited basis. Out of 11, 4 species *viz. F. chiloensis, F. daltoniana, F. nilgerrensis* and *F. vesca*, have been reported to be grown in different parts of the world. On the basis of the ch romosome number, the different species have been classified into 4 groups. The other species of *Fragaria* like *F. moschata* (the musky-flavoured strawberry), *F. virdis* (the green strawberry), *F. alba* and *F. sylvestris* have also been described. At the end of the fifteenth century, the two cultivated strawberries in the gardens were "wood strawberry" (*F. vesca*), and the "musky–flavoured strawberry (*F. moschata*), both characterized by their small and distinctly flavoured fruits.

Botany

The strawberry plant is a low-creeping perennial herb in which stem is compressed into a resettled crown with 2 mm long inter-nodes. The axillary buds in the leaf nodes of the crown either remain dormant, or develop into branch crowns, or stolons (runners), depending on the prevailing environment. Inflorescence is terminal, and vegetative growth continues by the uppermost axillary bud of the crown, resulting in a sympodial growth habit. Although, it is oftenly referred to as a herbaceous perennial, aging results in lignification of the crown, producing a hard woody tissue. With time, extensive branched crown development produces a structure resembling a highly compacted crown scaffold. However, vegetative and reproductive growths of strawberry are more sensitive to photoperiod and temperature than most of the other fruit-crops. Strawberry has two types of roots, the large primary ones, which originate in crown and small secondary lateral roots that make up the mass of the root system, which arise from the primary roots. There are usually 20 to 35 roots but these can be up to 100 or more primary roots and thousands of small rootlets in a good root system. In general, strawberry roots grow in a year and die the next, during the fruiting season but when all flower clusters are removed from a mature plant, most of its roots do not die at fruiting time. The leaves are arranged in a 2/5 spiral, and each 6th leaf is just above the first, that is for maximum light exposure. In reality, the strawberry crown is very shortened stem of the plant. Sometimes, it may become as much as 2 feets in length and nodes several inches apart (as in sand-dunes of Pacific Coast). During frost injury, the crown is the first to get damaged.

Many investigators have found a certain type of fungus (Endogne) in strawberry roots, and have suggested that it is mycorrhizal in nature *i.e.* the fungus may be furnishing nutrients to the host plant The commonly associated mycorrhizal fungi are *Rhizophagus* and/or *Phycomycetous*.

IMPORTANT VARIETIES

Strawberries are adapted to different climates, moderate, mediteranean, subtropical and can even be grown at high altitudes under tropical climate. Purposeful breeding has played leading role in the development of modern cultivars. Most important varieties of strawberry are: Chandler, Camarosa, Sweet Charlie, Tioga, Torrey, Etna, Fern, Catskill, Cambridge Vigour, Belrubi, Canoga. Gorella, Jewel, Mars, Pajaro, Ofra, Primetime, Premier, Selva, Sparkle, Senga Sengana *etc*. Due to development of area-specific varieties, there has been tremendous increase in strawberry production. Several cultivars have been developed in different countries and were evaluated under local conditions. In general, Senga Sengana is most suited to septentrional conditions while Tioga does well in meridional culture. Nowadays, day-neutral strawberries are becoming favourite in the world because these are not at all affected by the day length, and thus can be grown throughout the year.

Chief characteristics of some of the varieties are as under.

- ☆ **Chandler:** This is a strawberry with bright red color and strong flavor. Chandler strawberries are resistant to damages caused by rain and plants are higly resistant to viral diseases. These strawberry fruits are equally good for raw consumption and processing. Chandler strawberry is very large and each fruit weighs approx.15 to to 20 g. Nutritional value of Chandler strawberry is very high with 12 per cent TSS, 0.85 per cent of acidity, 55.5mg/100g of Vitamin C and 6.1 per cent of sugar.

- ☆ **Tioga:** Tioga is an early variety of strawberry. Tioga strawberries are large and juicy with firm skin. Hence these are suitable for dessert and processing purposes. As far as nutritional values are concerned, Tioga strawberries contain 12.2 per cent TSS, 0.98 - 1 per cent acidity and 6.2 per cent sugar content. Each Tioga strawberry weighs about 9-10 g.

☆ **Torrey**: Torrey strawberry plants are very tough and highly resistant to plant viruses. They produce large good quality fruits which are suitable for both dessert and peocessing purposes. Each fruit weighs approx. 7 g. Torrey strawberries contain 12 per cent TSS, 0.98 - 1 per cent acidity and 6.2 per cent sugar content.

☆ **Fern**: Fern is a day-neutral cultivar of strawberry. It is an early ripening variety. Fern strawberries are large, bright red with excellent flavor and aroma. Fruit skin is firm and fruits are fleshy. Each Fern strawberry weighs around 20-25 g. Fruits contain 11 per cent TSS; 0.88 per cent acidity and sugar content 6 per cent.

☆ **Belrubi**: Belrubi strawberries are large and conical in shape. Each fruit weighs around 15 g. Fruits contain 11.8 per cent TSS; 0.98 per cent acidity and sugar content 6 per cent.

☆ **Selva:** Selva is also a day-neutral variety of strawberry. Selva strawberry plants produce berries even during off-seasons. It produces large fruits with firm skin and good dessert quality which can endure long distance transportaion. Hence these strawberries are best suited for exports. Each individual Selva strawberry weighs around 15 to 20 g and contains TSS 11 per cent, acidity 1 per cent and sugar content 5.5 per cent.

☆ **Pajaro**: Pajaro strawberries are susceptible to rain damage, resistant to plant viruses and good to grow under summer system. Each fruit weighs approx. 7.5g, with TSS content of 12.2 per cent, acidity 0.97 per cent and sugar content of 5.5 per cent.

☆ **Camarosa**: Plant growth is similar to 'Chandler'. Compared to Chandler, leaf colour is distinctly lighter on the underside. Individual leaflets are larger, somewhat longer and narrower than Chandler. Leaves (including petioles) are similar in length to Chandler, but are much broader. It produces fruit early and has good storage properties.

☆ **Winter Down**: Highly productive. Fruits are medium to large in size and moderately resistant to *Botrytis* and anthracnose.

☆ **Festival**: It is a short day strawberry variety. Average height and width of mature plants is 23 and 30 cm, respectively. Average petiole length and diameter is 120 mm and 3.5 mm, respectively. Petioles have medium pubescence. Average length and breadth of terminal leaflets is 78 and 73 mm, respectively. Leaflet margins are crenate and average 21 serrations per terminal leaflet, and 26 serrations per secondary leaflet. The upper leaf surface is dark grey green colour, and the lower leaf surface is a light grey green in colour. Petiole colour is medium yellow green.

☆ **Sweet Charlie**: Plants of Sweet Charlie are smaller and more compact when compared to Camarosa. Leaves are generally slightly cupped, medium to dark green and semi glossy.

Other commercial cultivars of strawberry are Bangalore, Florida 90, Katrain Sweet, Pusa Early Dwarf (especially suitable for North Indian plains), Premier, Jutogh Special, Shimla Delicious, Red Coat, Local Jeolikot, Dilpasand, Blakemore, Olympus, Hood and Shuksan. Olympus, Hood and Shuksan are best suited for ice-cream making, whereas Midway, Midland, Cardinal, Hood, Redchief and Beauty are quite suitable for processing.

FLOWERING, POLLINATION AND FRUIT-SET

Light is the major environmental factor regulating growth and development of *Fragaria* species. The optimal day-length for floral induction in short-day cultivars appears to be between 8 and 11 hours. Most strawberry cultivars produce hermaphrodite flowers and are self-fertile. However, some also produce male or staminate, imperfect, and female or pistillate flowers. Hermaphrodite flowers are self-fertile and pistillate flowers require cross pollination for fruitful production. The presence of the pollinators and pollinizer in the vicinity of the strawberry planting is important for cultivars whether self-fertile or self-sterile. In self-sterile cultivars, fruit-set improves considerably as cross-pollination is required for them. In the absence of a suitable pollinator and/or pollinizer, or insufficient pollination, percentage of malformed or misshapen fruits increase. Honey-bees are chief insect pollinators of the strawberries in different countries. Other pollinators are blow-flies, and bumble. Four strong honey-bee colonies should be placed in strawberry field/ha for proper pollination.

Even under proper management, most of the varieties do not give 100 per cent fruit-set. Increased fruit-set can be obtained with nitrogenous fertilizers, auxins and with insect-pollinators, especially, honey-bees, in the vicinity of the strawberry plantations.

SOIL AND CLIMATIC REQUIREMENTS

Location and Site

Selecting a good site with a suitable soil is the first step in growing strawberries successfully. Factors of prime importance in selecting a site for commercial strawberry plantations are (a) accessibility to markets, (b) transportation facilities, (c) adequate labour availability (d) community interest, and (e) appropriate climate. If strawberries are to be grown for general market, it is usually best to select a site, which is nearer to a town or city. Strawberry can be grown on a variety of soils, ranging from heavy clay to gravelly soils. Strawberry has a fibrous root system and most of its roots are confined to the top 15-20 cm layer of soil, and thus it grows best in light porous soils that are rich in humus. However, in light soils, frequent irrigation is required for proper establishment of runners and to maintain berry-size and quality. In heavy soils, development and penetration of roots of runners is inhibited adversely. Though, strawberry is not sensitive to soil reaction (pH), it however, prefers slight acidic soils with pH of 5.8-6.5. Strawberry suffers badly in poorly drained soils. Waterlogging, even for a limited period, may kill plants. Poorly drained soils also encourage fruit-rot diseases, affect plant growth, fruit yield, fruit quality adversely.

Climate

Cultivated strawberry can be grown profitably under extremely different climates. Among the different climatic factors, temperature, and day-length affect considerably the growth and yield of strawberry. However, a precise control on light and temperature is must.

On the basis of photoperiodic requirement for floral induction, strawberry cultivars have been categorized as short-day (SD), long-day (LD) or day-neutral (DN). In short-day cultivars, floral induction occurs with photoperiods of less than 14 hours, though these cultivars flower continuously regardless of day-length but environmental temperature should be less than 16°C. The day-neutral cultivars, usually called as everbearers, flower continuously, regardless of day-length, although temperature also modifies photoperiodic response but day-neutral cultivars are sensitive to lesser extent to high temperature than short-day cultivars. The third group is between short day and day-neutrals in which flowering takes place under long-day conditions.

PLANT PROPAGATION

Strawberry can be propagated through sexual (seed) and asexual (vegetative) means. However, propagation by seed is not considered as a viable method of propagation for cultivated strawberry as they do not come true-to-type. Thus, strawberry is usually propagated through runners. Nowadays, a large-scale

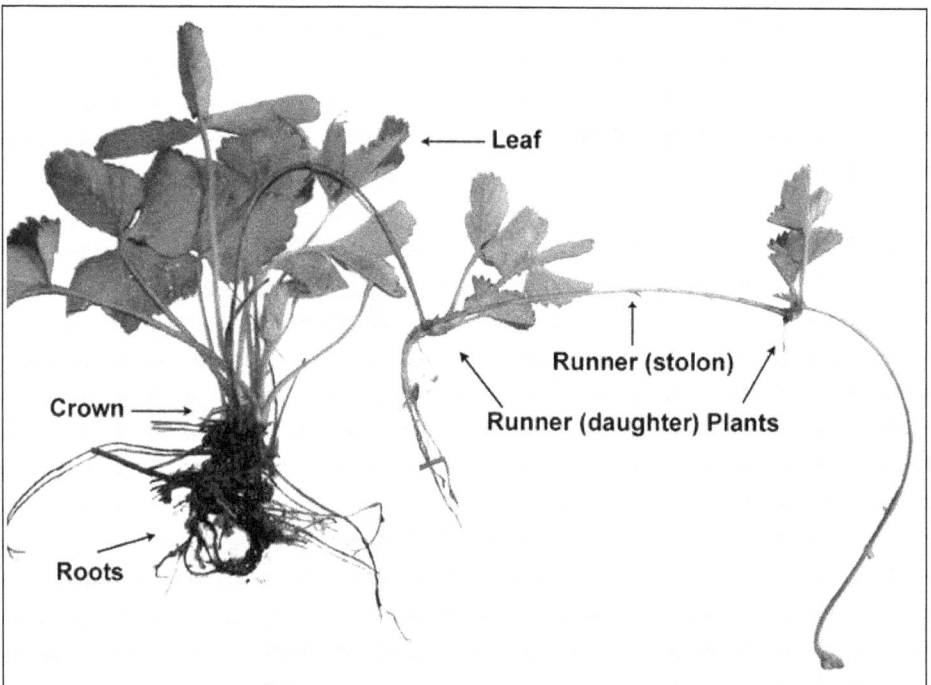

Figure 20.1: Runner Production in Strawberry.

commercial propagation by tissue culture has also been used widely in strawberry industry.

Vegetative Propagation

The stolon, a creeping stalk, is produced in the leaf axil of the plant and grows-out from the parent plant during summer. At the second node, a runner plant is formed and a new stolon arises on the runner plant to continue runner train that often branches. Initially, runner plant produces fewer roots but thereafter makes excessive fibrous roots. Thus, the plant naturally propagates itself by vegetative method of runner production. After the runner plant has acquired sufficient growth and roots, it is separated from mother-plant, and be planted elsewhere.

Micropropagation

Strawberry is perhaps the first fruit-crop in which micropropagation technique has been standardized, and now its large-scale commercial propagation is being done through tissue culture. This fruit-plant requires very high number of plants/ha (50-60 thousand) and such a high demand can be easily met through micropropagation.

PLANTING AND ORCHARD ESTABLISHMENT

Planting

Planting can be done either on furrows or on raised beds in a particular system of planting. In India, raised beds are usually preferred. The planting may be done by hand or by machine but care should be taken to prevent damage and drying of roots of runners. After removing runners from the soil, they may be planted directly in the field or placed in a some shadow before transplanting. It is better if runners are kept in moist sphagnum moss grass. It will prevent drying of roots and can keep runners in a fresh condition for longer period. Planting should be done in early or late evening hours. It should, however, be avoided in sunny day and under wet or frosty conditions. Immediately after planting, a light irrigation is essential.

Systems of Planting

Three planting systems are, in general, used in strawberry; (a) the hill, (b) the spaced row, and (c) the matted row system. Further, single hedge, double-hedge row, single row or double row hill are also followed and in backyard plantings, barrel or pyramids can be used.

1. **Hill system**: This system of planting is commonly followed in cultivars, which produce few runners. In this system, plants are grown either in single or double rows on raised beds. The beds are usually made 15-20 cm high and runners are set 20-25 cm apart in twin rows, 30-35 cm apart and distance of 90-120 cm is kept between twin rows. With this system, a small garden tractor or a field tractor can be used for tillage; greatly reducing the expense of hand labour. In home garden, where hand labour is used, rows can be spaced closer, or about 45-50 cm apart. Sometimes, twin rows are set 40-60 cm apart, then a wide space is left and another two rows are set. In some cases, triple rows are set. However, these are only

modifications of hill system with the plants set at the same distance apart in the row in each case. This system demands a much higher economic input and a large number of runners are set per unit area. As a result of crowding, removing of runners become difficult in this system.

2. **Spaced row system**: This system of planting is used for varieties, which are moderate to weak in producing runners and thereby daughter plants. In this, runners are usually set at 30-50 cm apart in rows with a spacing of 90-100 cm between rows. The rows may be wider than 90 cm in rich soils. It helps in easy-picking of fruits by moving between rows easily.

3. **Matted row system**: This system is adopted in areas where crown injury occurs in winter either due to freezing or temperature fluctuations. Runners are planted along the row at about 90 cm apart and at a space of 45 cm between plants. The width of mats is maintained between 40-45 cm. In this, more number of plants can be accommodated per unit area, which may give higher yield under suitable conditions.

Although planting can be done at any time between July to April but early planting ensures good crop. In temperate humid climate, planting is usually done in spring. If weather is undesirable, runners can be stored in poly-bags at 0°C till conditions become favourable. Planting during July-August, however needs adequate care, particularly from strong sunshine or dry soil conditions. During this period, irrigation should also be given frequently otherwise leaves may desiccate and defoliate, which may result in delayed fruiting. However, planting in October and November usually does not require much special care for establishment. The planting season vary widely with elevation of planting site and nature of cultivars. For example, Chandler is sown during March under hilly tracts (Shimla) of India and the best time for it under subtropical zones is the last week of October or 1st week of November

INTERCULTURAL OPERATIONS

Nutrition Management

Practically, all strawberry plantings benefit from manures and fertilizers application, but fertillizers are not, as assumed, a cure-all, because several other factors like, fertility and physical conditions of soil, careful planting, thrifty disease tested plants, suitable cultivars, adequate moisture, the quantity, quality and time of applying mulch, and weed and insect-pest control, may be as important as the application of fertilizers. Certain cultivars are inherently more productive than others in a given region. Fertilizers do not normally contribute to cause yield of a low-yielding or a disease-susceptible variety to equal to a high-yielding and disease-resistant cultivar. Similarly, fertilizer will not make a low chilling cultivar to succeed in warm areas and *vice versa*.

Water Management

Due to an evergreen, semi-herbaceous growth habit, rapid growth rates, high fruit productivity and relatively shallow root system, growth and development of strawberry plant are sensitive to variation in soil moisture. Strawberry responds to irrigation in a more profitable manner than any other fruit crop. Irrigation is useful in utilizing nutrients in the soil, preventing blossoms from frost damage, hastening growth of runners and increasing bearing size and yield.

Strawberry is a shallow-rooted plant and roots are confined to the upper 25-30 cm of the soil surface. Due to shallow root system, irrigation is required more frequently than any other fruit-crop. Shortage or excess of water should be avoided to harvest a good crop of quality fruits. Irrigation at the time of active vegetative growth and during flowering is necessary. It should, however, be avoided during ripening, as it may result in softening of fruits.

The most commonly used irrigation systems are the surface and drip or trickle system. However, nowadays drip or trickle irrigation systems are gaining popularity. This method is more convenient to regulate water as well as to minimize amount of water. On an average, water used is 30-40 per cent less for trickle irrigation as compared to furrow method.

Mulching and Winter Protection

Mulching is recognized as one of the most beneficial intercultural practices in strawberry as it helps in better conservation of soil moisture, minimizes winter and frost injury, suppresses weed growth, reduces soil erosion and avoids contact of berries with soil thereby reducing number of dirty and rot infected berries. Mulching can be done by using black or white plastic sheet, or paddy straw.

PLANT PROTECTION

Major Insect-Pests and their Control

- ☆ **Red spider mite** (*Tetranychus urticae*): It is the most serious pest of strawberry. Damage is caused both by nymphs and adults by sucking sap from leaves, causing a rusty brown colour. Plants are stunted and show reduced yields. Dicofol 18.5 EC @ 2.5 ml/L is effective for controlling red spidermite.

- ☆ **Blossom weevil** (*Anthonomus rabi*): Blossom weevil is also called as elephant weevil. These cause damage by making holes in flowering-stalks and buds, and causes whole truss to wilt and die. Systemic pyrethroids Deltamethrin (0.05 per cent) and Sumicidin (0.025 per cent) sprays just before flowering saves plants from blossom.

- ☆ **Root weevil** (*Otiorhynchus rugostriatus*): This is a wingless weevil and its young ones (grubs) are more harmful than adults. Small white legless grubs feed on small rootlets in winter and spring and make deep tunnels in the crown base and finally plants collapse. Adults feed on foliage at night but damage is less severe than grubs. Application of carbaryl or malathion reduce pest population and damage.

- ☆ **White grubs**: In some places white-grubs cause severe damage. Its larvae are dirty white with brown heads, usually coiled and with distinct legs.

They eat off roots and kill plants, usually between planting time and runner formation. Drenching of soil with Chlorpyriphos (0.02 per cent) reduces damage of white grubs. Chickens or hogs may also destroy grubs.

☆ **Thrips** (*Thrips atratus*): Nymphs and adults cause damage by sucking sap from growing points, leaves and fruits. The damage is more severe in late-season strawberries. Severe infestations may result in fruit distortion, leading to down-grading and finally financial losses to farmers. Frequent sprays of synthetic pyrethroids, Deltamethrin (2.25 g a.i./ha) and Cypermethrin (16.8 g a.i./ha), reduce thrips incidence and number of distorted fruits.

☆ **Cut-worms and army-worms**: Several species of cut-worms and army-worms infest strawberries. Large, stout, stripped mottled or grey caterpillars eat on fruits and make large holes. They usually feed at night and hide during day. They cut off young plants at the ground level, and may eat leaves and berries of established plants. Drenching soil with Chlorpyriphos (0.01 per cent) during March is effective for cut-worms. Similarly 4-5 deep ploughings during land preparation also controls cut-worms to some extent.

☆ **Tarnished plant bug** (*Lygus lineolaris*): Adult bug is coppery-brown with piercing and sucking mouth parts. It causes damage by puncturing berries, causing uneven ripening or "catfacing". It also feeds on individual achenes (seeds) and destroys their contents. Spraying with Imidachloprid (0.05 per cent) controls it effectively.

☆ **Nematodes:** Sting nematode (*Belonolamus* sp.) is more destructive to strawberries than any other nematode, and it occurs mainly in light soils. It feeds on the surface of roots and is usually dislodged when plants are dug. Root-knot nematode (*Meloidogyne* sp.) penetrates and feeds on small roots and causes knot-like enlargements. Affected plants weaken and produce few runners and fruits. Nematodes that enter roots can stay alive even when plants are dug, stored and shipped. It has been reported to infest all strawberry cultivars in North India. (Lesion nematode (*Pratylenchus* sp.) also attacks roots and initially causes small spots that are amber to dark-brown. Roots may decay and may be much like caused by root-rot diseases. Length of leaf stalks of the affected plants shortens, and leaves become yellow, and plant growth is reduced. This nematode may increase chances of *Verticillium* wilt. Bud nematode (*Aphalenchoides* sp.) lives in most succulent tissues of buds and very small leaves. Flower buds are often killed by feeding, resulting in flowerless plants. It causes bud disease 'spring crimp' or 'red plant'. Stem nematode (*Ditylenchus dipsaci*) lives in plant stems and causes damage by sucking sap. It results in shortening and thickening of stem and leaf stalks.

All soil-inhabitating nematodes can be checked by fumigating soil with nematicides, methyl bromide or by micro granular nematicides, Thimet or Temik. Root-knot nematode can be controlled by Basamid (dazomet) and solar-heat treatment. Nematodes in roots, leaves and buds can be killed by hot-water dip of dormant plants at 52.8°C for 2-3 minutes.

☆ **Birds**. Sparrows, parrots and peacocks cause heavy losses to strawberry at the time of berry-ripening. The damage by birds is more alarming than the one caused by insect-pests or diseases. It can be reduced by using bird scarer, reflecting ribbons and by beating drums. Covering field with nylon-netting is very useful. The birds are very clever and get acclimatized to the scaring measure used, and hence no scaring device should be continuously used for a longer time.

Major Diseases and their Control

☆ **Powdery mildrew**: Mildew may affect strawberries in nearly all regions during dry periods. It is a fungal disease caused by *Sphaerotheca macularis*. Its characteristic appearance is an upward curling of leaves and a cobweb-like mold on the lower surface as the fungus destroys surface layers of leaves. Catskill, Dunlop, Klondike, Marshall and Sparkle are notable resistant varieties. Baytelon or Karathane (0.05 per cent) at 10-15 days interval between May to September is quite effective for controlling powdery mildew.

☆ **Red stele**: It is the most destructive disease of the plants in cool and moist soils and pooerly drained soils in late winter and early spring. It is also called as 'red core' or 'brown core'. It is caused by a fungus, *Phytophthora fragariae*. The fungus enters central parts of roots, the stele, which turns reddish while its cortex appears normal. If the season is dry, infected plants may die before blossoming, and if the season is wet, plants may blossom, but they die before ripening of fruits. Affected plants at first show dull, bluish green colour, which soon wilt and may die in a few days. In well-drained sandy soils, it is rarely serious. The disease is spread by drainage water, machines, tools and by infected plants. Improvement in drainage system, ridge planting, crop rotation, covering plants with cloches and use of resistant varieties are important cultural techniques to reduce red stele incidence. Soil drenching with disinfectant Ridomil and treatment of planting material before planting with bordeaux mixture (1 per cent) or copper oxychloride (0.1 per cent) for 15-20 minutes reduce the occurrence of red stele in field.

☆ **Verticillium wilt**: It is caused by a fungus, *Verticillium albo-atrum*, which is active in cool weather. Fungus invades roots and interferes with movement of water to leaves. Disease symptoms include internal browning of vascular tissues at the crown base. Outer older tissues wilt and dry at margins and become reddish to dark brown. In affected plants, only a few leaves develop, which tend to be stunted, and may curl up along the mid vein. The central young leaves are small, new roots from the crown are short with blackish tips, and plants look dry and may suddenly collapse. Soil fumigation is the best way to keep disease under control.

☆ **Grey mould or botrytis rot**: It is a serious disease of strawberry in field and storage. It is caused by a fungus, *Botrytis cinerea*. The disease appears as a light-brown soft spots on the green and ripening fruits. The berries dry out, become tough and get covered by a dusty fungal growth. Infection is common on raw and ripe fruits resulting in complete rotting of fruits. Fruit loss of 50 per cent in strawberry may only be due to *Botrytis* rot. A single infected fruit may spread rot to all fruits during transit or storage. Following control measure should be followed for reducing its incidence (i). Don't allow fruits to touch the soil; (ii) provide suitable mulch; (iii) avoid excessive application of N fertilizers; (iv) avoid excessive irrigation; (v) ensure proper drainage; (vi) follow proper sanitary measures; (vii) spray carbendazim (0.05 per cent) or captan (0.05 per cent 0 or thiram (0.05 per cent just after the opening of flowers) (viii) remove all infected fruits before storage or transportation.

Viral Diseases

Several viral diseases damage strawberry plants. They may produce mottling, mild yellow edge, crinkle, vein chlorosis, leaf stunting or dwarfing. Viruses are transmitted by leaf hoppers and nematodes. Control of viral diseases is possible only if insect vectors are kept under control, following clean cultivation and by crop rotation. Initially, virus-free planting material should be used and infected plants may be uprooted and burnt to avoid further spread

Physiological Disorders and their Management

The causes and controls of the important disorders are as follows.

☆ **Fruit malformation:** Malformed fruits can be commonly seen even in recognized varieties of strawberry. In this, common shape of berry is changed. Usually, the primary and secondary flowers produce more malformed fruits due to undeveloped achenes at the distal end of the receptacle than the tertiary and quaternary ones. Many factors have been found associated with malformation of fruits. However, planting of more vigorous runners and high N levels and insufficient pollination and lack of growth promoting substances are the ones chiefly associated with the production of malformed fruits. Provision of adequate pollinizer cultivars in commerical plantings of the main cultivar and honey-bee colonies reduce fruit malformation. Similarly, planting of young and less vigorous plants and avoidance of excessive N fertilizers also reduce it to a greater extent.

☆ **Albinism:** It is the most serious disorder of strawberry fruit, occurring primarily at the time of ripening. It has attained an alarming situation in the USA, Belgium and the Netherlands. Fruits suffering from albinism appear bloated, and these develop white or pink areas on their surface; the pulp remains pale. These fruits have poor flavour and tend to be acidic. Affected fruits develop normally but do not ripen uniformly and show waxy appearance and are liable to severe damage during harvesting and are highly susceptible to fruit rot during storage. This disorder tends to develop in densely growing crops, dense plantings or due to excessive fertilizer application. It generally develops in crops grown in sandy, low pH soils and soils with high N, K, Ca contents and sometimes P content also. Albinism is more severe in closed tunnels (with little ventilation). Similarly, its incidence is more in crops mulched with black film than those with white films, probably due to high N mineralization under black film, which generates higher temperature. Selection of a suitable variety, avoiding dense planting and excessive application of fertilizers particularly N and K etc are some measures to reduce its menace.

MATURITY, HARVESTING AND YIELD

Maturity

The crop is ready for harvesting in March-April, but in cool climates, it matures by mid-June. The fruits are usually harvested when $^2/_3$ to ¾ portion of the fruit peel has developed colour. For distant markets, berries are sometimes harvested while they are still green, or white and are hard. However, a little delay in picking may increase percentage of over-rotted berries. Berries should be harvested in shallow containers as they are highly perishable and damage early if deep and bulky containers are used.

Harvesting

Usually, berries should be picked up daily in warm weather and 2-3 times a week in cool weather. As far as possible, berries should be picked in early and late hours of the day. Keep picked berries in a shade and a dust-free place. A shelter bed, even a structure with simply a roof and a wall on one side of the prevailing wind is advantageous to reduce farm heat of berries and other adverse effects. Pick-your-own (PYO) method of harvesting strawberries is the most popular method of direct marketing in several advanced countries of the world. In this, strawberry farms are established near road-sides and consumers come and pick the wanted fruits in small baskets, get it weighed and pays money to the growers. This concept of PYO farms began when harvesting labour became scarce and increasingly expensive. The advantages of PYO farms include no direct harvest labour and transportation costs, improved quality, increased product availability for consumers, and reduced packing costs.

Yield

Yield is influenced by soil and climate, variety, planting system, planting density, number of crowns/plant, number of fruits/plant, fruit size and various management practices. In general, primary and secondary flowers bear large-sized fruits, as compared to tertiary and quaternary flowers, thus affecting yield. Average strawberry yield is about 5-7 tones/ha.

POSTHARVEST HANDLING

Pre-cooling

Strawberries reach markets in a better condition if pre-cooled at 4.4° C within 2 hours and kept at this temperature. This is especially true during warm weather when fruit is picked during the heat of the day and fruit temperature is 26.4°C or more. At higher temperature, rate of respiration goes high, which result in further reduction in postharvest life of berries.

Grading and Packing

After harvesting, fruits should be graded according to size, shape and colour. Fruits of uniform size and colour should be selected for A grade packs and slightly damaged ones may be packed and sold out as B grade. Before grading, remove all cull, diseased or rotten fruits. In India, sorting is usually done manually. In some advance countries, pan-method of sorting is followed for grading. The grading pan is 25 cm long, 2.5 cm wide and 1.5 cm deep. Commercial growers may run fruits over a belt, grade it and then jumble back into cardboard trays of different sizes, which can conveniently be stacked for pre-cooling and marketing. Standardized grades promote honest and fair-dealing, and discourage careless and unscrupulous packing.

Strawberries may be packed in small baskets, tins, polyethylene over-wrapped baskets and crates. Nowadays, special packing cases of plastics, commonly called as 'plastic punnets', have been developed for packing strawberries. About 200-250 g fruits can be safely packed in one punnet. These packs provide proper ventilation for strawberries because they have many perforations (10-12 holes/punnet). After packing fruits in punnets, these are placed in corrugated fibre-trays. Punnet-filled trays should be kept under shade or shelter to reduce heat and water-loss from berries. In recent development, packing of strawberries in cellophane (cellulose film and chitosan films has been recommended for better postharvest life of strawberry.

Properly filled baskets should neither be slack nor so full that berries are crushed by covers or dividers. Packs should be full enough to look attractive and maintain a well-filled appearance for the consumer. Similarly, the berries of one cultivar should be packed in one container and may not be mixed with the other. The pack should be properly labeled with name of variety, garden and date of packing.

Storage

The strawberry fruit is highly perishable. It should be marketed soon after harvesting. Due to perishable nature, refrigerated transportation and short-term storage of strawberries can give good money to grower. Further, strawberries are available only for a short time (15-20 days) and thus their proper storage can extend their availability.

Many techniques like pre-harvest application of fungicides and other chemicals have been tried to prolong postharvest life of strawberries, but with limited success. The better way of storage of strawberries is the controlled atmospheric (CA) storage. The requirement for temperature and concentration of CO_2 and O_2 in controlled atmosphe varies with cultivar but, in general, they can be stored for 10-12 days at

2 per cent O_2 and 5 per cent CO_2 at 3°C. Some investigators have suggested that long-term storage of strawberries at 1 per cent or lower concentration of O_2 may lead to off-flavour, and thus higher concentration of CO_2 (10 per cent or more) should be restricted in storage for up to a week only, where adequate refrigeration is not available. The postharvest life of berries in storage can further be increased if plants are sprayed with Bavistin (0.1 per cent) during flowering and then storing of harvested berries in 20 per cent CO_2 and 0-3 per cent O_2.

Value Addition

In addition to table and dessert purposes, various value added products can be prepared from strawberry. Berries can be used for canning, candy, jam and jelly-making and for flavouring ice-creams. Juice is also made from strawberries, which is considered as a refreshing and soothing drink in summer. Some developing countries also make wine and freeze products from strawberry. In our country, strawberry is consumed as a fresh fruit. However, it is also used for flavouring ice-cream.

21

Walnut

INTRODUCTION

Walnut is the leading nut crop of temperate regions of the world. After cashewnut, walnut is considered as the most important nut fruit crop in India. Walnuts have innumerable health benefits. Walnuts are delicious nuts recognized since ancient times as the symbol of intellectuality. The nuts are rich source of health benefiting omega-3 fatty acids.

COMPOSITION AND USES

Composition

Kernel is the edible portion of walnut, which is good source of energy, protein, fat, and dietary fiber. The protein in walnuts provides many essential amino acids, which have beneficial health promoting effects. Its kernel contains high amount of B complex vitamins especially B-6, and vitamin E (Table 21.1). Its kernels contain appreciable amount of essential minerals such as iron, zinc, copper, magnesium, phosphorus and potassium. English walnuts contain 4.5 times Omega-3 fatty acid content than black walnuts. Unlike most nuts that are high in monounsaturated fatty acids, walnut oil is composed largely of polyunsaturated fatty acids (47.2 per cent), particularly α-linolenic acid and linoleic acid. Raw walnuts contain antioxidants. They also contain triglycerides of the n-3 fatty acid alpha-linolenic acid which is effective in reducing heart risk. Green hulls and immature fruits are rich source of ascorbic acid. Both kernels and oil reduce serum chloestrol and triglycerides level in the blood.

Table 21.1: Nutritional Composition of Walnuts per 100g of Meat

Attribute	Contents	Attribute	Contents
Energy	654 kcal	Vitamin B$_6$	0.537 mg
Carbohydrates	13.71	Pantothenic acid (B$_5$)	0.570 mg
Sugars	2.61	Vitamin A	20 IU
Dietary fiber	6.7	Vitamin E	0.7 mg
Fat	65.21	Calcium	98 mg
Saturated fats	6.126	Iron	2.91 mg
Monounsaturated	8.933	Magnesium	158 mg
Polyunsaturated	47.174	Phosphorus	346 mg
Protein	15.23	Potassium	441 mg
Thiamine (B$_1$)	0.341 mg	Zinc	3.09 mg
Niacin (B$_3$)	1.125 mg		

Source: USDA Nutrient Database.

Uses

All parts of walnut are utilized one or the other way. Dry kernels are used as table fruit. Nuts consumed fresh, roasted, or salted, used in confectioneries, pastries, and for flavoring. The shells may be used as antiskid agents for tires, blasting grit, and in the preparation of activated carbon. Activated charcoal and fructose have recently been suggested to foil the alcohol "breathalizer". Ground nut shells used as adulterant of spices. Crushed leaves, or a decoction used as insect repellant and as a tea. Outer fleshy part of fruit very rich in Vitamin C and produces a yellow dye. Fruit, when dry pressed, yields valuable oil used in paints and in soap-making; when cold pressed, light yellow edible oil used in foods as flavoring. Young fruits made into pickles, also used as fish poison. Twigs and leaves lopped for fodder in India. Decoction of leaves, bark, and husks used with alum for staining wool brown. Wood hard, durable, close-grained, heavy, used for furniture and gun-stocks. Tree often grown as ornamental. The husk of the black walnut is used to make an ink and a brown dye for fabrics. Historically, walnut oil was prescribed for colic, to soothe intestines, and to relieve diarrhea and hemorrhoids. Further folk uses include treating rickets, frostbite, and glandular disturbances, and as an astringent, tonic restorative, and disinfectant. Blisters, ulcers, itchy scalp/dandruff, sunburn, and perspiration are some of the conditions treated with various walnut preparations.

TAXONOMY AND BOTANICAL DESCRIPTIONS

Taxonomy

Walnuts are part of the tree nut family (Juglandaceae), pecans and genus *Juglans* having basic chromosome number 16 and somatic chromosome number 32. The genus *Juglans* includes 21 species. All the *Juglans* species are edible but English walnut or Persian walnut (*Juglans regia* L.) is the most widely cultivated in several countries of the world. Other important walnut species are *J. nigra*

(American Black walnut), *J. hindsii* (California Black walnut), *J. cinera* (Butternut), *J. sieboldiana* (Japanese walnut), *J. major* (Arizona Black walnut) and *J. microcarpa* (Texas Black walnut). The Latin name *Juglans* derives from *Jovis glans*, "Jupiters acorn": figuratively, a nut fit for a god. The word walnut derives from Old English wealhhnutu, literally "foreign nut", wealh meaning "foreign" (wealh is akin to the terms Welsh and Vlach). The walnut was so called because it was introduced from Gaul and Italy. The previous Latin name for the walnut was nux Gallica, *i.e.*, "Gallic nut".

The best-known member of the genus is the Persian Walnut (*Juglans regia*), native from the Balkans in southeast Europe, southwest and central Asia to the Himalaya and southwest China. The Black walnut (*Juglans nigra*) is a common species in its native eastern North America, and is also widely cultivated elsewhere. The nuts are edible, but have a smaller kernel and an extremely tough shell, and they are not widely grown for nut production. The Butternut (*Juglans cinerea*) is also native to eastern North America, where it is currently endangered by an introduced disease, butternut canker, caused by the fungus *Sirococcus clavigignenti*. Its leaves are 40-60 cm long, and the nuts oval. The Japanese walnut (*Juglans ailantifolia*) is similar to Butternut, distinguished by the larger leaves up to 90 cm long, and round (not oval) nuts.

Some interesting interspecific hybrids like Paradox (*Juglans hindsii* x *J. regia*) and Royal (*J. nigra* x *J. hindsii*) have been developed but are primarily used as rootstock.

Botany

Juglans regia is a large, deciduous tree attaining heights of 25–35 m, with a short trunk and broad crown. It is a light-demanding species, requiring full sun to grow well. The bark is smooth, olive-brown when young and silvery-grey on older branches. The pith of the twigs contains air spaces. The leaves are alternately arranged, having 5–9 leaflets, paired alternately with one terminal leaflet. The male flowers are in drooping catkins, and the female flowers are terminal. The developing fruit is green, with semi-fleshy husk and a brown, corrugated nut. The whole fruit, including the husk, falls in autumn. The seed is large, with a relatively thin shell, and edible, with a rich flavour.

ORIGIN, HISTORY AND DISTRIBUTION

The walnut is considered to be the native of North America, South America and south east Europe to East Asia. The English walnut (*J. regia*) is the native of Caucasus region.

The worldwide production of walnuts has been increasing rapidly in recent years, with the largest increase coming from Asia. At present, China is the world's largest producer of walnuts, with a total harvest of 1.7 million metric tonnes. The other major producers of walnuts are Iran, United States, Turkey, Ukraine, Mexico, Romania, India, France and Chile (Table 21.2). Of these countries, The USA is the world's largest exporter of walnuts. In India walnut remained confined to Kashmir and spread to Himachal Pradesh and Uttarakhand during the last century only and is yet to gain popularity in east Himalayan ranges and hills of South India.

Table 21.2: Walnut Growing Countries of the World and their Walnut Production

Country	Production (tones)	Country	Production (tones)
China	1,700,000	Ukraine	96,900
Iran	450,000	India	40,000
United States	425,820	Chile	38,000
Turkey	194,298	France	36,425
Mexico	110,605	Romania	30,546

Source: FAO Production Year Book- 2012.

SOIL AND CLIMATIC REQUIREMENTS

Soil

For profitable cultivation, the top soil should be fertile and well drained with a pH of 6.7. Besides, the subsoil should be free from hard pan, impervious clay or gravel layer up to 3 m. In general, walnut can be grown profitably in well drained, silt loam soils rich in organic matter. Thus, coarse and sandy soils with hard pan should be avoided for walnut plantation. Fluctuating water table is also not suitable for walnut cultivation. Walnuts are also most sensitive to alkaline soils.

Climate

Main climatic limitations of walnut cultivation are spring frost, extreme summer heat and insufficient winter chilling. Walnut requires more than 1000 chilling hours during winters. An annual rainfall of 80 cm is considered sufficient for the cultivation of walnut. High temperatures in spring or summers above 40 °C accompanied by humidity may cause sunburn of exposed tissues. Such conditions early in the season may cause blank nuts, but if occurs later, kernels may be shriveled. In warm areas, walnuts do not receive adequate chilling causing delayed bud break and flowering resulting in poor fruit set.

ROOTSTOCKS AND PROPAGATION

The open pollinated walnut seedlings are generally used as rootstocks. The seedling stocks are vigorous in growth and susceptible to oak root fungus but make smooth graft union free from constriction with scion varieties. Some of the important rootstocks suitable for walnut are Manregian seedlings, *J. hindsii*, Royal and Paradox. Walnut is usually propagated by side veneer grafting, patch budding, annular budding and chip budding.

PLANTING AND ORCHARD ESTABLISHMENT

Planting of walnut is done any time from December to March during dormancy but an early planting is advisable for successful field establishment of plants. In warmer areas without irrigation facilities, walnuts should be planted during rainy season with earth ball or seedlings raised in polythene bags to ensure better establishment of the plants. For *in situ* grafting, 3 to 4 seeds are sown 10 to 15 cm deep during winters at one location and the desirable scion is grafted on the selected seedling during July-August. Since the walnut plants on seedling rootstock are vigorous in growth, plant to plant spacing of 8 m is suitable. Pits of 1 x 1 x 1 m size are dug about one month before planting. The grafting point should be kept at least 15-20 cm above the ground level at the time of planting.

IMPORTANT VARIETIES

In India, all walnut trees are of seedling origin except that few seedlings have been named locally or some cultivars have been introduced from other countries. However, for profitable cultivation, proper cultivar should be selected on the basis of microclimate of the site, soil fertility, bearing habit, availability of chilling hours and fruit *etc*. Some commercially grown varieties of walnut are Eureka, Sunland, Placentia, Wilson Wonder, Prolific, Franquette, Howard, Tehama, Vina, Pedro, Chandler, Hartley, Fernor, chico, Payne and Serr. The promising varieties of walnut in India are Govind, Kashmir Budded, Eureka, Pratap, Placentia, Wilson and Franquette. Recently, ICAR-IARI, New Delhi has released Pusa Khor and ICAR- CITH, Srinagar has also developed and released some interesting selections.

Chief characteristics of some varieties of walnut are described hereunder.

Chandler

Harvest mid-season; large smooth oval nut, with good shell seal and high quality kernel. Kernel colour is excellent, light grade consistently 90 per cent or better. Has potentially high fruitfulness with 80 to 90 per cent of lateral buds fruitful. Medium size tree is a moderately vigorous and semi-upright, highly productive tree. Pollinizers are Cisco and Scharsch Franquette.

Chico

Small, upright, early harvest, highly productive tree. Nut size is small with excellent kernel quality. Due to smaller size trees and a very high percentage of lateral pistil late; bloom 90 to 100 per cent. It is well suited for high-density plantings. Pollinizers are Payne, Serr or Sunland.

Cisco

Its main attribute is as a pollinizer for Chandler and Howard. In growth habits Cisco is semi-upright and a small tree.

Eureka

Tree is very large, somewhat spreading growth habit. Harvest early to mid-season. Nut is medium size elongated with a good shell seal.

Franquette

A late leafing Californian variety which is good for areas with late spring frosts. Unfortunately, its late flowering also means it misses the pollen shed by other varieties, so nut set and yields are often poor. It may be worth trying 'Mayrick,' also late flowering, as a pollinizer or 'Rex.' Most Californian varieties are susceptible to walnut blight, and are therefore poorly suited to wet and humid areas, but Franquette seems to have some degree of blight resistance. Franquette is a terminal bearer. The nuts are large, and attractive. Crack out is around 31 per cent. Franquette reputedly also has very high quality timber.

Hartley

Percentage of light kernels nearly 90 per cent. Tree size is moderate to large, moderately spreading with good vigor on fertile soil. Hartley needs 40 to 45 foot spacing for mature tree. Most widely planted walnut variety in California. Acceptable pollinizers are the late blooming Amigo and Scharsch Franquette.

Howard

Harvest mid-season; nut is large, round and smooth with a good seal. Kernel quality is excellent at 90 per cent light, and kernel percentage is 50 per cent. Tree size is small to medium and semi-upright with moderate vigor, which makes it a good candidate for high-density plantings. Pollinizers are Cisco or Scharsch Franquette.

Payne

Harvested early in the season. Payne nut size is medium to small. Shell seal is very good. Nuts average 48 per cent kernel with approximately 50 per cent light. Yield potential is high to very high. Approximately 80 per cent to 90 per cent of the lateral buds on shoots are fruitful. Very productive. Tree is medium round shaped. Heavy pruning is required when trees are young to avoid overbearing.

Pedro

A particularly desirable walnut for the home gardener because it is a relatively 'small' tree at about nine metres/30 feet high. It is self fertile, needing 400 hours of winter chill (not suited to areas with late frost), and the nut is both well sealed and particularly liked when tested in consumer taste panels.

Serr

Harvest is early to mid-season. Nut size is large, with a fair to good shell seal. Kernel is 60 per cent light. Percentage of kernel is high at 59 per cent. Serr planted on shallower, heavier, or less fertile soil seems to bear better. Serr tree size is large and requires a spacing of at least 40 feet. Shape is moderately spreading and vigor is good to excessive. Suitable pollinizers include Chico and Tehama.

Tulare

Harvest mid-season. The nut and kernel are large, with a well-sealed, nearly round nut. The Tulare requires no pollinizer. Tree has upright growth habit, moderately vigor, suitable for hedgerow and other high-density planting systems.

Vina

Harvest early to mid-season; medium size pointed nut, with a good shell seal. Kernel color is good at 60 per cent light with 48 per cent kernel. Tree size is small to medium; vigor is moderate to good and highly productive. Pollinizers are Chico, Chandler, Howard and Tehama.

Hamdan

Trees are vigourous and spreading, mid to late blooming, protoandrous, terminal bearing, nuts are medium to large in size, oblong to ovate in shape, kernels are medium creamy light brown, soft shelled and mid-season maturity.

Sulaiman

Trees are spreading and vigourous, mid blooming and protoandrous, terminal bearing, nuts are medium to large in size, roundish oblong in shape, kernels are medium creamy light brown, soft shelled and mid-season maturity.

CITH Walnut – 1

Suitable for export as well as domestic market, having light kernel colour, nuts are bold, kernels are large in size, good kernel recovery (47 per cent), light shell colour, long trapezoidal in shape, easy to remove kernel halves.

CITH Walnut - 2

Nuts are large in size, ovate in shape, medium shell texture, medium shell colour, strong shell seal, intermediate shell strength, complete shell integrity, satisfactory kernel flavour, well filled kernel, plumy, easy to remove kernel halves and light kernel colour.

CITH Walnut - 3

Nuts are large in size, round in shape, medium shell texture, medium shell colour, strong shell seal, strong shell strength, complete shell integrity, satisfactory kernel flavour, well filled kernel, plumy, difficult to remove kernel halves and light kernel colour.

Jammu and Kashmir, CITH Walnut - 4

Nuts are large, ovate, rough shell texture, light shell colour, strong shell seal, intermediate shell strength, complete shell integrity, thin, satisfactory kernel flavour, well filled kernel, moderately plumy, very easy to remove kernel halves and light kernel colour.

CITH Walnut - 5

High yielder, having extra light kernel colour, suitable for export market, bigger nut (19 g) and kernel (9.5 g) size, good kernel recovery (48.9 per cent), light shell colour, ovate in shape, moderate to remove the full kernel halves.

CITH Walnut - 6

Nuts are large in size, ovate in shape, shell colour medium, intermediate shell seal, Intermediate shell strength, satisfactory kernel flavour, well filled kernel, moderate plumy and easy to remove kernel halves.

CITH Walnut - 7

Matures in 155-160 days after full bloom, nuts are medium in size, ovate in shape, medium shell texture, medium coloured shell, intermediate shell seal, intermediate shell strength, satisfactory kernel flavour, well filled kernel, plumy, moderate removal of kernel halves.

CITH Walnut - 8

Nuts are having light kernel colour, nut weight (20.4 g), and kernel weight (11.01 g), good kernel recovery (54 per cent), light shell colour, long trapezoidal in shape, very easy to remove kernel halves, rough shell texture, strong shell seal and strong shell strength.

CITH Walnut - 9

Nuts are medium in size, round in shape, light in colour, strong shell seal, intermediate shell strength, well filled kernel, plumy, moderate to remove the kernel halves.

CITH Walnut - 10

Heavy bearing, nuts are small in size, round in shape, smooth shell texture, medium colour shell, intermediate shell seal, intermediate shell strength, satisfactory flavour, well filled kernel, plumy and easy to remove kernel halves.

Partap

Trees are large in vigour and heavy bearer, nuts are large, oblong, smooth, light amber in colour, kernels are light in colour and very easy to remove, semi soft shelled.

Kotkhai Selection - 1

Trees are vigourous and good yielder, nuts are medium in size, smooth light weight, kernels are light in colour, well filled, thin shelled and early maturity.

Chakrata Selection

Trees are vigorous and spreading, mid bloomer and protoandrous, nuts are small to medium in size, round ovate or short trapezoid, thin shelled and early maturity.

Pusa Khor

It is a semi-vigorous variety and bears terminally as well as laterally. The fruits are self-cracking in nature which indicates its maturity. The fruit weight with husk was 59.99 g whereas, without husk (nut) weight recorded 23.67g.The length of the fruit with husk (nut) it was 61.58 mm whereas, without husk (nut) length recorded 43.87 mm. The dry nut weight is above 12 g.

FLOWERING, POLLINATION AND FRUIT SET

The seedling trees start bearing after 8-10 years whereas grafted trees start bearing in about 4-5 years. Walnut is a monoecious plant and bear male and female flowers separately but on the same tree. The male flowers (catkins) are borne on lateral buds of last year growth, whereas the female flowers (spikes) are borne terminally on current season's growth. Some of the varieties are lateral fruiting as well having high yield potential like Sunland, Chico, Howard and Tehama, Walnut is a wind pollinated fruit (Anemophilous) and most of the varieties are self-compatible and self-fruitful. But the plantation of different varieties is useful as the anthesis of male and female flowers may not synchronize in one tree due to dichogamy. Pedro is considered as a good pollenizing variety.

INTERCULTURAL OPERATIONS

Training and Pruning

The traditional training system is the goblet, however, modified central leader system is now commonly followed for training walnuts in which 5 to 6 laterals are allowed to develop around the stem. After the initial frame of the tree is developed, walnut plants require minimal pruning. Some thinning out of branches is desirable in heavy bearing varieties.

Nutrition Management

For fruitful production, annual application of nitrogenous and phosphatic fertilizers is required for proper growth, flowering and regular fruiting. In practice, the walnut plantations are scattered and seldom manured resulting in alternate bearing habit. The fertilizer doses depend on age, yield, size and the fertility of soil. Usually 60 kg/ha each of N, P and K is sufficient for fully grown up trees. Walnut trees are sensitive to zinc deficiency which can be corrected by foliar application of 0.4 per cent zinc and sulphur.

PLANT PROTECTION

Major Insect-Pests and their Management

Many insect-pests cause damage to walnut but the following pests cause serious damage:

☆ **Indian gypsy moth** (*Lymantria obfuscata*): This pest causes considerable damage to walnuts and found in all walnut growing areas of the world. Its caterpillars feed on foliage during night and defoliate the whole tree. Caterpillars eat away the whole lamina and leave behind only the hard veins. For its control, it is advisable to collect its eggs and destroy them and spray Imidachloprid (0.05 per cent) during March-April.

☆ **Walnut weevil** (*Alcidodes porrectirostris*): It is found in all walnut growing areas of the world. Weevils lay eggs on fruits, which on hatching bore deeper and feed on kernels. They also excavate circular feeding holes on the fruit surface. Grubs inflict the damage by boring into the fruits and reduce the kernels into a black mass, as a result, the attacked fruits drop prematurely with grubs inside. For its control, the fallen fruits should be collected and destroyed. This method is efficient only if tree owners carry out this operation of collection and destruction of infested fruits on campaign basis. Spray carbaryl (0.2 per cent) twice at 7 days interval as and when first attack of weevil is visible.

☆ **Walnut husk fly** (*Rhagoletis cingulata*): It causes extensive damage to late season cultivars by feeding on fleshy tissues of the husk. The feeding process produces a substance that penetrates the shell of the nut and stains the kernel. Off-coloured kernels are not only unattractive but are also off-flavour. As a result of extensive feeding, nutshell get stained, which renders shell unfit for sale. Sometimes, it also attacks kernels, which get shriveled and turn brown. Pick up and remove infested walnuts from the plantation as soon as possible after they fall from the trees. In addition, this pest can be controlled by applying sticky trap spheres or spraying Imidachloprid (0.05 per cent) before the expected attack of fly. Sprays should be done between mid-August to mid-September.

☆ **Ambrosia beetles**: Two species of Ambrosia beetles attack black walnut trees but the most serious is *Xylosandrus germanus*. This beetle occurs throughout most of the walnut growing regions of the world. Young walnut trees are most often attacked by this beetle. Female beetle may introduce a *Fusarium* fungus into the tree as she excavates her tunnel into wood. This fungus causes a cankered area in the wood, usually causing top dieback and resprouting from the base of the tree. Cankering, however, is not always apparent. In some plantations, dieback in 1 year due to ambrosia beetle/*Fusarium* canker attack has been reported on 30 to 40 per cent of the trees. The attack of this beetle is usually not detected until there is profuse sprouting from the base of the trees or until the trees are dead. Close examination is necessary to locate the tiny pinholes in the lower stem area or in small, low-hanging branches. For its control, cut and remove dead or *Fusarium*-canker infected tree tops and branches and burn, if possible.

☆ **Walnut caterpillar**: The most obvious insect pest on walnut is the walnut caterpillar. They feed in clusters and cause noticeable defoliation. However, it is not of great importance because it usually does not occur until late summer when defoliation is limited to a few limbs, and actual damage is relatively light. The small caterpillars are reddish-brown and raise their heads and tails when disturbed. They seldom appear on trees before late July or early August. Spot treatment of infested limbs is sufficient to check the feeding of this insect.

☆ **Walnut curculio**: Two species of curculios attack walnuts, but *Conotrachelus retentus* is more frequent to occur in different walnut growing countries. In some areas, walnut curculio can cause 60 per cent or more loss of nuts. Damage is caused by adult female curculio by laying eggs in young nuts during May-July. The larvae bore into the developing nuts and cause great losses during the so-called 'June drop' of walnuts. Walnut curculio larvae also cause the meager filling of walnuts that remain on the tree. A small exit hole in the side of a fallen nut is evidence that walnut curculio larvae have been present. If only few trees are affected, immediately pick up and destroy the immature nuts that drop prematurely during the growing season.

☆ **Codling moth** (*Cydia pomonella*): In addition to apple, codling moth causes serious damage to walnut. The larvae of the pest enters the fruit from any point on surface and tunnels upon seeds and cause excessive damage in core region. The pest is found both on apple and pear. The control strategy for this problematic pest includes mass pheromone trapping (25 traps/ha), collection and destruction of over wintered pupae in cocoons during April to June, deep burying of fallen fruits during August in addition to 2 sprays of DDVP or phosphorus (0.04 per cent) during June-July at an interval of 2-3 weeks. Use of egg parasitoids *Trichogramma* spp. has shown great potential against this dreadful pest.

Major Diseases and their Management

Several diseases infest walnut but the diseases causing extensive harm have been described briefly hereunder:

☆ **Walnut blight:** Walnut blight is caused by a bacterium, *Xanthomonas campestris* pv. *juglandis*). It is mainly a disease of the leaves, new shoots, nuts, and catkins of walnut. It is considered as the most destructive disease of walnuts throughout the world. Walnut trees are particularly susceptible at flowering, especially during wet weather in Spring and early Summer. Initially, small, angular black spots, appear on the leaves, leaf petioles, and succulent new shoots. The spots can become so numerous as to run together. While the disease appears destructive on these plant parts, it does not kill the tree, though it may cause some shoot dieback and some loss of leaves. Walnut blight can be destructive when it attacks the nuts. On nuts, the initial symptoms are small, round, elevated black spots. Early infection causes nut drop while the late infection results in shriveling and discolouration of the nuts. Affected nuts/kernels becomes unfit for consumption.

Since water is essential for the infection, hence, rains, fog, dew and sprinkler irrigation aggravate the problem. All varieties are susceptible to blight but early suffer severe damage than late flowering ones. The disease can be managed by spraying the trees with copper-based sprays and Bordeaux mixture and by planting late-flowering cultivars to avoid the worst infection period.

☆ *Phytophthora* **root rot:** *Phytophthora* root rot is also a major disease of walnut the world over. Three species affect nut trees: *Phytophthora cinnamomii*, *P.cactorum* and *P.citricola*. These fungi are present in most orchard soils and spread quickly by mobile spores when the soil is saturated, especially in warm weather. Infected leaves turn yellow and drop and the trees may die within a few years. Careful soil management and attention to irrigation and drainage will reduce the risk of infection.

☆ **Anthracnose**: This is a widespread and destructive disease of certain walnut species in different walnut growing areas. In India, it is a serious disease of walnuts in Kashmir valley. The disease is caused by a fungus that has both a sexual state (*Gnomonia leptostyla*) and an asexual (*Marssonina juglandis*). Both stages are involved in the disease. It attacks all aerial parts of walnut and cause severe loss to the growers. Initially, there is development of small circular light brown spots on leaves, which enlarge with the time. However, when climatic conditions are favourable, the disease rapidly becomes epidemic and black walnut trees may become prematurely defoliated. Premature loss of leaves results in poorly-filled, low-quality, darkened kernels. The amount of rain and the duration of wet periods from the time of pollination up to September hold the key as to the seriousness of anthracnose.

It is advisable to collect and burn the fallen leaves and fruits to avoid further spread of the disease and spray zineb or captan (0.01 per cent) twice or thrice at fortnightly interval as and when the infestation of the disease is noticed.

☆ **Armillaria root rot:** This disease is caused by *Armillaria mellea*, and its symptoms are almost similar to *Phytophthora* root rot. It is also called as mushroom root rot or shoestring root rot. To avoid its attack, it is advisable to fumigate the soil with methyl bromide and use of *J. hindsii* instead of *J. regia*.

☆ **Thousand cankers disease:** In certain walnut growing regions, thousand cankers disease can cause serious losses. This disease was first reported from Colorado (North America) in 2001. This is caused by a fungus, *Geosmithia morbida*, which is vectored by the twig beetle, *Pityophthorus juglandis*. When the beetles form galleries beneath the bark of walnut trees, they carry the fungus with them. The fungus forms dark cankers in the phloem around the beetle galleries. The number of beetles that attack an individual tree is enormous and the number of cankers that form is correspondingly large. On black walnut, the disease is lethal, causing cankers that coalesce and eventually girdle the trunk and branches. English walnut (*Juglans regia*), the species primarily responsible for commercial nut production, seems to be resistant.

Despite the graphic name of this disease, cankers are not the most obvious symptom. Initial symptoms of infection may be subtle. Leaves may flush in spring, but then suddenly wilt. Gradually the upper branches die back. Cankers are hidden beneath the bark and can only be seen in the early stages of disease when a thin layer of bark is cut away. A dark brown stain is apparent in the phloem just beneath the outer bark. The discoloration does not extend into the xylem (the wood), so care should be taken to avoid cutting too deeply when examining trees for cankers. Beetle galleries are also present in affected phloem tissue and tiny exit holes may be present, especially in branch crevices. Beetles are approximately 2 mm long, so exit holes are very small and may be hard to see. On smooth barked branches the exit holes are easier to see than on branches that have developed rough bark. Trees typically die about two years after the first symptoms are noticed, but this is thought to be many years after the initial infection actually occurred. Trees may resprout from the base, but sprouts are also infected and killed.

At present, there are no known control measures for thousand cankers disease. Because transport of logs and/or firewood is one of the main avenues for spread of the disease, several states in the US have enacted quarantines restricting the movement of black walnut logs. The movement of timber and nursery stock is also restricted by these quarantines. Education of the public about the need to prevent the transport of logs and firewood is also of paramount importance.

☆ **Viral diseases:** The blackline and walnut mosaic are most serious viral diseases of walnut. All walnut varieties budded/grafted on Paradox or California Black rootstocks are susceptible to blackline. It is a serious disease in different parts of the world but occurs very rarely in India. In this disease, there is development of small yellow drooping leaves, profuse suckering and small holes and cracks in the bark at graft union. The affected plants may break up from graft union. This disease can be controlled by using *J. regia* rootstock instead of *J. nigra* for Persian walnut, virus free scion and uprooting and destroying the infected plants.

MATURITY, HARVESTING AND YIELD

Maturity

For the production of quality nuts, harvesting at proper maturity is of prime importance. Any delay in picking after maturity of kernels deteriorates quality and increases the incidence of diseases and pests. The different maturity indices are used to for determining the harvest date for walnuts. However, when the hulls are cracked and exposed the nut, it is the right time for its picking. Similarly, depending upon climate, harvesting should be done after about 2 weeks of browning or kernel or the walnuts should be harvested when 10-15 per cent fruits have dropped themselves, indicating fruit maturity. A spray of ethephon (200-500 ppm) on kernel maturity advances maturity by about a week and makes harvest of entire crop at one time and promotes hull splitting.

Harvesting

For harvesting, knock the walnuts from the tree by striking the branches and small shoots with a long, stout pole. For small trees, we only need a pole but for larger trees, use a long pole with a large hook affixed to one end to enhance shaking. Pick the walnuts up immediately after harvest. Nuts that lie on the ground are very susceptible to mold infections, and may get darkened.

Yields

Yield of walnut trees varies with the variety, age and size. However, a full bearing walnut tree can yield about 40 kg nut/tree.

POSTHARVEST HANDLING

Hulling

Nuts have an inedible outer *hull* (also called a *husk*) that should be removed promptly after harvest so the nuts can be dried properly. The longer the hulls remain on nuts after harvest, the more the nut quality deteriorates. After harvesting and hulling, nuts are dried properly to reduce kernel moisture. Undried or improperly dried nuts are more likely to develop molds and a disagreeable flavour (rancidity), and have a shorter storage life. The harvested nuts are dehulled by keeping in a heap under wet leaves. Several methods are used to remove hulls from the kernels of walnut. Where only a few bushels of nuts are involved, they can be tramped off, placing the nut with husk on a hard surface and rolling it under a heavy foot. Light hammer blows will remove the husk; the nut can even be hammered through a hole in a piece of wood forcing the hull off.

Bleaching

Bleaching of nuts is a practice to give attractive appearance by dipping nuts for 5-10 seconds in a mixture of 8 kg each of Salsoda (sodium carbonate) and lime dissolved in 227 litre of water.

Drying

To dry walnuts, spread the hulled in-shell nuts in a single layer on a smooth, flat surface in a shady area where the air can circulate freely. Stir the nuts daily. If chances of the rains are there, cover the nuts with some material. If necessary, cover the nuts with screen or plastic netting to prevent theft by scrub jays and other birds.

At normal temperatures, walnuts will dry adequately in 3 to 4 days. Walnuts are considered adequately dried when they have brittle kernels and brittle and moisture content reaches 8-10 per cent. Rubbery kernels require further drying. Inadequately dried walnuts are susceptible to mold and quickly become rancid.

Storage

Walnuts, like other tree nuts, must be processed and stored properly. Poor storage makes walnuts susceptible to insect and fungal mold infestations; the latter produces aflatoxin—a potent carcinogen. A mold-infested walnut batch should be entirely discarded. Slow development of rancidity and kernel darkening. Similarly, proper storage prevents mold growth, stop insect damage, and maintains uniform nut moisture content, preventing rancidity. The ideal temperature for longest possible storage of walnuts is -3 to 0 °C and low humidity for industrial and home storage. However, such refrigeration technologies are unavailable in several countries.

References

Aggrawal, R.S. (2004). *Jamun*. Manoj Publications, India.

Anonymous. 2006. Jackfruit (*Artocarpus heterophyllus*). Chichester, England, UK: Southampton Center for Underutilized Crops, UK

APAARI. 2012. Jackfruit Improvement in the Asia-Pacific Region – A Status Report. Asia-Pacific Association of Agricultural Research Institutions, Bangkok, Thailand. 182 p.

Avocado culture in India, California Avocado Society 1967 Yearbook 51: 97-106.

Bender, Gary S. Avocado Flowering and Pollination.

Board, N. (2004). Persimmon. In: Cultivation of fruits, vegetables and flowers. National Institute of Industrial Research, India, pp.224-230.

Campbell, C.W. 1984. Tropical fruits and nuts. In: CRC Handbook of Tropical Food Crops, (FW Martin, ed.). Boca Raton Florida: CRC Press, Inc.

Chatterjee, P (2012). The Jamun Tree and Other Stories on the Environment. The world Bank Group.

Childer, N.F. (2003). The strawberry: For growers and other. NF Childers Publications, USA.

Chundawat, B.S. (1990) Arid Fruit Culture. Oxford and IBH Publishing Co. Pvt. Ltd., New Delhi.

Condit I.J. (1947). The Fig. Chronica Botanica Co., Massachusetts, USA.

Crane, J. H. (2001). The carambola (star fruit). University of Florida Cooperative Extension Service, Institute of Food and Agriculture Sciences, EDIS.

Das, M. R., Hossain, T., Mia, M. B., Ahmed, J. U., Karim, A. S., and Hossain, M. M. (2013). Blooming pattern of passion fruit flower (*Passiflora edulis* Sims.) under diversified flashes. *American Journal of Agricultural and Biological Sciences*, 8(3), 173.

Das, M. R., Hossain, T., Mia, M. B., Ahmed, J. U., Kariman, A. S., and Hossain, M. M. (2013). Fruit Setting Behaviour of Passion Fruit. *American Journal of Plant Sciences*, 4, 1066-1073.

Dasgupta, P., Chakraborty, P., and Bala, N. N. (2013). Averrhoa carambola: an updated review. *International Journal of Pharma Research and Review*, 2, 54-63.

Galilee, Agricultural R and D. Western, and Agricultural Experiment Farm. "Avocado pollination–A review."

Geroge, A.P. and Nissen, R.J. (1985). The persimmon as a subtropical fruit crop. *Queensland Agricultural Journal*, May/June. pp., 133–140.

Hancock, J.F. (1999). Strawberries. CABI, New York, USA.

Hofshi, R., and Arpaia, M. L. (2002). Avocado fruit abnormalities and defects revisited. Calif Avocado Soc Yearb, 86, 147-162.

Ish-Am, G. (2005). Avocado pollination: A review. New Zealand and Australia Avocado Grower's Conference '05. 20-22 September 2005. Tauranga, New Zealand. Session 7. Flowering, fruit set and yield. 9 pages.

Kassim, A., Workneh, T. S., and Bezuidenhout, C. N. (2013). A review on postharvest handling of avocado fruit. *Afr. J. Agric. Res*, 8, 2385-2402.

Kitagawa, A.H. and Glucina, P.G. (1984). Persimmon culture in New Zealand. New Zealand Department of Science and Industrial Research Information Service series No. 159, Science information Publishing Centre, Wellington, New Zealand.

Knott, S. Benefits of the Avocado in India.

Kore, V. T., Devi, H. L. and Kabir, J. (2013). Packaging, storage and value addition of *aonla*, an underutilized fruit, in India. *Fruits*, 68(3), 255.

Kumar, L., Gupta, V.K., Jha, B.K., Singh, I.S., Bhatt, B.P. and Singh, A.K. (2011). Status of Makhana (*Euryale ferox* Salisb.) cultivation in India. *Technical Bulletin*. ICAR Research Complex for Eastern Region, Patna.

Kumar, S. (2014). A Critical Review on Loquat, *Eriobotrya japonica* Thunb Lindl. *International Journal of Pharmaceutical and Biological Archive*, 5(2): 1-7.

Kundu, M., Schmidt, L. H. (Ed.), and Jørgensen, M. J. (Ed.) (2012). *Emblica officinalis* Gaertn. *Seed Leaflets*, (154), 1-2.

Lanz, H. (2012). Grow your own strawberries. Sea-to-Sea Publications, USA.

Manda, H., Vyas, K., Pandya, A. and Singhal, G. (2012). A complete review on: *Averrhoa carambola*. *World Journal of Pharmacy and Pharmaceutical Sciences*, 1, 17-33.

Mishra, R.K., Jha, V. and Dehadrai, P.V. (2008). *Makhana*. DKMA, ICAR Publications.

Morton, J. (1987). Fruits of Warm Climates. Julia F. Mor ton, Miami, Florida.

Morton, J. 1987. Fig. In: Fruits of warm climates. Julia F. Morton, Miami, FL. p. 47–50.

Morton, J. F. (1987). Fruits of Warm Climates. Creative Resource Systems, Inc. Winterville, N. C., USA, 503 p.

Mowat, A.D., Collins, R.J. and George, A.P. (1995). Cultivation of persimmon (Diospyros kaki l.) under tropical conditions. *Acta Hort. (ISHS)* **409**:141-150

Murli, T.P. and Balachandran, M. (2009) Fig. In: A textbook on pomology: Subtropical fruits. Chattopadhyay, T.K. (Ed.); Kalyani Publishers; New Delhi; pp. 158-169.

Orwa, C., Mutua, A., Kindt, R., Jamnadass, R., Simons, A. (2009). Agroforestree Database: A tree reference and selection guide version 4.0 (http://www.worldagroforestry.org/ af/treedb/)

Pathak, R. K. (2003). Status Report on Genetic Resources of Indian Gooseberry–*Aonla* (*Emblica officinalis* Gaertn.) in South and Southeast Asia. IPGRI Office for South Asia, New Delhi, India.

Pushpakumara, D. K. N. G., and Heenkenda, H. M. S. (2007). Chapter 6: *Nelli* (*Amla*) (*Phyllanthus emblica* L.). Pushpakumara, DKNG, Gunasena, HPM and Singh, VP (2007).(eds) Underutilized fruit trees in Sri Lanka. World Agroforestry Centre, South Asia Office, New Delhi, India, 180-221.

Riotte, L. (1998). Grow the best strawberries. Storey Books, USA.

Roy, S. K (1990a). *Bael*. In Fruits: Tropical and Subtropical. (eds.) T. K Bose and S. K Mitra. Naya Prakash 206 Bodhan Sarani, Calcutta, pp. 740-745.

Sanewski, G. (1991). Custard apples: cultivation and crop protection (No. Ed. 2). Queensland Department of Primary Industries.

Sanyal, D. and Das, P.C. (2009) Jackfruit. In: A textbook on Pomology. Chattopadhyay, T.K. (Ed.); Kalyani Publishers; New Delhi; pp. 120-124.

Saúco, V. G., Tindall, H. D., and Menini, U. G. (1993). Carambola cultivation (No. 108). Food and Agriculture Org.

Sharma, R.R. (2002). Growing strawberries. International Book Distributing Co., Lucknow, India.

Sharma, V,P. and Sharma, R.R. (2004) The strawberry. DMKA, ICAR, New Delhi, India.

Sharpe, R. H. (2010). Loquat: botany and horticulture. *Horticultural Reviews*, 23, 233.

Shivanna, K. R. (2012). Reproductive assurance through unusual autogamy in the absence of pollinators in *Passiflora edulis* (passion fruit). *Current Science*, 103(9), 1091-1096.

Singh, A. K. and Chaurasiya, A. K. (2014). Postharvest Management and value addition in *Bael* (*Aegle marmelos* Corr.). *International J. Interdisciplinary and Multidisciplinary Studies*, 9(1), 65-71.

Singh, S. and Hoda, M.K. (2009). *Makhana* (Gorgon Nut). In A textbook on pomology. Chattopadhyay, T.K. (Ed.); Kalyani Publishers; New Delhi; pp. 296-301.

Singh, G. (2012). Checklist of Commercial Varieties of Fruits. Department of Agriculture and Co-operation Ministry of Agriculture, Krishi Bhawan, New Delhi. P. 144.

Srivastava, S. S. (1996). Suskha phalodyaniki (Dryland fruit culture). Central Book House, Raipur, India, pp. 275-278, 278-280 and 339-342.

Storey W.B., Enderund J.E., Saleeb W.S. and Nauer E.M. (1977). The Fig (*Ficus carica* Linnaeus). Its Biology, History, Culture and Utilisation. Jurupam Mountains Cultural Centre, California USA.

Sweedman R. (1981). Fig Growing. Division of Plant Industries Bulletin H3.1.19. NSW Agriculture, Orange.

Tripathi, S.N. and Choubey, L.K. (2009). Bael. In A textbook on pomology. Chattopadhyay, T.K. (Ed.); Kalyani Publishers; New Delhi; pp. 120-124.

Von-Carlowitz, P. G. (1991). Multipurpose trees and shrubs-sources of seeds and inoculants. ICRAF, Nairobi, Kenya, 328 pp.

Uniyal, S. and Misra, K. K. (2015). Effect of plant growth regulators on fruit drop and quality of Bael under Tarai conditions of Uttarakhand. *Indian Journal of Horticulture*, 72(1), 126-129.

Vyas, K., Manda, H., Sharma, R. K., and Singhal, G. An update review on Annona squamosa. *International Journal of Pharmacy and Therapeutics*, 3(2), 2012, 107-118.

Index

B

Bacterial canker 39, 69, 96, 251

Bacterial gummosis 10, 203

Bacterial spot 39, 96, 204, 252

Badama 139

Bael 63

Baggugosha 215

Balanagar 104

Bark eating caterpillar 9, 26, 169, 220

Bartlett 213

Belrubi 262

Bhado 139

Bigarreau Noir Grossa 91

Bing 90

Black cherry aphid (*Myzus cerasi*) 95

Black heart 90

Black knot 252

Bleaching 288

Blossom blight 221

Blossom thrips 250

Blossom weevil (*Anthonomus rabi*) 267

Bosc 214

Botany 3, 19, 87, 160, 227, 260

Botrytis rot 270

Brown rot 10, 39, 96, 203, 252

Brown spot 185

Bruising 97

Bruno 148

C

California advance 162

California paper shell 4

Calyx cavity 236

Camarosa 262

Canning varieties 194

Capri fig 116

Carambola 73

Carapace spot 61

Carbohydrate 3, 30, 100, 124, 130

Cercospora leaf spot (*Cercospora kaki*) 235

Cercospora spot (*Cercospora purpurea*) 59

Chakaiya 21

Chakrata selection 282

Chandler 261, 279

Charmagz 33

Cherry 85

Cherry fruit fly (*Rhagoletis cerasi*) 96

Chico 279

Chilling injury 42, 60, 82, 171

Circular leaf spot (*Mycosphaerella* spp.) 235

Cisco 279

CITH cherry 91

CITH walnut 281

Cleft sutures 206

Climate 66, 78, 161, 278

Cling stone 193

Clonal rootstock 93

Codling moth (*Cydia pomonella*) 285

Collar rot 10, 186

Colletotrichum acutatum 11

Colletotrichum gloeosporioides 11

Comice 214

Composition 2, 73, 100

Concorde 214

Conference 214

Cordia myxa 24

Crest heaven 194

Crick-side 61

Crop regulation 106

Cross-pollination 53

Crown and root gall 97

Crown gall 10, 235

Crown rot (*Phytophthora* sp.) 169

Cryptolaemous montrouzieri 110